Mathematics
UNLIMITED

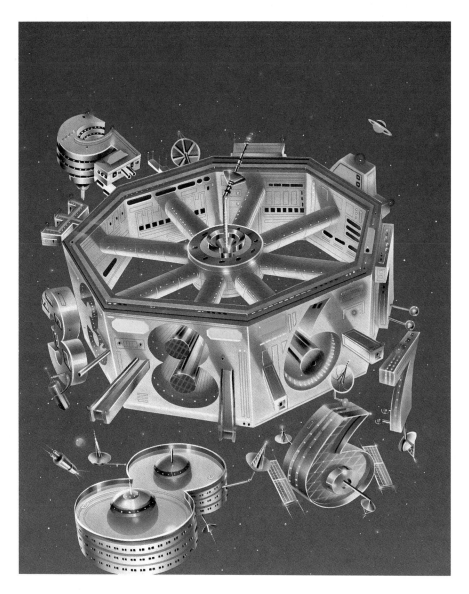

HOLT, RINEHART and WINSTON, Publishers
New York • Toronto • Mexico City • London • Sydney • Tokyo

AUTHORS

Francis "Skip" Fennell
Chairman, Education Department
Associate Professor of Education
Western Maryland College
Westminster, Maryland

Barbara J. Reys
Assistant Professor of Curriculum
and Instruction
University of Missouri, Columbia, Missouri
Formerly Junior High Mathematics Teacher
Oakland Junior High, Columbia, Missouri

Robert E. Reys
Professor of Mathematics Education
University of Missouri
Columbia, Missouri

Arnold W. Webb
Senior Research Associate
Research for Better Schools
Philadelphia, Pennsylvania
Formerly Asst. Commissioner of Education
New Jersey State Education Department

ILLUSTRATION

Bob Aiese: 32–3, 44, 78, 79, 110, 138–9, 180, 181, 254–5, 320–1, 344, 358–9, 392, 393 • Donald Crews: 206, 292, 294, 295, 388 • Nancy Didion: 14, 15, 140, 141, 216, 226, 227, 322, 323 • Hovik Dilakiam: 196, 368 • Betsy Feeney: 152 • Edwin Figueroa: 369 • Mark Giglio: 22–3, 102, 121, 130, 230–1, 326–7 • Jim Ludtke: 18, 37, 67, 116, 168, 169, 217, 301, 346, 364 • Linda Miyamoto: 398, 399 • Deidre Newman: 4, 5, 56, 86, 166, 167, 236–7, 165, 174–5, 314, 331, 374, 375 • Michael O'Reilly: 16, 106, 260–61, 286 • Dixon Scott: 46–7, 76, 172, 187, 190, 304, 308, 330, 373 • Marti Shohet: 2–3, 28, 384, 385 • Arthur Thompson: 8, 132–3, 146, 159, 377 • Paul Vaccarello: 92, 93, 240–1, 309 • Mike Van Ralte: 228–9 • Fred Winkowski: 88, 89, 256, 257, 272, 273 • Nina Winters: 6, 20, 37, 182, 183, 210–11, 198, 199, 248, 249 • Lane Yerkes: 114, 154–5, 200, 246–7, 335.

Cover Illustration: Jeannette Adams   Chapter Opener Illustrations: Jim Owens: 1, 43, 75, 105, 129, 165, 195, 225, 271, 307, 339, 383.

PHOTOGRAPHY

American Heritage/Duncan: p. 94; Frost: p. 94 • Black Star/Fred Ward: p. 184 • Dennis Brack: p. 176 • DPI/Ron Sefton: p. 136; Linda K. Moore: p. 316 • Bruce Coleman, Inc./Jeff Foott: p. 26 • DRK Photo/Tom Bledsoe: p. 84; J. Wengle: p. 84 • Duomo/Tony Duffy: p. 62 • Focus on Sports: pp. 51, 64, 96 • Focus West/Dave Black: p. 58 • Michal Heron: pp. 108, 134, 135, 156 • The Image Bank/Grafton M. Smith: p. 80; Merrell Wood: p. 109; Earl Roberge: p. 118; Jay Freis: p. 394 • Imagery: p. 296 • International Stock Photo: p. 288 • Linda K. Moore: p. 262 • NASA: pp. 348, 370 • National Center for Atmospheric Research/National Science Foundation: pp. 212 • Marvin E. Newman: pp. 174 top, 175 • Omni-Photo Communications, Inc./Ken Karp: pp. 48–9, 54–55, 148–9, 204–5, 282, 284, 324–5, 396, 397; John Lei: pp. 10–11, 25, 82–3, 144, 242–3, 252–3, 276–7, 278, 280, 290–1, 328 • Photo Researchers, Inc.: pp. 238, 352, 360, 366; Wesley Bocxe: p. 174; Richard Hutchinson: p. 297; Carleton Ray: p. 19 • Rainbow/Coco McCoy: p. 112 • The Stock Market: p. 310, 239; p. 310; Stan Tess: p. 390 • Taurus Photos: p. 311 • Woodfin Camp & Associates/Craig Aurness: p. 27; Sissie Brimberg: p. 328; Robert Frerck: p. 208; David Alan Harvey: p. 26; Mike Maple: pp. 202, 214; Jeff Lowenthal: p. 312; Robert McElroy: p. 142; Wally McNamee: pp. 150–1; Chuck Nicklin: p. 12; Mike S. Yamashita: p. 12.

ISBN 0-03-006437-6

6 7 8 9 0 032 9 8 7 6 5 4 3 2 1

CONTENTS

RATIO AND PERCENT

GEOMETRY

STATISTICS AND PROBABILITY

The world's oceans are teeming with life. How many different kinds of creatures can you name? Could you put them in order by length, weight, or speed?

1 PLACE VALUE, ADDITION AND SUBTRACTION
Whole Numbers

Numbers to Hundred Thousands

One hundred twenty-five thousand, four hundred six people have contributed money to help save the whales. Write a number that shows how many people contributed.

125,406 people contributed money.

Thousands			Ones		
hundred thousands	ten thousands	thousands	hundreds	tens	ones
1	2	5,	4	0	6

A comma is used to separate large numbers into groups of three digits.

In 125,406:

The value of the digit 1, in the hundred thousands place, is 100,000.
The value of the digit 2, in the ten thousands place, is 20,000.
The value of the digit 5, in the thousands place, is 5,000.
The value of the digit 4, in the hundreds place, is 400.
The value of the digit 0, in the tens place, is 0.
The value of the digit 6, in the ones place, is 6.

Standard form: 125,406
Expanded form: 100,000 + 20,000 + 5,000 + 400 + 6

Checkpoint Write the letter of the correct answer.

What is the value of the blue digit?

1. 386,846

a. 8
b. 80
c. 800
d. 846

2. 930,834

a. 3
b. 300
c. 3,834
d. 30,000

3. 864,020

a. 0
b. 20
c. 100
d. 200

What is the value of the blue digit?

1. 456,782
2. 385,621
3. 598,364
4. 786,320
5. 976,841

6. 304,562
7. 343,754
8. 600,032
9. 750,401
10. 806,150

11. 596,321
12. 846,329
13. 795,423
14. 134,769
15. 612,439

16. 316,030
17. 453,230
18. 985,063
19. 321,119
20. 980,020

21. 409,265
22. 763,518
23. 547,028
24. 993,457
25. 821,593

Write in standard form.

26. 80,000 + 7,000 + 900 + 20 + 6

27. 200,000 + 40,000 + 6,000 + 200 + 30 + 1

28. 600,000 + 20,000 + 1,000 + 500 + 90 + 2

29. 50,000 + 3,000 + 500 + 60 + 8

30. 40,000 + 5,000 + 200 + 80 + 5

31. 800,000 + 9,000 + 800 + 7

Write in expanded form.

32. 1,238
33. 27,569
34. 438,451
35. 998,915

36. 342,671
37. 357,954
38. 529,346
39. 82,165

40. 3,972
41. 34,602
42. 15,045
★43. 707,399

FOCUS: MENTAL MATH

You can sort a list of numbers by using the value of the front digit of each number as a guide.

134

987

This number is close to 100.

This number is close to 1,000.

Sort the list of numbers into two groups:

125; 88; 1,013; 91; 970; 922; 53; 1,321; 988; 879; 85; 1,279; 1,009; 73; 103; 898

1. those close to 100.

2. those close to 1,000.

Numbers to Hundred Billions

A. Most of Earth's surface is covered with water. Four oceans—the Pacific, the Atlantic, the Indian, and the Arctic—cover about one hundred twenty-seven million, three hundred forty-eight thousand square miles. Write this number in standard form.

Each group of three digits is called a **period.** Periods simplify the reading and writing of large numbers.

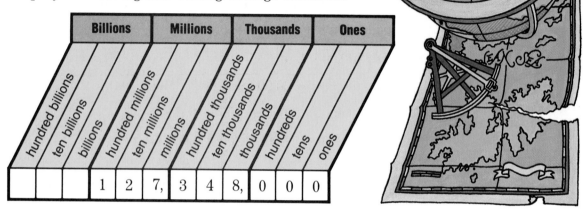

Billions			Millions			Thousands			Ones		
hundred billions	ten billions	billions	hundred millions	ten millions	millions	hundred thousands	ten thousands	thousands	hundreds	tens	ones
			1	2	7,	3	4	8,	0	0	0

In 127,348,000:

The value of the digit 1, in the hundred millions place, is 100,000,000.
The value of the digit 2, in the ten millions place, is 20,000,000.
The value of the digit 7, in the millions place, is 7,000,000.
The value of the digit 3, in the hundred thousands place, is 300,000.
The value of the digit 4, in the ten thousands place, is 40,000.
The value of the digit 8, in the thousands place, is 8,000.
The value of the digit 0, in the hundreds place, is 0.
The value of the digit 0, in the tens place, is 0.
The value of the digit 0, in the ones place, is 0.

Write: 127,348,000.

Read: one hundred twenty-seven million, three hundred forty-eight thousand.

B. You can write short word names for large numbers by writing the digits in each period, followed by the name of each period.

127,348,000   27,315,656,423

127 million, 348 thousand   27 billion, 315 million, 656 thousand, 423

What is the value of the blue digit?

1. 228,603,000

2. 755,049,001

3. 100,004,522

4. 591,440,700,000

5. 807,611,882,000

6. 780,049,467,795

7. 492,332,079

8. 561,978,445

9. 893,556,423,009

10. 133,567,238,000

11. 475,506,231,005

★12. 9,000,054,723,976

For 583,247,612,741, write the digits in the

13. billions period.

14. millions period.

15. thousands period.

Write in standard form.

16. 600 million, 29 thousand 500

17. 140 million

18. 845 million, 349 thousand, 1 hundred 1

19. 200 billion, 67 million, 1 thousand, 9 hundred 5

Write the short word name for each.

20. 42,799,928,725

21. 164,989,100,000

22. 608,010,002,700

23. 301,499,500

24. 992,456,204,089

25. 757,322,468,271

CHALLENGE

Ancient Egyptians used *hieroglyphics*, a form of picture writing, to write numbers. They did not use place value.

$\mathring{8}$ = 1,000	9 = 100
$\cap$ = 10	$/$ = 1

The Egyptians would write 3,762 like this:

$\mathring{8}\mathring{8}\mathring{8}$ 9999999 $\cap\cap\cap\cap\cap\cap$ //

Write each number.

1. 999 $\cap\cap$ ////

2. $\mathring{8}$ 9999 $\cap\cap$ ///

Write each number in hieroglyphics.

3. 5,192   **4.** 4,628   **5.** 8,325   **6.** 3,531

Comparing and Ordering

A. A **number line** can be used to compare numbers.

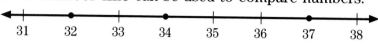

31 32 33 34 35 36 37 38

32 is to the left of 34.
So, 32 *is less than* 34.

Write: 32 < 34.

37 is to the right of 34.
So, 37 *is greater than* 34.

Write: 37 > 34.

You can compare to decide whether two numbers are equal.

$$34 = 34 \qquad 34 \neq 37$$

≠ means not equal to

B. You can compare without using a number line. Compare the average depths of the Atlantic and the Pacific oceans. Which has a greater average depth?

Atlantic: 3,575 meters   Pacific: 3,940 meters

Compare 3,575 and 3,940.

Line up the digits.	Begin to compare digits at the left.	Continue comparing.
3,575 3,940	3,575 3,940 3 = 3	3,575 3,940 5 < 9

So, 3,575 < 3,940.

The Pacific Ocean has a greater average depth.

C. You can order 4,219; 867; 911; and 298 by comparing.

Line up the digits.	Begin to compare at the left.	Compare the remaining numbers.
4,219 867 911 298	4,219 867 911 298	911 > 867 867 > 298

Only number with a thousands place; it is the greatest.

From the greatest to the least: 4,219; 911; 867; 298
From the least to the greatest: 298; 867; 911; 4,219

Compare. Use >, <, or = for ●.

1. 47 ● 17
2. 105 ● 204
3. 62 ● 620

4. 80 ● 802
5. 5,468 ● 4,599
6. 11,301 ● 9,098

7. 27,687 ● 21,688
8. 62,546 ● 101,829
9. 878,450 ● 678,405

Order from the least to the greatest.

10. 105; 744; 298; 741
11. 9,056; 821; 1,751; 1,052

12. 301; 6,981; 3,010; 7,059
13. 5,422; 6,001; 542; 512

14. 97; 809; 3,840; 3,048
15. 723; 737; 1,737; 373

Order from the greatest to the least.

16. 588; 198; 258
17. 1,419; 5,712; 2,576
18. 9,082; 482; 9,544

19. 8,222; 3,079; 8,041
20. 22,098; 22,911; 23,004; 21,992

21. 6,776; 6,767; 7,667; 799
22. 54,698; 5,499; 75,001; 57,698

Solve. Use the chart.

23. Which sea is deeper, the Mediterranean Sea or the Caribbean Sea?

24. Scientists have learned that some whales dive to a depth of 3,609 feet. Which seas on the chart have a depth that is greater than the depth to which the whales can dive?

★25. Copy the chart and arrange the seas in order from the shallowest to the deepest.

Sea	Depth
Mediterranean	4,902 feet
Black	3,826 feet
Baltic	282 feet
Caribbean	8,173 feet
North	308 feet

ANOTHER LOOK

Write the number in expanded form.

1. 9,506
2. 89,682
3. 75,897

4. 869,708
5. 456,789,123
6. 3,509,000

PROBLEM SOLVING
Using the Help File

If you are having trouble solving a problem, try using the Help File on pages 411–414 of this book. It has suggestions that may help you. To use this file, you should follow these steps.

1. Try to decide why you are having trouble. Then, go to the part of the Help File that can help you.

2. Look over the ideas in the part of the file you chose. Find one that you think will help you. Remember that there is no rule about which idea you should select. Different people may choose different ways to help them solve a problem.

3. Try to solve the problem by using the idea you chose.

4. If you still cannot solve the problem, look for another idea in the Help File.

Read the problem. Then choose the part of the Help File each student should use. Write the letter of the correct answer.

Ann Marie spent two weeks studying the life in a coral reef located off the coast of Jamaica. In the first week, she spent 38 hours diving. In the second week, she dived for a total of 25 hours. For how many more hours did she dive during the first week than the second week?

1. Bill read the problem. Then he read it again. He read it one more time. Then he said, "I do not even know what I am looking for."

   Where should Bill look in the Help File?
   a. Tools   **b.** Questions
   c. Solutions   **d.** Checks

2. Cathy read the problem. "I see," she said. "Ann Marie spent 38 hours diving in the first week. She spent 25 hours diving in the second week. But how do I find how many more hours she spent diving during the first week?"

   Where should Cathy look in the Help File?
   a. Solutions   **b.** Questions
   c. Checks   **d.** Tools

3. Francis knew he had to subtract to solve the problem. But he had trouble with subtraction. "What do I do now?" he wondered.

   Where should Francis look in the Help File?
   a. Checks   **b.** Tools
   c. Solutions   **d.** Questions

4. Mona answered the question. "Ann Marie spent 13 more hours diving in the first week than in the second week," she said. "Now, I had better check the answer."

   Where should Mona look in the Help File?
   a. Tools   **b.** Questions
   c. Checks   **d.** Solutions

Properties of Addition

A. Many people are interested in the plant and animal life of a coral reef. John Sparrow takes people out to the reef in his glass-bottom boat. He completes 7 trips Saturday and 8 trips Sunday. How many trips does he complete?

To find how many, you can add.

Addition can be shown in two ways.

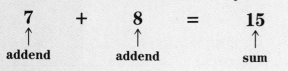

$$7 \quad + \quad 8 \quad = \quad 15$$

addend   addend   sum

$$
\begin{array}{r}
7 \quad \longleftarrow \text{ addend} \\
+\,8 \quad \longleftarrow \text{ addend} \\
\hline
15 \quad \longleftarrow \text{ sum}
\end{array}
$$

John completes 15 trips.

B. Addition has special properties.

Commutative Property If the order of the addends is changed, the sum remains the same.	$3 + 8 = 8 + 3$ $11 = 11$
Zero Property If one of the addends is zero, the sum is equal to the other addend.	$4 + 0 = 4$ $0 + 6 = 6$
Associative Property If the grouping of the addends is changed, the sum remains the same.	$(2 + 4) + 5 = 2 + (4 + 5)$ $6 \quad + 5 = 2 + \quad 9$ $11 = 11$

Complete. Identify the property used.

1. $9 + 3 = 3 + \blacksquare$ **2.** $4 + 7 = \blacksquare + 4$ **3.** $6 + 5 = 5 + \blacksquare$

4. $(2 + 5) + 3 = 2 + (\blacksquare + 3)$ **5.** $6 + (3 + 2) = (\blacksquare + 3) + 2$

6. $4 + (9 + 2) = (\blacksquare + 9) + 2$ **7.** $(3 + 6) + 5 = 3 + (\blacksquare + 5)$

Add.

8. 4
  + 7

9. 6
  + 9

10. 5
   + 8

11. 9
   + 8

12. 6
   + 7

13. 8
   + 9

14. 7
   + 5

15. 2
   + 9

16. 8
   + 4

17. 0
   + 6

18. $7 + 8$

19. $4 + 9$

20. $5 + 8$

21. $8 + 7$

22. $8 + 6$

23. $9 + 6$

24. $0 + 9$

25. $6 + 8$

26. $(2 + 2) + 8$

27. $(3 + 4) + 9$

28. $(7 + 1) + 3$

Solve.

★**29.** $0 + \blacksquare = 9$

★**30.** $4 + \blacksquare = 5$

★**31.** $7 + \blacksquare = 11$

★**32.** $6 + \blacksquare = 15$

★**33.** $(379 + \blacksquare) + 685 = 379 + (471 + 685)$

★**34.** $5,438 + \blacksquare = 3,481 + 5,438$

★**35.** $562 + (468 + 750) = (562 + \blacksquare) + 750$

★**36.** $13,541 + 25,403 = \blacksquare + 13,541$

Solve.

35. On one trip, the passengers spot 7 angelfish. Then they see 6 more angelfish. How many angelfish do the passengers see?

36. Robert and Sarah saw the same number of fish. Robert saw 8 fish in the morning and 9 fish in the afternoon. Sarah saw 9 fish in the morning. How many fish did she see in the afternoon?

37. Look at Ray's chart. How many fish did he spot on each outing? On which outing did he spot more fish?

TROPICAL FISH

Fish	Outing 1	Outing 2
Trunkfish	7	2
Glassfish	4	6

FOCUS: MENTAL MATH

You can use doubles as an addition shortcut.

Add $6 + 5$.   **Think:** 5 is 1 less than 6.
                  $6 + 6 = 12$ So, $6 + 5 = 11$.

Compute mentally.

1. $7 + 8$

2. $5 + 4$

3. $8 + 9$

4. $8 + 7$

5. $6 + 7$

6. $5 + 6$

Related Facts

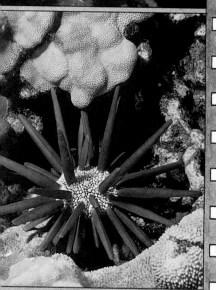

A. At Buck Island Reef National Monument, snorkelers swim through miles of fragile coral gardens. Nan sees 6 cardinal fish. Then she sees 3 more cardinal fish. How many cardinal fish does Nan see?

To find how many, you can add.

6 + 3 = 9 ⟵ sum

Nan sees 9 cardinal fish.
If 3 cardinal fish swim away, how many are left?
To find how many are left, you can subtract.

9 − 3 = 6 ⟵ difference

There are 6 cardinal fish left.

B. You can use the numbers 9, 6, and 3 to write a family of facts.

6 + 3 = 9	9 − 3 = 6
3 + 6 = 9	9 − 6 = 3

You can use a family of facts to solve a subtraction problem by writing a related addition problem.

$14 - n = 5$ n stands for the missing number.

Think: 9 + 5 = 14. So, 14 − 9 = 5.

C. As you subtract, remember:

If 0 is subtracted from a number, the difference is equal to that number.	7 − 0 = 7
If a number is subtracted from itself, the difference is 0.	8 − 8 = 0
Subtraction is not commutative.	9 − 4 = 5 4 − 9 ≠ 5

Copy and complete.

1. 7 + 8 = 15
    15 − ▨ = 8

2. 9 + 4 = 13
    13 − ▨ = 9

3. 6 + 5 = 11
    11 − ▨ = 6

4. 3 + 9 = 12
    12 − ▨ = 9

5. 7 + 6 = 13
    13 − ▨ = 7

6. 8 + 2 = 10
    10 − ▨ = 2

Subtract.

7. $\begin{array}{r} 11 \\ -\ 8 \\ \hline \end{array}$ **8.** $\begin{array}{r} 17 \\ -\ 8 \\ \hline \end{array}$ **9.** $\begin{array}{r} 12 \\ -\ 4 \\ \hline \end{array}$ **10.** $\begin{array}{r} 6 \\ -5 \\ \hline \end{array}$ **11.** $\begin{array}{r} 8 \\ -4 \\ \hline \end{array}$

12. $\begin{array}{r} 14 \\ -\ 7 \\ \hline \end{array}$ **13.** $\begin{array}{r} 5 \\ -1 \\ \hline \end{array}$ **14.** $\begin{array}{r} 9 \\ -2 \\ \hline \end{array}$ **15.** $\begin{array}{r} 7 \\ -2 \\ \hline \end{array}$ **16.** $\begin{array}{r} 15 \\ -15 \\ \hline \end{array}$

17. $12 - 12$ **18.** $10 - 4$ **19.** $15 - 8$ **20.** $6 - 4$

21. $8 - 0$ **22.** $11 - 5$ **23.** $5 - 2$ **24.** $4 - 2$

Solve for n.

★**25.** $54 + 32 = 86$ ★**26.** $63 + 34 = 97$ ★**27.** $107 + 132 = 239$ ★**28.** $425 + 85 = 510$

$86 - n = 54$ $\qquad$ $97 - n = 34$ $\qquad$ $239 - n = 107$ $\qquad$ $510 - n = 425$

Write a family of facts for each group of numbers.

29. 4, 9, 13 $\qquad$ **30.** 17, 9, 8 $\qquad$ **31.** 13, 6, 7 $\qquad$ **32.** 15, 7, 8

33. 6, 8, 14 $\qquad$ **34.** 5, 3, 2 $\qquad$ **35.** 11, 6, 5 $\qquad$ **36.** 12, 8, 4

Solve for n. Use related facts.

37. $14 - n = 7$ **38.** $n - 3 = 5$ **39.** $5 - n = 1$ **40.** $n - 9 = 9$

Solve.

41. On her snorkeling trip, Rachel took 18 photographs with her underwater camera. When the film was developed, 9 of the photographs were too dark. How many photographs were of good quality?

42. Use the information below to write several families of facts.

Type of eel	Mon.	Tues.	Wed.
Wolf eel	4	6	7
Moray eel	6	9	5
Total number of eels	10	15	12

CHALLENGE

Copy and complete.

Add or subtract to find each missing number.

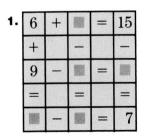

13

Front-End Estimation

A. Grouping pairs of numbers to sums of 100 or 1,000 can help you estimate sums.

Group numbers in this box whose sum is about 100.

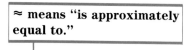

≈ means "is approximately equal to."

Think: $45 \approx 4$ tens, $53 \approx 5$ tens.
So, $45 + 53 \approx 100$
$19 + 83 \approx 100$, $75 + 36 \approx 100$.

4 tens	45	83	75
	19	36	53 — **5 tens**

Group numbers in this circle whose sum is about 1,000.
$129 + 837 \approx 1,000$
$376 + 649 \approx 1,000$ $489 + 465 \approx 1,000$

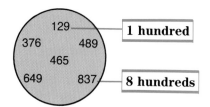

129 — **1 hundred**
376   489
465
649   837 — **8 hundreds**

B. Look at the chart of fish tagged by volunteers at the Oceanographic Institute. Their goal was to tag 600 fish daily. Estimate to find whether or not the goal was reached.

Estimate: $115 + 86 + 113 + 175$.

You can use front-end estimation.

FISH TAGGED ON TUESDAY

Erica	115
Sean	86
Sally	113
Kevin	175

Add the numbers in the greatest place.

$$\begin{array}{r} 115 \\ 86 \\ 113 \\ +175 \\ \hline 3 \end{array}$$

Adjust by grouping the other amounts.

$\left.\begin{array}{r} 115 \\ 86 \end{array}\right\}$ about 100

$\left.\begin{array}{r} 113 \\ 175 \end{array}\right\}$ about 100

Adjustment: $100 + 100 = 200$.

Rough estimate: 300.
Adjusted estimate: $300 + 200 = 500$.
$115 + 86 + 113 + 175 < 600$
So, they did not tag enough fish to meet their goal.

Another example:

$$\begin{array}{r} \$4.67 \\ 3.42 \quad \text{about } \$1 \\ + \ 0.85 \quad \text{about } \$1 \\ \hline \$7 + \$1 + \$1 = \$9 \end{array}$$

14

Write the two numbers whose sum is about

1. 100.   85   45   13   35

2. 100.   48   75   53   86

3. 50.   17   57   25   35

4. 50.   24   49   40   27

5. 1,000.   450   275   560   110

6. 1,000.   285   700   100   580

Estimate. Write > or < for ●.

7. 57 + 69 ● 100

8. 86 + 97 ● 200

9. 46 + 74 ● 100

10. 157 + 294 ● 300

11. 219 + 689 ● 1,000

12. 895 + 129 ● 900

13. 465 + 789 + 921 ● 2,000

14. 389 + 471 + 59 ● 1,000

15. $8.75 + $0.89 ● $10.00

16. $5.37 + $1.76 + $1.98 ● $10.00

★17. 8,957 + 85 + 125 ● 10,000

★18. 9,875 + 4,327 + 2,756 ● 20,000

Estimate. First write your rough estimate.
Then write your adjusted estimate.

19.
```
   54
  487
  215
+ 149
```

20.
```
  345
  159
   95
+ 220
```

21.
```
4,276
  345
5,729
+ 3,287
```

22.
```
9,217
6,029
5,788
+ 955
```

23.
```
$9.57
 8.39
 0.95
+ 9.08
```

Solve.

24. On a field trip to the Gulf of Mexico, a marine biologist identifies 73 types of coral, 61 types of fish, 22 types of plankton, and 38 types of crustaceans. Did the biologist identify more than 200 types of ocean life?

25. Mary stocks the Oceanographic Institute's exhibit of saltwater fish. She stocks 38 chimera, 69 herring, 117 dogfish, and 478 flounder. About how many saltwater fish is that?

FOCUS: REASONING

Look at the shapes below. Each shape differs in only one way from the shapes next to it.

Rearrange the shapes so that each shape differs in *two* ways from the shapes next to it.

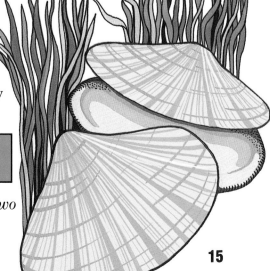

Rounding and Estimating Sums

A. Australia's coral Great Barrier Reef is about 1,250 miles long. To the nearest thousand miles, how long is the reef?

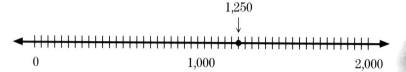

1,250 is between 1,000 and 2,000. It is closer to 1,000. So, to the nearest thousand miles, the reef is 1,000 miles long.

B. Sometimes you have to take a closer look at the digits of a number when rounding. To the nearest hundred miles, how long is the reef?

- Find the place to which you are rounding.   **1,250**
- If the digit to the right is 5 or greater, round up.
- If the digit to the right is less than 5, round down. The digit to the right is 5. Round up.   **1,250**

So, to the nearest hundred miles, the reef is 1,300 miles long.

C. Rounding can help you estimate sums.

Estimate 4,857 + 1,426.

The place to which you round depends upon the size of the numbers and how easily you can mentally compute the rounded numbers.

Round each addend to the nearest thousand.	Add the rounded numbers.	Round each addend to the nearest hundred.	Add the rounded numbers.
$4,857 \rightarrow 5,000$ $+\ 1,426 \rightarrow 1,000$	$5,000$ $+\ 1,000$ $\overline{6,000}$	$4,857 \rightarrow 4,900$ $+\ 1,426 \rightarrow 1,400$	$4,900$ $+\ 1,400$ $\overline{6,300}$

Both 6,000 and 6,300 are good estimates.

Round each number to the nearest hundred, thousand, and ten thousand.

1. 15,326   2. 24,856   3. 17,909   4. 21,998   5. 36,111   6. 32,762

7. 47,283   8. 32,931   9. 76,270   10. 81,635   11. 57,469   12. 64,983

Round to the nearest ten dollars.

13. $75.75   14. $12.13   15. $19.99   16. $18.90   17. $49.99   18. $16.75

19. $34.08   20. $18.98   21. $32.95   22. $23.90   23. $93.46   24. $29.01

Estimate. Write > or < for ●

25. 723 + 184 ● 1,000   26. 8,519 + 2,426 ● 10,000   27. 4,326 + 2,148 ● 7,000

28. 423 + 136 + 214 ● 700   29. 8,329 + 4,916 + 2,463 ● 17,000

Estimate.

30. 436 + 538   31. 568 + 273   32. 8,916 + 7,228   33. 15,642 + 23,984

34. 426 + 298 + 137   35. 5,326 + 4,216 + 5,593   36. 14,316 + 12,149 + 27,823

Solve.

37. A scientist in a minisub dives 2,345 feet into the Coral Sea to do some experiments. Then he descends another 1,672 feet. About how deep did the scientist dive?

★38. A commercial diver collected 1,467 pounds of white coral, 1,687 pounds of pink coral, and 2,632 pounds of red coral. About how much coral did the diver collect?

MIDCHAPTER REVIEW

What is the value of the blue digit?

1. 786,213   2. 34,077   3. 299,493   4. 137,244   5. 347,214

Order from the least to the greatest.

6. 744, 323, 536   7. 2,519; 5,912; 5,291   8. 63,246; 63,426; 63,446

Complete.   Round to the nearest hundred.

9. 26 + ■ = 29   10. 0 + ■ = 93   11. 16,924   12. 901,099

Estimating Differences—Rounding

A. The Aquafest hoped that the exhibition of Biff the Sea Lion would increase daily attendance by 1,000 people. Last year's daily attendance was 2,516. This year it was 4,326. Estimate whether the goal was reached.

Round to estimate 4,326 − 2,516.
The place to which you round depends upon the size of the numbers and how easily you can compute the rounded numbers.

Round each number to the nearest thousand.	Subtract the rounded numbers.
$4,326 \longrightarrow 4,000$ $-\ 2,516 \longrightarrow 3,000$	$\begin{array}{r} 4,000 \\ -\ 3,000 \\ \hline 1,000 \end{array}$

The goal of about 1,000 was reached.

B. Estimate 2,746 − 593.

Round each number to the nearest hundred.	Subtract the rounded numbers.
$2,746 \longrightarrow 2,700$ $-\ \ \ 593 \longrightarrow \ \ \ 600$	$\begin{array}{r} 2,700 \\ -\ \ \ 600 \\ \hline 2,100 \end{array}$

The difference of 2,746 and 593 is about 2,100.

Estimate by rounding.

1. $\begin{array}{r} 8,427 \\ -\ 3,916 \end{array}$	**2.** $\begin{array}{r} 5,684 \\ -\ 3,196 \end{array}$	**3.** $\begin{array}{r} 7,426 \\ -\ \ \ 783 \end{array}$	**4.** $\begin{array}{r} 847 \\ -\ 638 \end{array}$	**5.** $\begin{array}{r} 3,237 \\ -\ \ \ 759 \end{array}$

6. 23,416 − 20,817   **7.** 93,426 − 79,846   **8.** 34,816 − 7,323

Solve.

9. Biff is trained to shoot baskets into a hoop with his nose. In June, Biff made 847 baskets. By July, he had improved enough to make 1,397 baskets. About how great was his improvement?

18

PROBLEM SOLVING
Estimation

In some situations, underestimating the answer makes the most sense.

The Oceanographic Museum will give a school lecture series if at least 600 students are interested. The local school boards are asked to estimate the number of students that will attend. Each school made a tally and gave the school board this chart.

STUDENTS INTERESTED IN ATTENDING LECTURES

Faye School	Rapp School	Gray School	Day School
156	236	224	328

To quickly find the number of interested students, the school board estimates. To be sure they have the *minimum* number required, *underestimation* is the best method. Front-end estimation gives an underestimate.

$$
\begin{array}{r}
156 \\
236 \\
224 \\
+\ 328 \\
\hline
8
\end{array}
$$
  Rough estimate: 800

At least 800 students are interested. The museum will sponsor the lecture series.

Solve. Use front-end estimation.

1. The lectures are held in a 4-section auditorium. At first only the first 112 seats are opened, but the lectures have become so popular that the museum must open the second section of 207 seats, the third section of 234 seats, and then the fourth section of 125 seats. Estimate the minimum number of seats available to the audience.

2. One lecture topic was a project that studied underwater volcanoes. At least 700 samples had to be recorded to make the project a success. The first dive recorded 207 items, the second dive found 218 samples, and the last dive recorded 326 items. Estimate to find whether the project recorded at least the number of samples necessary for success. Was the project a success?

Addition of 2- and 3-Digit Numbers

A. Students from the Webb School visit the Seaquarium as part of their Science Week. They go in two groups. One group has 158 students. The other group has 135 students. How many students visit the Seaquarium?

Find 158 + 135.

First estimate the sum.

$$
\begin{array}{r}
158 \longrightarrow 200 \\
+\ 135 \longrightarrow +\ 100 \\
\hline
300
\end{array}
$$

Add the ones. Regroup if necessary.	Add the tens. Regroup if necessary.	Add the hundreds. Regroup if necessary.
$\overset{1}{1}58$	$1\overset{1}{5}8$	158
$+135$	$+135$	$+135$
3	93	293

293 students visit the Seaquarium.
The answer is reasonably close to the estimate.

B. You add money the same way you add whole numbers. Remember to write the dollar sign and the cents point.

$\overset{1\ 1}{\$3.66}$	$\overset{1\ 1}{\$2.69}$	$\overset{1\ 1}{\$4.67}$
$+\ \ 2.75$	$+\ \ 0.83$	$+\ \ 5.48$
$\$6.41$	$\$3.52$	$\$10.15$

Checkpoint Write the letter of the correct answer.

Add.

1. $\begin{array}{r} 56 \\ +\ 28 \\ \hline \end{array}$

   a. 28
   b. 74
   c. 84
   d. 114

2. $\begin{array}{r} 862 \\ +\ 239 \\ \hline \end{array}$

   a. 111
   b. 1,001
   c. 1,091
   d. 1,101

3. $6.65 + 3.76$

   a. $9.31
   b. $9.41
   c. $10.31
   d. $10.41

4. $647 + 84$

   a. 621
   b. 631
   c. 731
   d. 1,531

Add.

1. 84
   + 13

2. 31
   + 56

3. 23
   + 53

4. 35
   + 34

5. $0.66
   + 0.43

6. 538
   + 145

7. 643
   + 309

8. 866
   + 424

9. 107
   + 773

10. $9.39
   + 2.23

11. 27
   + 796

12. 83
   + 129

13. 92
   + 428

14. 285
   + 57

15. $1.58
   + 0.77

16. 753
   + 288

17. 47
   + 38

18. 856
   + 939

19. 584
   + 463

20. $8.06
   + 2.29

21. 18 + 17

22. 151 + 12

23. 77 + 362

24. $1.08 + $8.87

25. 739 + 113

26. 119 + 78

27. 273 + 98

28. $1.19 + $0.37

Solve.

29. At one Seaquarium tank, the students see 67 different kinds of fish from Hawaii. In another tank, they see 49 different kinds of fish from Florida. How many kinds of fish do the students see?

30. At the Seaquarium Book Shop, Jo buys two books about the fish she saw. The book about Hawaiian fish costs $2.85. The Florida fish book costs $1.95. How much does Jo spend for the books?

31. The Seaquarium is involved in a special breeding program to help save endangered species. Use the information in the table to write and solve your own addition problems.

NUMBER OF MAMMALS AT THE SEAQUARIUM

Mammal	Total Number (1985)	Number born (1986)
Dolphin	28	5
Manatee	12	3
Whale	7	1

CHALLENGE

Copy and complete each subtraction problem. Use only the digits 3, 7, and 8.

1. ■■
   − ■■
   ────
    5 5

2. ■■
   − ■■
   ────
    4 9

3. ■■
   − ■■
   ────
    3 6

4. ■■■
   − ■■
   ────
    2 8 6

Addition of Larger Numbers

Every winter, gray whales migrate from their feeding grounds in arctic waters to Baja California. Whale-watchers along the coast first spot the whales 3,295 miles from the feeding grounds. The whales must swim another 1,725 miles before they reach Baja. How far will the whales travel?

Add 3,295 + 1,725.

Add the ones. Regroup if necessary.	Add the tens. Regroup if necessary.	Add the hundreds. Regroup if necessary.	Add the thousands.
$\begin{array}{r} 1 \\ 3{,}2\,9\,5 \\ +\,1{,}7\,2\,5 \\ \hline 0 \end{array}$	$\begin{array}{r} 1\ 1 \\ 3{,}2\,9\,5 \\ +\,1{,}7\,2\,5 \\ \hline 2\,0 \end{array}$	$\begin{array}{r} 1\ 1\ 1 \\ 3{,}2\,9\,5 \\ +\,1{,}7\,2\,5 \\ \hline 0\,2\,0 \end{array}$	$\begin{array}{r} 1\ 1\ 1 \\ 3{,}2\,9\,5 \\ +\,1{,}7\,2\,5 \\ \hline 5{,}0\,2\,0 \end{array}$

The whales will travel 5,020 miles.

Other examples:

$$\begin{array}{r} 1\ 1\ 1 \\ 5\,6{,}8\,4\,3 \\ +\ \ \ 5{,}6\,7\,2 \\ \hline 6\,2{,}5\,1\,5 \end{array} \qquad \begin{array}{r} 1\ 1\ 1\ 1\ 1 \\ \$1{,}6\,0\,3.8\,6 \\ +\ \ \ \ 4\,9\,7.6\,4 \\ \hline \$2{,}1\,0\,1.5\,0 \end{array}$$

Checkpoint   Write the letter of the correct answer.

Add.

1. $\begin{array}{r} 8{,}078 \\ +\ 1{,}546 \\ \hline \end{array}$

2. $\begin{array}{r} 47{,}954 \\ +\ 31{,}676 \\ \hline \end{array}$

3. $6{,}725.58 + \$298.54$

a. 9,514
b. 9,524
c. 9,624
d. 9,651

a. 78,520
b. 78,530
c. 78,630
d. 79,630

a. $6,024.12
b. $6,913.02
c. $7,024.12
d. $70.2412

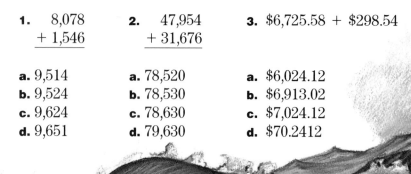

Find the sum.

1.	1,556 + 1,377	**2.**	3,251 + 5,699	**3.**	6,847 + 7,068	**4.**	5,188 + 1,784	**5.**	$43.48 + 5.82
6.	69,994 + 22,953	**7.**	41,923 + 9,884	**8.**	64,383 + 42,762	**9.**	$26,870 + 16,261	**10.**	$762.47 + 49.72
11.	58,659 + 27,343	**12.**	67,094 + 37,738	**13.**	40,653 + 37,438	**14.**	12,009 + 68,994	**15.**	$567.04 + 798.92

16. 229,775 + 50,426   **17.** 176,754 + 436,437   **18.** $758.25 + $238.47

19. $2,456.67 + $729.16   **20.** 721,423 + 53,724   **21.** $675.48 + $138.09

22. 3,485 + 7,947   **23.** 8,967 + 36,775   **24.** 96,783 + 4,563

Solve.

25. A whale named Gigi is tagged to track her migration. She swam 1,957 miles during July and 2,443 miles during August. How many miles did Gigi swim during the two months?

26. Whale watchers observe a group of killer whales in Canadian waters. They spend $2,567.85 to rent a ship. They spend $3,853.78 on equipment. How much do they spend in all?

FOCUS: ESTIMATION

You can estimate differences by using the front digits of numbers.

	Subtract the front digits.	Write zeros in the other places.
8,369 − 3,629	8 − 3 = 5	8,369 − 3,629 ⟶ 5,000

5,000 is a rough estimate of the difference.

Estimate.

1.	6,243 − 1,427	**2.**	7,846 − 3,598	**3.**	4,276 − 3,569	**4.**	67,416 − 28,497

More Practice, page 421

Column Addition

A. Michael uses a minisubmarine to study the plant and animal life of the coral reef. So far, he has noted 132 different types of coral and 47 different types of sea plants. He has also recorded 348 kinds of fish. How many types of coral-reef life has Michael seen?

Add 132 + 47 + 348.

Line up the numbers so that the ones are in a column.

Add the ones. Regroup if necessary.	Add the tens. Regroup if necessary.	Add the hundreds.
132 47 +348 ⎯⎯⎯ 7	132 47 +348 ⎯⎯⎯ 27	1 132 47 +348 ⎯⎯⎯ 527

Michael has seen 527 types of coral-reef life.

B. You can check your answer by adding up.

```
 1 1
 132 ↑
  47 |
+348 |
―――――
 527
```

Checkpoint Write the letter of the correct answer.

Add.

1.　46
　　　　39
　　　+ 17

2.　$12.56
　　　　2.01
　　　+　3.15

3. 5,329 + 2,438 + 59 + 47,601

1.	2.	3.
a. 82	**a.** $17.62	**a.** 44,397
b. 92	**b.** $17.72	**b.** 55,407
c. 102	**c.** $18.72	**c.** 55,427
d. 822	**d.** $28.72	**d.** 184,071

Add. Check by adding up.

1. 29
   31
   + 28

2. 31
   11
   + 13

3. 12
   29
   + 69

4. 44
   29
   + 69

5. 10
   22
   + 58

6. $3.15
   5.21
   + 1.54

7. 620
   107
   + 162

8. 491
   11
   + 274

9. $2.81
   1.14
   + 0.04

10. 117
    604
    + 229

11. 6,192
    1,420
    1,600
    + 3,837

12. 2,519
    1,228
    3,927
    + 190

13. $14.27
    42.54
    5.17
    + 4.30

14. 6,711
    2,633
    1,225
    + 1,347

15. $80.88
    94.16
    56.53
    + 71.92

16. 4,246 + 874 + 840

17. $1.59 + $29.97

18. 353 + 466 + 2,886

19. 327 + 517 + 59 + 73

20. 149 + 902 + 65 + 281

21. 970 + 784 + 29 + 197

Solve.

22. Patricia made three dives during one day. On her first dive, she spent 37 minutes underwater. Her second dive was 54 minutes long, and her third dive lasted 107 minutes. How much time did Patricia spend underwater that day?

★23. Pat bought the following supplies for her diving trip: flippers and a wet suit cost $358.97; goggles and an air tank cost $257.99; an underwater camera cost $287.35. The boat trip and all other expenses amounted to $455.30. What was the cost of Pat's trip?

ANOTHER LOOK

Write >, <, or = for ●.

1. 2,562 ● 2,065

2. 85,689 ● 9,995

3. 6,038 ● 6,039

4. 26,785 ● 2,683

5. 247,850 ● 25,787

6. 478,934 ● 479,734

PROBLEM SOLVING
Using Outside Sources Including the Infobank

Sometimes you may have to look in outside resources for the information you need to solve a problem.

Some of the sources in which you can find information are books, magazines, catalogs, and newspapers. You can also obtain information by contacting government agencies, museums, companies, and organizations.

Information from many different sources has been gathered into an Infobank, which you will find on pages 415–420 of this book. You can use this information to solve many problems in this textbook.

Read the following problem.

> Grace needs some equipment for her whale-watching trip. She orders a compass, a canteen, and a flashlight from Explorer's Outfitters, Inc. How much does this order cost?

You need to know the cost of each item that Grace orders. You could find this information by calling Explorer's Outfitters, Inc. You might find an advertisement for this company. Or you could look in the company's catalog.

A part of their catalog appears in the Infobank. Once you find it, you can use the information to solve the problem.

Item	Price
Compass	$11.25
Canteen	3.20
Flashlight	+ 8.75
	$23.20

Grace's order costs $23.20.

Read each problem. Then choose the item in the Infobank that provides the missing information. Write the item.

1. Betsy reads that a man rode on a roller coaster for 458 hours. She wonders if that time is as long as the record-setting ride. Where should she look?

2. The stratosphere is 7,000 meters above Earth. Where can you look to find how much farther away from Earth the mesosphere is than the stratosphere?

Use the Infobank to solve.

3. The numbers of blue, humpback, and Bryde's whales differ today from what they were before the whaling boom. Put the current populations of these whales in order from the least to the greatest.

4. The world record for the most words typed in an hour was set by Margaret Hanna. To the nearest thousand, how many words did she type?

5. There are box kites, parafoil kites, delta kites, and flat kites. If these four kites without their tails were laid end to end, how many inches long would this row of kites be?

6. Some people are interested in comparing the physical features of United States Presidents. Which President since Calvin Coolidge weighed the most?

7. Scientists and oceanographers try to keep track of the whale population. They have discovered that there are more Bryde's whales today than there are blue whales. Estimate how many more Bryde's whales there are than blue whales.

8. Bill reads in the newspaper that the Campabout Store is having a sale on sleeping bags. Each sleeping bag sells for $79.53. Bill finds the same sleeping bag in the Explorer's Outfitters, Inc., catalog. Which store has the better price?

Subtraction of 2- and 3-Digit Numbers

A. Scientists begin a two-day trip to follow dolphins along the California coast. The scientists will travel a total of 85 miles. On the first day, they travel 58 miles. How many more miles do they need to travel on the second day?

To find how many miles are needed to complete the trip, you can subtract. Find 85 − 58.

There are not enough ones.	Regroup. 1 ten 5 ones = 15 ones	Subtract the ones.	Subtract the tens.
$\begin{array}{r} 8\,5 \\ -\,5\,8 \\ \hline \end{array}$	$\begin{array}{r} {\scriptstyle 7\ 15} \\ 8\!\!\!/\,5\!\!\!/ \\ -\,5\,8 \\ \hline \end{array}$	$\begin{array}{r} {\scriptstyle 7\ 15} \\ 8\!\!\!/\,5\!\!\!/ \\ -\,5\,8 \\ \hline 7 \end{array}$	$\begin{array}{r} {\scriptstyle 7\ 15} \\ 8\!\!\!/\,5\!\!\!/ \\ -\,5\,8 \\ \hline 2\,7 \end{array}$

The scientists need to travel 27 miles on the second day.

Add to check your answer.
$$\begin{array}{r} 58 \\ +\,27 \\ \hline 85 \end{array}$$

B. You can subtract money the same way you subtract whole numbers. Remember to write the dollar sign and the cents point and the zero when necessary.

$$\begin{array}{r} {\scriptstyle 15} \\ {\scriptstyle 0\ 5\ 10} \\ \$1.6\,0 \\ -\ 0.9\,7 \\ \hline \$0.6\,3 \end{array}$$

You can use addition to check your answer.

$0.63 + \$0.97 = \$1.60$

Checkpoint Write the letter of the correct answer.

Subtract.

1. $\begin{array}{r} 74 \\ -\,38 \\ \hline \end{array}$

a. 36
b. 44
c. 46
d. 112

2. $\begin{array}{r} \$8.50 \\ -\ 1.48 \\ \hline \end{array}$

a. $7.02
b. $7.11
c. $7.19
d. $9.99

3. 350 − 76

a. 184
b. 274
c. 322
d. 426

Subtract. Check your answer.

1.	46 − 13	**2.**	72 − 22	**3.**	84 − 39	**4.**	80 − 65	**5.**	92 − 17
6.	295 − 47	**7.**	430 − 125	**8.**	615 − 407	**9.**	848 − 629	**10.**	$9.81 − 8.43
11.	371 − 294	**12.**	545 − 367	**13.**	915 − 19	**14.**	250 − 78	**15.**	$8.82 − 6.96

16. $345 - 67$ **17.** $298 - 109$ **18.** $814 - 585$ **19.** $9.62 - $0.79

20. $345 - 72$ **21.** $918 - 678$ **22.** $764 - 378$ **23.** $4.57 - $3.88

Find n.

24. $112 - n = 88$ **25.** $226 - n = 17$ **26.** $114 - n = 91$ **27.** $8.72 - n = $1.61

Solve.

28. Marine biologists were studying a group of 117 dolphins. They were able to tag 108 of them. How many dolphins were not tagged?

29. Scientists were studying a group of 93 dolphins. They discovered that 27 of the dolphins were adults. How many of the dolphins were not adults?

30. The Miami Seaquarium is famous for its performing whales and dolphins. The whale show attracted 118 people one day. The next day, 211 people saw the dolphin show. During the two days, how many people watched the shows?

★31. Some scientists teach human words to dolphins. A dolphin named Elvar took 32 weeks to learn five words. Another dolphin, Chee Chee, learned to say the same words in only 23 weeks. How much longer did it take Elvar to learn the words?

CHALLENGE

Copy the number sentence. Write $+$ or $-$ for ● to make the number sentence true.

1. $458 ● 529 ● 268 = 719$

2. $846 ● 124 ● 512 = 1,234$

3. $654 ● 235 ● 752 ● 3,456 = 4,627$

4. $269 ● 322 ● 312 ● 411 = 690$

Subtracting Larger Numbers

When he was young, Jack read *20,000 Leagues Under the Sea*, a fantastic story of a submarine voyage. Now, Jack is an underwater explorer. His deepest dive was 4,526 feet in a bathyscaphe. Jack also spent one week in a submarine at a depth of 1,987 feet. How much deeper did Jack go in the bathyscaphe?

You can subtract to compare two numbers.

Find 4,526 − 1,987.

First estimate the difference.

```
  4,526  ⟶    5,000
− 1,987  ⟶  − 2,000
            ───────
              3,000
```

Regroup. Subtract the ones.	Regroup. Subtract the tens.	Regroup. Subtract the hundreds.	Subtract the thousands.
$\begin{array}{r}\;^{1}\;^{16}\\4,5\,2\,6\\-\,1,9\,8\,7\\\hline 9\end{array}$	$\begin{array}{r}^{4}\;^{11}\;^{16}\\4,5\,2\,6\\-\,1,9\,8\,7\\\hline 3\,9\end{array}$	$\begin{array}{r}^{3}\;^{14}\,^{11}\,^{16}\\4,5\,2\,6\\-\,1,9\,8\,7\\\hline 5\,3\,9\end{array}$	$\begin{array}{r}^{3}\;^{14}\,^{11}\,^{16}\\4,5\,2\,6\\-\,1,9\,8\,7\\\hline 2,5\,3\,9\end{array}$

The bathyscaphe went 2,539 feet deeper than the submarine.

The answer is reasonably close to the estimate.

Other examples:

```
  813 15                12 15              11 13 12
  8,9 4 5            4 2 5 10            5 X 3 2 12
               1 3 5,3 6 0          $8,6 2 4.3 2
− 3,2 6 7      −   1 2,4 7 5         −  4,3 5 8.6 7
─────────      ───────────          ─────────────
  5,6 7 8        1 2 2,8 8 5           $4,2 6 5.6 5
```

Checkpoint Write the letter of the correct answer.

Subtract.

1.   5,684
  − 3,275

2.   7,508
  − 246

3. $6,557.44 − $2,239.66

a. 1,419	**a.** 7,262	**a.** $4,317.78
b. 2,409	**b.** 7,342	**b.** $4,230.00
c. 2,419	**c.** 7,362	**c.** $4,238.88
d. 23,109	**d.** 7,754	**d.** $4,322.22

Find the difference.

1. 423
  − 264

2. 345
  − 178

3. 743
  − 287

4. 605
  − 283

5. $8.35
  − 5.49

6. 6,832
  − 2,754

7. 3,512
  − 1,674

8. 4,320
  − 2,753

9. 7,543
  − 4,786

10. $36.74
  − 16.98

11. 27,684
  − 16,895

12. 75,074
  − 56,847

13. 32,761
  − 18,775

14. 42,635
  − 24,067

15. $678.41
  − 489.53

16. 689,315
  − 149,878

17. 961,332
  − 710,556

18. 781,121
  − 526,573

19. 864,218
  − 439,629

20. $5,472.31
  − 1,385.85

21. 47,083
  − 6,529

22. 13,529
  − 9,468

23. 129,863
  − 49,127

24. 457,208
  − 70,651

25. $294.63
  − 97.25

26. 3,123 − 373

27. 58,166 − 2,628

28. 516,915 − 376,314

29. 59,656 − 2,468

30. 764 − 598

31. $245.72 − $88.73

Solve.

32. Jack read that in 1952, a submarine made an unsuccessful attempt to sail under the North Pole. That feat was completed in 1958 by the submarine U.S.S. *Nautilus*. How many years after the first attempt were people able to sail under the North Pole?

33. Jack reads a report that during the last ten years, the population of Bryde's whales has decreased by 6,363. Only 33,637 Bryde's whales remain. How many Bryde's whales were there ten years ago?

CHALLENGE

Copy these dots on a sheet of paper. Without lifting your pencil, draw four straight lines that pass through all nine dots.

Subtracting Across Zeros

Scientists keep track of whale populations. The right whale is a protected species. In 1976, scientists counted 3,073 right whales in the world's oceans. Recently, they counted 4,102. By how much did the count increase?

Subtract 4,102 − 3,073.

Regroup tens. There are no tens to regroup. So, regroup hundreds.	Regroup tens.	Subtract.
$\begin{array}{r} {}^{0\ 10}\ 4{,}1\not{0}\,2 \\ -\ 3{,}0\,7\,3 \\ \hline \end{array}$	$\begin{array}{r} {}^{9}\ {}^{0\ \not{10}\,12}\ 4{,}1\not{0}\,\not{2} \\ -\ 3{,}0\,7\,3 \\ \hline \end{array}$	$\begin{array}{r} {}^{9}\ {}^{0\ \not{10}\,12}\ 4{,}1\not{0}\,\not{2} \\ -\ 3{,}0\,7\,3 \\ \hline 1{,}0\,2\,9 \end{array}$

The right whale population has increased by 1,029.

Add to check your answer.

$$\begin{array}{r} 1{,}029 \\ +\ 3{,}073 \\ \hline 4{,}102 \end{array}$$

Other examples:

$$\begin{array}{r} {}^{9}\ {}^{2\ \not{10}\,14}\ \not{3}\not{0}\,4 \\ -\ \ \ 2\,6 \\ \hline 2\,7\,8 \end{array} \qquad \begin{array}{r} {}^{9\ 9}\ {}^{3\ \not{10}\,\not{10}\,17}\ \$\not{4}\not{0}.\not{0}\,7 \\ -\ \ \ 2\,8.3\,9 \\ \hline \$1\,1\,.6\,8 \end{array} \qquad \begin{array}{r} {}^{13\ 9}\ {}^{8\ \not{3}\ \not{10}\,10}\ 5\not{9}\,4{,}\not{0}\not{0}\,8 \\ -\ 2\,4\,6{,}2\,2\,7 \\ \hline 3\,4\,7{,}7\,8\,1 \end{array}$$

Checkpoint Write the letter of the correct answer.

Subtract.

1. $\begin{array}{r} 500 \\ -\ 376 \\ \hline \end{array}$	2. $\begin{array}{r} \$70.05 \\ -\ \ 36.79 \\ \hline \end{array}$	3. $\begin{array}{r} 6{,}400 \\ -\ 3{,}725 \\ \hline \end{array}$	4. $\begin{array}{r} 460{,}052 \\ -\ 235{,}766 \\ \hline \end{array}$
a. 124	**a.** $14.36	**a.** 2,675	**a.** 214,286
b. 134	**b.** $33.26	**b.** 2,685	**b.** 224,286
c. 200	**c.** $34.26	**c.** 2,700	**c.** 234,286
d. 224	**d.** $43.26	**d.** 3,675	**d.** 235,714

Find the difference.

1.	2.	3.	4.	5.
700 − 212	906 − 527	700 − 27	800 − 327	$3.00 − 0.87

6.	7.	8.	9.	10.
4,704 − 2,439	51,007 − 7,668	27,103 − 17,195	$150.00 − 147.19	$400.60 − 384.78

11.	12.	13.	14.	15.
623,400 − 551,982	700,000 − 632,757	$6,012.00 − 2,759.28	$2,010.01 − 1,754.68	260,005 − 199,278

16. 500 − 483   17. 5,304 − 418   18. 20,055 − 16,739   19. 340,781 − 228,627

20. 278,005 − 119,259   21. $7,050.70 − $2,193.84   ★22. $9,012.21 − $64.79

★23. 340,051 − 26,834   ★24. $4,010.25 − $38.49   ★25. 700,085 − 24,296

Solve. For Problem 27, use the Infobank.

26. Blue whales are the largest creatures on Earth. A blue-whale calf can weigh 2,873 pounds at birth. It weighs 4,000 pounds at the end of one week. How much weight does it gain in one week?

27. Use the information on page 415 to write and solve two of your own word problems.

★28. Scientists know that a whale's favorite food is squid. A 60,000-pound whale will often dive 3,000 feet to find squid. If a giant squid weighs 446 pounds, how much heavier is the whale?

CHALLENGE

What are the next two numbers in the pattern?

1.
5	34	63		

2.
7	33	59		

3.
14	47	80		

4.
320	345	370		

PROBLEM SOLVING
Choosing the Operation

If you read a problem carefully, you may get clues to help you solve the problem. The clues will help you decide whether you should add or subtract.

> There were 4 marine biologists in the research group studying the Beluga whales. When a new grant was approved, 5 more scientists were hired. How many scientists are there in the research group?

Hints:

If you know	and you want to find	you can
• how many there are in two or more groups	how many in all	add.
• how many there are in one group • how many join it	the total number	add.
• how many there are in one group • how many there are in a second group	how many there are in a second group	add.
• how many there are in one group	how many are left	subtract.
• the number taken away	how much larger one group is than the other	subtract.
• how many there are in each of two groups • how many there are in one group • how many there are in part of the group	how many there are in the remaining part of the group	subtract.

Once you have decided, you can solve the problem.

how many are in one group		how many join it		the total number
4	+	5	=	9

There are 9 scientists in the research group.

Write the letter of the operation you would use to solve the problem.

1. A Greek ship had 60 oars. The captain wanted it to go faster. Builders added 34 more oars. How many oars did the ship have then?

a. add   **b.** subtract

2. A British ship arrived in Tahiti with 45 crew members. Some of the crew members liked the island so much that they stayed. The ship left with 28 crew members. How many stayed in Tahiti?

a. add   **b.** subtract

Solve.

3. A Roman grain ship could haul 800 tons of grain. If workers loaded the ship with 340 tons of grain in one day, how many more tons would they have to load to fill the ship?

4. The *Windjammer* sails into an island bay in the Caribbean. The chain for its anchor is 98 feet long, but the anchor cannot reach the ocean floor. The crew adds 53 more feet of chain to the anchor. How long is the chain?

5. In 1816, passenger ships like the *Blue Star* crossed the Atlantic from Liverpool, England, to New York in 40 days. In 1860, the ship *Andrew Jackson* completed the trip in 15 days. How many more days did it take the *Blue Star* to make the trip?

6. The ancient Egyptians sailed the Nile in boats made of reeds. Some of the boats were 50 feet long. The Vikings sailed the seas in boats as much as 80 feet long. How much larger were the Viking boats than those sailed by the Egyptians?

★7. A ship leaves Calcutta, India, carrying 635 crates of spice. The ship stops to trade in ports in Africa. When it reaches Lagos, Nigeria, it has sold 389 crates of spice. It also takes on 245 more crates of spice. How many crates of spice are on board when the ship sails from Lagos?

★8. A certain kind of boat rides on a cushion of air above the water. It has room for up to 254 passengers and 30 automobiles. If there are 134 passengers and 25 automobiles already on board the boat, how many more autos and passengers can it carry?

CALCULATOR

Study the following example. Write each digit in the correct place in the numeral. Then press the correct calculator keys and read the display upside down. What word do you see?

Example: 7 hundreds, 0 ones, 1 tens   **Numeral:** 710

Calculator Keys: [7] [1] [0]

Display: OIL

Copy and complete the table.

	Numeral	Calculator Keys	Display
1. 5 ones 3 hundreds 7 thousands 4 tens 7 ten-thousands	⬛⬛,⬛⬛⬛	☐☐☐☐☐	⬛
2. 3 ten-thousand 8 ones 3 millions 9 hundred-thousand 3 tens 5 hundreds 1 thousand	⬛,⬛⬛⬛,⬛⬛⬛	☐☐☐☐☐☐☐	⬛

You can also use a calculator to find sums and differences that spell out simple sentences. Follow the example then complete the table.

		Calculator Keys	Display
EXAMPLE	518,067 + 12,867 530,934	[5][1][8][0][6][7] [+][1][2][8][6][7][=]	**HE GOES**
3.	55,145,632 + 2,589,713	☐☐☐☐☐☐☐☐ ☐☐☐☐☐☐☐☐☐	⬛
4.	95,746,215 − 18,594,870	☐☐☐☐☐☐☐☐ ☐☐☐☐☐☐☐☐☐☐	⬛

GROUP PROJECT

"And the Winner is . . ."

The problem: You won the Lucky Travelers Contest. You have a choice of spending a week on a tropical island or in a big city of your choice, with all travel and housing expenses paid. To help you make this decision, draw up for each place a schedule that lists what you would do in one day.

Key Facts

*Things to do
on a tropical island*

- swimming
- sailing
- snorkeling
- collecting seashells
- lying on the beach
- fishing

*Things to do
in a big city*

- sightseeing
- going to museums
- eating at fancy restaurants
- going to sporting events
- seeing plays
- shopping

SCHEDULE

Island Paradise		Dream City	
8:00		8:00	
9:00		9:00	
10:00		10:00	
11:00		11:00	
12:00		12:00	
1:00		1:00	
2:00		2:00	
3:00		3:00	
4:00		4:00	
5:00		5:00	
6:00		6:00	
7:00		7:00	
8:00		8:00	

CHAPTER TEST

Write in expanded form. (page 2)

1. 243,758

Write in standard form. (pages 2 and 4)

2. 50,000 + 2,000 + 900 + 80 + 2

3. 432 million, 348 thousand, 300

Write the short word name for each. (page 4)

4. 37,453,295,432

Write the value of the blue digit. (page 4)

5. 375,060　　　　**6.** 221,039　　　　**7.** 42,304,676,532　　　　**8.** 328,753,021

Compare. Use >, <, or = for ●. (page 6)

9. 323,650 ● 232,567

10. 6,327 ● 7,632

Order from the least to the greatest. (page 6)

11. 3,057,339; 3,958; 3,032

Order from the greatest to the least. (page 6)

12. 2,473; 22,468; 22,472; 2,508

Estimate using front-end estimation. (page 14)

13.
```
    34
   793
   157
 + 124
```

14.
```
 $6.23
  5.49
  0.75
+ 3.04
```

Estimate. Use rounding. (page 16)

15.
```
  7,615
+ 6,523
```

16.
```
  4,623
− 2,157
```

Round to the nearest hundred, thousand, and ten thousand. (page 16)

17. 25,639　　　　**18.** 32,898

Round to the nearest ten dollars. (page 16)

19. $65.80　　　　**20.** $83.08

Add. (pages 20, 22, and 24)

21.
```
  $5.62
+  7.39
```

22.
```
  4,382
+ 2,634
```

23.
```
  52,757
+ 29,039
```

24.
```
  5,778
  3,239
+    87
```

Subtract. (pages 28 and 30)

25.
```
  85
− 73
```

26.
```
  403
− 217
```

27.
```
 $7.43
− 0.65
```

28.
```
  3,200
− 1,576
```

29.
```
  239,865
− 177,598
```

Write the letter of the operation you would use to solve the problem. (pages 34 and 35)

30. Al buys a model of a blue whale for $5.25. Tom buys one for $3.72 more than that. How much does Tom spend?

 a. addition   **b.** subtraction

31. Al works 155 hours at the Aquarium. Tom works there 200 hours. How many more hours does Tom work than Al?

 a. addition   **b.** subtraction

Estimate using front-end estimation. (page 19)

32. At the aquarium, Jake counted 273 blowfish, 157 eels, and 439 anglefish. Estimate the total number of fish that Jake counted.

33. Pablo dives for oysters with his friends. They collect 239 oysters on Monday, 168 on Tuesday, 353 on Wednesday, and 411 on Thursday. About how many oysters do they collect?

BONUS

This place-value chart shows numbers to hundred trillions.

Trillions			Billions			Millions			Thousands			Ones		
hundred trillions	ten trillions	trillions	hundred billions	ten billions	billions	hundred millions	ten millions	millions	hundred thousands	ten thousands	thousands	hundreds	tens	ones
3	5	4	6	0	1	2	4	4	9	8	0	1	2	1

Read: 354 trillion, 601 billion, 244 million, 980 thousand, 121.
Write: 354,601,244,980,121.

Add or subtract. Then compare the answers.
Write >, <, or = for ●.

1.   2,375,987,020,505
   + 4,769,301,512,737

2.   9,301,586,000,121
   + 7,200,392,412,030

3.   357,680,294,027,598
   − 323,455,686,342,000

RETEACHING

When you subtract from a number that has zeros, you need to regroup before subtracting.

Subtract 6,104 − 2,066.

Regroup tens.
There are no tens to regroup.
So, regroup hundreds.

$$
\begin{array}{r}
^{0\ 10}\ \ \\
6,1\cancel{0}4 \\
-2,066 \\
\hline
\end{array}
$$

Regroup tens.

$$
\begin{array}{r}
^{9}\ \ \\
^{0}\ \cancel{10}14\ \\
6,1\cancel{0}4 \\
-2,066 \\
\hline
\end{array}
$$

Subtract.

$$
\begin{array}{r}
^{9}\ \ \\
^{0}\ \cancel{10}14\ \\
6,1\cancel{0}4 \\
-2,066 \\
\hline
4,038
\end{array}
$$

Other examples:

$$
\begin{array}{r}
^{4}\ ^{10}18\ \\
\cancel{5}\cancel{0}\cancel{8} \\
-459 \\
\hline
49
\end{array}
\qquad
\begin{array}{r}
^{9\ 9}\ \\
^{6}\ \cancel{10}\cancel{10}15\ \\
\$\cancel{7}\cancel{0}.\cancel{0}\cancel{5} \\
-19.27 \\
\hline
\$50.78
\end{array}
\qquad
\begin{array}{r}
^{12\ 9}\ \\
7,2\ ^{10}10\ \\
4\cancel{8}\cancel{3},\cancel{0}\cancel{0}7 \\
-135,116 \\
\hline
347,891
\end{array}
\qquad
\begin{array}{r}
^{8\ 101\ 10}\ \\
9\cancel{0},2\cancel{0}7 \\
-32,162 \\
\hline
58,045
\end{array}
$$

Subtract.

1. 600 − 289	**2.** 401 − 302	**3.** 300 − 284	**4.** 700 − 309	**5.** $4.00 − 2.20
6. 3,700 − 1,029	**7.** 5,006 − 2,249	**8.** 7,408 − 4,409	**9.** 1,000 − 804	**10.** $708.06 − 707.98
11. 57,000 − 20,600	**12.** 20,009 − 17,576	**13.** 70,209 − 70,199	**14.** 62,044 − 24,398	**15.** $800.70 − 532.47
16. 700,903 − 384,565	**17.** 423,000 − 403,823	**18.** 750,001 − 453,628	**19.** $2,000.60 − 970.44	

20. 8,000 − 2,345

21. 300 − 176

22. 62,000 − 42,551

23. 680,005 − 217,586

24. 33,005 − 999

25. $4,000.00 − $3,947.00

26. 1,204 − 376

27. 80,006 − 32,007

28. 402,600 − 393,691

ENRICHMENT

Roman Numerals

Hundreds of years ago, the Romans did not use numbers like those we use today. Instead, they used letters to represent amounts. Here are the Roman numerals for 1, 5, 10, 50, 100, 500, and 1,000.

I	V	X	L	C	D	M
1	5	10	50	100	500	1,000

To find what number a Roman numeral names, add.

$LXVII = L + X + V + I + I = 50 + 10 + 5 + 1 + 1 = 67$
$MDCCI = M + D + C + C + I = 1,000 + 500 + 100 + 100 + 1 = 1,701$

The Romans almost never wrote more than three of the same numerals in a row. When the numeral for a smaller number comes before the numeral for a larger number, subtract.

$IV = V - I = 5 - 1 = 4$ $CD = D - C = 500 - 100 = 400$
$MMXL = M + M + (X - L) = 1,000 + 1,000 + (50 - 10) = 2,040$

Often you can group Roman numerals to make them easier to add.

Write the number that is named by each.

1. III
2. LXV
3. XVI
4. MDCCCI
5. MCLXX
6. CCLXVII
7. MMCCLIX
8. CDX
9. MDCCLXII
10. MMMDXCI
11. CDII
12. MDCCXLIV

Write the Roman numeral for each.

13. 27
14. 112
15. 64
16. 647
17. 3,417
18. 956
19. 1,001
20. 1,249
21. 2,228
22. 1,599
23. 2,110
24. 1,898

Write the Roman numeral for

25. the year in which you were born.
26. the year in which your school was built.
27. the year in which you will graduate from high school.
28. the year Columbus came to America.
29. the year the Declaration of Independence was signed.

CUMULATIVE REVIEW

Write the letter of the correct answer.

1. What is the value of the blue digit?
530,727,654,300

  a. 700   **b.** 70,000
  c. 700,000,000   **d.** not given

2. Write in standard form:
500,000,000 + 70,000,000 + 1,000,000 + 70,000 + 1,000 + 80 + 1.

  a. 5,071,981   **b.** 50,071,981
  c. 500,071,071,071   **d.** not given

3. Compare. Choose >, <, or = for ●.
549,781 ● 354,892

  a. =   **b.** <
  c. >   **d.** not given

4. Estimate: 7,469 − 5,188.

  a. 1,000   **b.** 2,000
  c. 3,000   **d.** not given

5. 4,567 + 3,841 + 287

  a. 7,585   **b.** 8,695
  c. 8,595   **d.** not given

6. Round to the nearest thousand:
6,539.

  a. 5,000   **b.** 6,000
  c. 7,000   **d.** not given

7. 4,003 − 2,987

  a. 1,016   **b.** 2,984
  c. 2,126   **d.** not given

8. Write the word name: 407,300,799.

  a. 407 billion, 300 thousand, 799
  b. 407 million, 300 thousand, 799
  c. 407 billion, 300 million, 799
  d. not given

9. $374.53 + $594.48

  a. $968.01   **b.** $968.91
  c. $969.01   **d.** not given

10. Order from the greatest to the least:
62,089; 75,300; 63,411; 57,980.

  a. 75,300; 62,089; 63,411; 57,980
  b. 75,300; 63,411; 62,089; 57,980
  c. 57,980; 62,089; 63,411; 75,300
  d. not given

11. The Halpern Chicken Farm sells 738 cartons of eggs in 6 months. The Whiting Farm sells 287 more cartons than that. How many cartons of eggs does the Whiting Farm sell?

  a. 451   **b.** 1,025
  c. 1,225   **d.** not given

12. Of 267 acres of farmland, 98 are planted with wheat. Choose the operation to find the number of acres not planted with wheat.

  a. add   **b.** subtract
  c. compare   **d.** not given

In sports, athletes continually break world records. What do you think the world record for running 1 mile will be by the year 2000? What about the world record for the 100-meter freestyle in swimming for the same year? Before you begin, find out how the record has changed for each of the two events during the past 30 years.

2 PLACE VALUE, ADDITION AND SUBTRACTION
Decimals

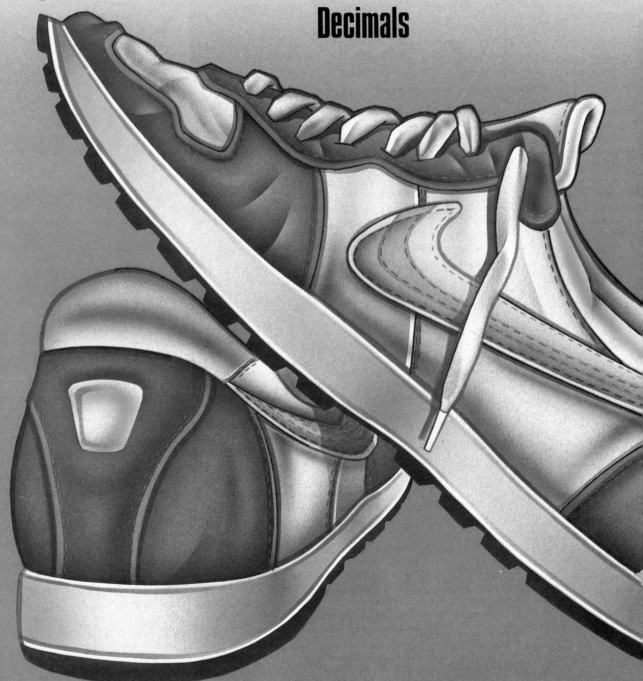

Tenths

A. There are ten rows of seats in one section at one Olympics event. Two of these rows are reserved for reporters and photographers. How can you write a number that describes the portion of rows that are reserved?

You can show the portion of reserved space in a picture. The square is divided into ten equal parts. Each part stands for one tenth. Two parts are shaded.

Ones	Tenths
0	2

Read: two tenths.
Write as a decimal: 0.2.
Write as a fraction: $\frac{2}{10}$.

Write 0 before the decimal point if there is no digit in the ones place.

Two tenths, or 0.2, of the rows are reserved.

B. You can write a decimal for a number greater than 1.

Ones	Tenths
1	4

Read: one and four tenths.
Write as a decimal: 1.4.
Write as a mixed number: $1\frac{4}{10}$.

Checkpoint Write the letter of the correct answer.

Complete.

1. Seven tenths is written as ▪.

   a. 0.07
   b. 0.7
   c. 0.71
   d. 7.10

2. Two and five tenths is written as ▪.

   a. 0.25
   b. 2.05
   c. 2.5
   d. 2 and 0.5

3. 90.1 is read as ▪.

   a. nine and one tenth
   b. ninety and one tenth
   c. ninety and one
   d. ninety-one

44

Write as a decimal.

1.

2.

3.

4. one tenth

5. one and six tenths

6. twelve and nine tenths

7. eight and one tenth

8. six tenths

9. three and four tenths

Copy this place-value chart. Write each decimal on the place-value chart.

10. 0.8   **11.** 1.3   **12.** 8.7   **13.** 2.3

14. 2.7   **15.** 4.5   **16.** 67.5   **17.** 15.4

18. 2.3   **19.** 34.7   **20.** 6.7   **21.** 16.8

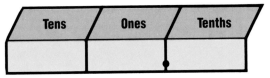

Write the word name for each decimal.

22. 0.3   **23.** 5.5   **24.** 10.3   **25.** 6.9   **26.** 11.1

27. 19.6   **28.** 8.2   **29.** 0.7   **30.** 17.6   **31.** 24.7

Solve.

32. Betty Cuthbert won the gold medal for the 200-meter dash in 1956. Her race time was twenty-three and four-tenths seconds. Write that number as a decimal.

33. In 1948, Mel Patton won the 200-meter dash in 20.1 seconds. If he had been one-tenth second faster, what would his time have been?

FOCUS: MENTAL MATH

Here is an addition shortcut that can help you add long columns of numbers.

$$\begin{array}{r} 34 \\ 25 \\ + 16 \\ \hline 75 \end{array}$$

Find numbers in the ones column that add up to 10 (4 + 6 = 10). Then finish adding the column: 10 + 5 = 15.

Compute. Look for tens.

1. 35 + 43 + 17   **2.** 28 + 56 + 82   **3.** 43 + 36 + 74   **4.** 15 + 39 + 85

5. 19 + 37 + 63   **6.** 51 + 32 + 98   **7.** 32 + 58 + 12   **8.** 24 + 63 + 23

Hundredths

A. There are 100 members on the Olympic team of one country. There are 48 women on the team. Write the number that describes the portion of women team members.

You can show this in a picture.
The square is divided into one hundred equal parts.
Each part stands for one hundredth.
Forty-eight parts are shaded.

Ones	Tenths	Hundredths
0	4	8

Read: forty-eight hundredths.
Write as a decimal: 0.48.   Write as a fraction: $\frac{48}{100}$.

The number that describes the portion of women team members is forty-eight hundredths, or 0.48.

B. You can write a decimal for a number greater than 1.

Hundreds	Tens	Ones	Tenths	Hundredths
1	2	3	5	8

Read: one hundred twenty-three and fifty-eight hundredths.
Write as a decimal: 123.58. Write as a mixed number: $123\frac{58}{100}$.

Checkpoint Write the letter of the correct answer.

Complete.

1. Twenty-nine hundredths is written as ▒.

   a. 0.029
   b. 0.209
   c. 0.29
   d. 29.100

2. 31.15 is read as ▒.

   a. thirty-one and fifteen hundredths
   b. thirty-one and fifty hundredths
   c. thirty-one and fifteen tenths
   d. thirty-one and fifteen

Write as a decimal.

1.

2.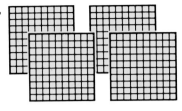

3.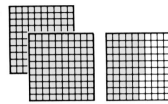

4. seventeen hundredths

5. six and two hundredths

6. forty and fifty-one hundredths

7. three and two hundredths

8. sixty-two hundredths

9. eleven and five hundredths

Copy this place-value chart. Write each decimal
on the place-value chart.

Hundreds	Tens	Ones	Tenths	Hundredths

10. 0.09

11. 2.19

12. 42.09

13. 14.12

14. 133.14

15. 320.08

16. 740.08

17. 69.73

18. 1.34

19. 87.03

20. 292.06

21. 100.97

Write the word name for each decimal.

22. 0.01

23. 0.38

24. 2.95

25. 17.63

26. 9.28

27. 3.05

28. 84.22

29. 532.74

Solve.

30. In 1972, Mark Spitz of the United States set the Olympic record for the men's 100-meter butterfly with a time of fifty-four and twenty-seven hundredths seconds. Write the record time as a decimal.

31. In 1984, Spitz's Olympic record in the 100-meter butterfly fell to West Germany's Michael Gross. The new record was fifty-three and eight-hundredths seconds. Write that as a decimal.

ANOTHER LOOK

Subtract.

1.   30
   − 17

2.   802
   − 105

3.   5,020
   − 2,528

4.   9,006
   − 2,528

5. $701.10
   − 565.98

Thousandths

A. A stopwatch can be used to measure the amount of time it takes to run a race. It can measure thousandths of a second. Look at the stopwatch. How would you write the word name for the decimal?

Ones	Tenths	Hundredths	Thousandths
0	7	7	3

Read: seven hundred seventy-three thousandths.
Write as a decimal: 0.773.
Write as a fraction: $\frac{773}{1000}$.

B. You can write a decimal for a number greater than 1.

Thousands	Hundreds	Tens	Ones	Tenths	Hundredths	Thousandths
3	8	2	1	8	7	5

Read: three thousand, eight hundred twenty-one and eight hundred seventy-five thousandths.
Write: 3,821.875.

C. You can write **equivalent** decimals that name the same number.

$$0.8 = 0.80 = 0.800 \qquad 2.6 = 2.60 = 2.600$$

Checkpoint Write the letter of the correct answer.

Complete.

1. Seven and forty-two thousandths is written as 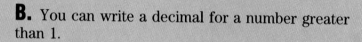.

2. 0.053 is read as ▪.

a. 0.742
b. 7.0042
c. 7.042
d. 7.420

a. fifty-three thousandths
b. fifty-three hundredths
c. fifty-three tenths
d. five and three thousandths

Write as a decimal.

1. six and thirty-one thousandths

2. twenty-three thousandths

3. six hundred forty-seven thousandths

4. ten and two thousandths

5. fifty-one and two thousandths

6. two hundred and one thousandth

Write the word name for each decimal.

7. 4.513   **8.** 0.606   **9.** 0.009   **10.** 0.112   **11.** 3.054

12. 0.031   **13.** 0.500   **14.** 225.620   **15.** 6.001   **16.** 1010.302

Write the value of the blue digit.

17. 0.005   **18.** 0.068   **19.** 0.321   **20.** 1.001   **21.** 0.809

22. 76.512   **23.** 100.298   **24.** 100.298   **25.** 600.606   **26.** 50.005

Solve.

27. Al Oerter won gold medals for the discus throw in four Olympics. His best throw was only 18 thousandths of a meter less than the world record. Write this as a decimal.

28. Each runner on a 400-meter relay team had an average time of 9.458 seconds. If each had run 0.002 seconds faster, what would be the average time of each runner?

FOCUS: MENTAL MATH

You can use what you know about place value to sort lists of decimals. The most important digit to look at is the digit to the right of the decimal point.

If the digit is 0 or 1, the decimal is close to 0.
If the digit is 4, 5, or 6, the decimal is close to half.
If the digit is 8 or 9, the decimal is close to 1.

The number of digits following this digit does not affect sorting.

Sort the decimals in the box into groups that are

0.94	0.436	0.51
	0.00046	
0.0987	0.11	0.8964

1. close to 0.   **2.** close to half.   **3.** close to 1.

More Practice, page 423

PROBLEM SOLVING
Using Broken-Line Graphs and Bar Graphs

A broken-line graph can show how something may have changed over a period of time.

The title states that this graph shows winning times for the men's 100-meter run.

The labels at the left tell you that the times are shown in seconds.

The labels at the bottom of the graph show you selected years of Olympic competition.

Each point on the graph shows the winning time for a selected year of competition.

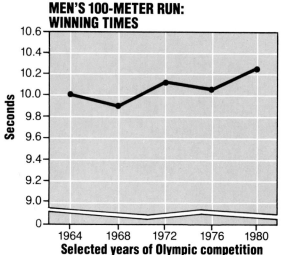

MEN'S 100-METER RUN: WINNING TIMES

A bar graph can help you compare information.

The title states that this graph compares women's 100-meter run times.

The labels at the left tell you that the winning times are shown in seconds. It is important to see that these numbers are not the same as those in the broken-line graph. Be careful when you compare information from different graphs.

The labels at the bottom show selected years of Olympic competition. It is important to see that the years for which information is given in the two graphs are not all the same.

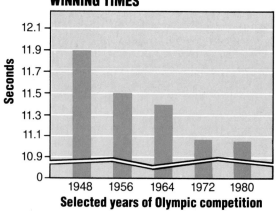

WOMEN'S 100-METER RUN: WINNING TIMES

A break in a graph means the information that would be there is unnecessary to complete the graph.

You can compare the information on the two graphs. The men's winning time in 1964 was 1.4 seconds faster than the women's winning time for that year.

Can you use the broken-line graph and the bar graph on page 50 to answer each question? Write *yes* or *no*.

1. What were the winning times for the men's and the women's 100-meter runs in 1952?

2. Whose winning time was faster in 1980, the men's or the women's?

Solve.

3. What was the first year in which the men's time was faster than 10.0 seconds?

4. What was the first year in which the women's time was less than 11.5 seconds?

5. Was the time run by the women's winner in 1972 faster than the time run by the men's winner in 1976?

6. Was the time run by the men's winner in 1976 fast enough to have won the race in 1968?

7. Was the time run by the women's winner in 1948 faster than the time run by the men's winner in the same year?

8. If the men's winning time in 1968 had been 0.1 second slower, what would it have been?

9. If the women's winner in 1964 had been 1 second faster, would her time have been faster than the men's winner in the same year?

10. In general, how have the times for both races changed over the years shown on the two graphs. Have they increased or decreased?

11. What are the differences between the information shown on the men's and the women's graphs?

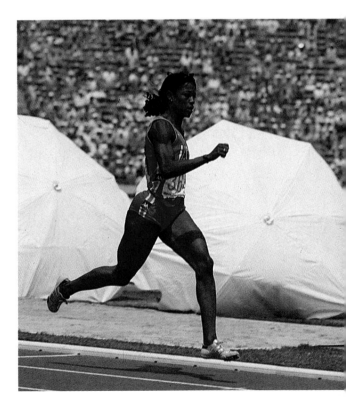

Comparing and Ordering Decimals

A. In the 1980 Olympics, Gerd Wessing of West Germany jumped 2.36 meters in the high-jump competition. Jörg Freimuth of West Germany jumped 2.31 meters. Which jump was higher?

A number line can help you compare decimals.

2.31 is to the left of 2.36 on the number line. So, 2.31 < 2.36.

B. You can compare decimals without using a number line.

Line up the decimal points.	Begin by comparing digits at the left.	Continue comparing.	Continue comparing.
2.36 **2.31**	**2.36** **2.31** 2 = 2	**2.36** **2.31** 3 = 3	**2.36** **2.31** 6 > 1

So, 2.36 > 2.31. The higher jump is 2.36 meters.

C. Sometimes you may need to write decimals as equivalent decimals before you compare.

Compare 0.4 and 0.43.   Think: 0.4 = 0.40.

Line up the decimal points.	Compare the tenths.	Compare the hundredths.
0.4 **0.43**	**0.40** **0.43** 4 = 4	**0.40** **0.43** 3 > 0

So, 0.43 > 0.4.

D. You can order the decimals 0.4, 1.04, and 1.024 by comparing them.

Line up the decimal points. Write equivalent decimals.	Begin to compare at the left.	Continue comparing.
0.400 **1.040** **1.024**	**0.400** **1.040** **1.024** Think: no ones. So, 0.400 is the smallest number.	**4 > 2** **1.040 > 1.024**

From the least to the greatest, the numbers are 0.4, 1.024, 1.04.
From the greatest to the least, the numbers are 1.04, 1.024, 0.4.

Compare. Write <, >, or = for ●.

1. 3.95 ● 0.946 **2.** 0.5 ● 0.352 **3.** 0.095 ● 0.14 **4.** 1.20 ● 1.2

5. 0.859 ● 0.806 **6.** 22.5 ● 22.9 **7.** 0.1 ● 0.10 **8.** 0.098 ● 0.31

9. 3.625 ● 3.025 **10.** 4.72 ● 7.72 **11.** 0.11 ● 0.011 **12.** 13.746 ● 1.376

13. 9.831 ● 91.8 **14.** 0.070 ● 0.700 **15.** 0.15 ● 0.150 **16.** 1.204 ● 1.240

Write in order from the least to the greatest.

17. 56.02; 0.56; 0.462 **18.** 2.582; 0.5; 2.52 **19.** 0.36; 6.31; 0.61

20. 5.9; 5.89; 6; 5.889 **21.** 0.6; 0.64; 1; 0.638 **22.** 1.079; 1.07; 1.6; 7

Write in order from the greatest to the least.

23. 5.98; 59.8; 0.673 **24.** 758.01; 75.801; 8.976 **25.** 0.325; 0.320; 0.239

26. 3.780; 3.761; 3.781 **27.** 67.563; 67.5; 67.82 **28.** 0.765; 7.659; 0.7659; 0.07

Solve.

29. June took 49.3 seconds to run the 400-meter race. Marcy ran the race in 47.9 seconds. Kate finished the race in 47.95 seconds. Who finished first? second? third?

30. Today the marathon is 42.195 kilometers. At one time it was 40.26 kilometers. Which distance is longer?

CHALLENGE

Who was the first American to break the sound barrier?

Copy the answer blanks and the decimals. Then match the decimals under them to those on the number line. Write the correct letter in the blank.

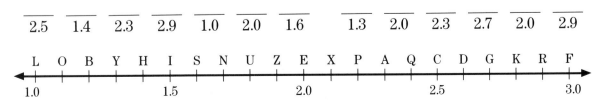

2.5   1.4   2.3   2.9   1.0   2.0   1.6   1.3   2.0   2.3   2.7   2.0   2.9

L   O   B   Y   H   I   S   N   U   Z   E   X   P   A   Q   C   D   G   K   R   F

1.0   1.5   2.0   2.5   3.0

Rounding Decimals

A. In the 1972 Olympics, two American women won the gold medal for the 100-meter freestyle swimming competition. They both swam the race in 55.92 seconds. What was their time to the nearest second?

Round 55.92 to the nearest whole number.

Look at the digit to the right of the ones place.
If the digit is 5 or greater, round up.
If the digit is less than 5, round down.   55.92 | 9 > 5 |
                                                   | Round up. |

55.92 rounded to the nearest whole number is 56.

Their time to the nearest second was 56 seconds.

B. To round to the nearest tenth, find the digit in the hundredths place, and compare it to 5.

Round 37.93 to the nearest tenth.

37.93 | 3 < 5 | So, 37.93 rounded to the
           | Round down. | nearest tenth is 37.9.

C. To round to the nearest hundredth, find the digit in the thousandths place, and compare it to 5.

Round 62.455 to the nearest hundredth.

62.455 | 5 = 5 | So, 62.455 rounded to the
           | Round up. | nearest hundredth is 62.46.

Round to the nearest whole number.

1. 4.54	**2.** 54.21	**3.** 19.71	**4.** 20.19	**5.** 38.5
6. 1.34	**7.** 13.759	**8.** 1.620	**9.** 346.04	**10.** 7.895
11. 12.87	**12.** 25.25	**13.** 49.49	**14.** 0.479	**15.** 0.765

Round to the nearest tenth.

16. 14.23	**17.** 0.39	**18.** 0.94	**19.** 34.11	**20.** 0.52
21. 0.567	**22.** 12.931	**23.** 23.257	**24.** 33.653	**25.** 0.150
26. 15.03	**27.** 34.5	**28.** 4.321	**29.** 0.845	**30.** 0.503

Round to the nearest hundredth.

31. 0.119	**32.** 12.501	**33.** 6.491	**34.** 15.455	**35.** 34.009
36. 12.915	**37.** 0.393	**38.** 45.563	**39.** 31.129	**40.** 0.014
41. 8.092	**42.** 7.931	**43.** 0.078	**44.** 0.706	**45.** 0.959

Solve. For Problem 49, use the Infobank.

46. In the 1972 Olympics, Mark Spitz won a gold medal for the United States in the 100-meter freestyle swimming event. His time was 51.22 seconds. What was his time to the nearest second?

47. In 1960, Robert Webster won a gold medal for the United States in high-board diving. His score was 165.56 points. In 1964, he won a second gold medal. His score was 148.58. In which year was his score higher?

48. In the 1980 Olympics, Martina Jäschke of West Germany won a gold medal in a diving event. Her score was 596.25. What was her score to the nearest tenth of a point? to the nearest whole number?

49. Use the information on page 415 to solve. Round each person's time to the nearest second and tenth of a second. Then make a chart that shows these results.

ANOTHER LOOK

Round to the thousands place. Estimate.

1. 4,565 + 3,465	**2.** 1,909 + 1,444	**3.** 9,858 − 6,983	**4.** 4,040 + 2,129	**5.** 9,575 − 9,281

Estimating Decimal Sums and Differences

A. As part of her training for the Olympic trials, Jennifer jogs four days each week. She records her distances on a training chart. Use Jennifer's training chart to estimate how many miles she jogged this week.

Day	Distance
Mon.	3.3 mi
Wed.	4.75 mi
Fri.	6.3 mi
Sun.	8.4 mi

Estimate 3.3 + 4.75 + 6.3 + 8.4.

You can estimate by rounding.

Round each number to the nearest whole number.

3.3 + 4.75 + 6.3 + 8.4
↓   ↓   ↓   ↓
3 + 5 + 6 + 8

Add the rounded numbers.

3 + 5 + 6 + 8 = 22

Jennifer jogged about 22 miles this week.

B. When you estimate, round to the place that allows you to mentally compute the rounded numbers.

Estimate $384.48 − $46.64.

Decide to which place value you will round.

To the nearest whole number:

$$\begin{array}{r} \$384.48 \rightarrow \$384 \\ -\quad 46.64 \rightarrow \quad 47 \\ \hline \end{array}$$

$384 − $47 is difficult to mentally compute; so, round to the nearest ten.

$$\begin{array}{r} \$384.48 \rightarrow \$380 \\ -\quad 46.64 \rightarrow -\ 50 \\ \hline \$330 \end{array}$$

You can see that $380 − $50 is easier to compute. So, the estimated difference of $384.48 − $46.64 is about $330.

Estimate. Write > or < for ●.

1. 6.38 + 1.96 ● 9
2. 15.75 + 4.39 ● 19
3. 2.764 + 1.09 ● 5
4. $15.89 + $1.97 ● $20
5. $39.88 + $7.89 ● $50
6. $13.36 + $5.55 ● $20
7. $8.59 − $2.67 ● $5
8. $20.00 − $14.79 ● $4
9. $34.86 − $26.99 ● $10
10. 11.275 − 1.87 ● 8
11. 26.75 − 15.891 ● 10
12. 1.295 − 1.198 ● 1
13. 8.6 + 4.273 + 9.01 + 3.75 ● 25
14. 19.835 + 5.6 + 4.9 + 27.61 ● 60

Estimate.

15.
```
    4.09
   14.315
 +  1.299
```

16.
```
    2.1794
   19.51
 + 36.453
```

17.
```
   93.694
 − 23.78
```

18.
```
   19.5
 − 8.876
```

Use the catalog of Olympic souvenirs and estimation to help you answer these questions.

19. Is $10 enough to buy B and D?
20. Is $20 enough to buy A, C, and D?
21. Is $10 enough to buy C and D?
22. Is $30 enough to buy C, D, and E?

OLYMPIC-SOUVENIRS CATALOG	
A. Medallion	$4.17
B. Bumper sticker	$5.49
C. Tote bag	$8.95
D. Socks	$4.49
E. Metal trophy	$15.85

Solve.

23. Olympic pole-vault records have risen in the past hundred years. In 1896, the best vault was 10.81 feet. In 1984, the best vault was 18.96 feet. About how many feet higher was the 1984 vault?

24. Sawao Kato scored 115.9 points at the 1968 Olympics in gymnastics. In 1972, he scored 114.65 points. About how many more points did he score in 1968 than in 1972?

MIDCHAPTER REVIEW

Write the value of the blue digit.

1. 7.012
2. 12.131
3. 6.793
4. 4.29
5. 0.704

Compare. Write >, <, or = for ●.

6. 3.625 ● 3.562
7. 0.6 ● 0.394

Round to the nearest hundredth.

8. 0.125
9. 3.006
10. 2.197

PROBLEM SOLVING
Estimation

Sometimes it is easier to solve a problem by estimating than by figuring an exact answer. First you must decide whether to overestimate or underestimate.

> To raise money for a new gym, the Linden School holds a "Sportathon." Tickets cost $2.75 for adults and $1.45 for children. Jean has $12.00. She wants to buy tickets for 1 adult and 1 child. She also wants to have at least $5.45 for lunch. Does she have enough money?

Since Jean wants to be sure she has enough money, she should overestimate her expenses. Round each value up to the next highest dollar.

Adult ticket: $2.75 $\longrightarrow$ $3.00
Child ticket: $1.45 $\longrightarrow$ $2.00
Lunch: $5.45 $\longrightarrow$ $6.00
                        $11.00 overestimated sum

Since $11 < $12, Jean has enough money.

When the gym committee feels they have earned the $1,750 they need, they will discount all prices. An hour before the Sportathon's end, the committee members will gather information about the amount of money earned. They need to know if they've earned enough so that they can begin to discount the prices.

Since they want to be sure that they've earned the money they need, they should underestimate their earnings. You would not need to calculate an exact sum. Round each value down to the next lowest hundred dollars.

School jackets: $931.70 $\longrightarrow$ $900
School shirts: $669.95 $\longrightarrow$ $600
School pennants: $281.50 $\longrightarrow$ $200
Program ads: $175.00 $\longrightarrow$ $100
                            $1,800 underestimated sum

Since $1,800 > $1,750, that is enough to begin to discount.

Explain why you would *underestimate* or *overestimate*.

1. Here is the coach's record for Doris.

   Day 1: 4.8 mi   Day 5: 4.2 mi
   Day 2: 3.6 mi   Day 6: 3.5 mi
   Day 3: 4.4 mi   Day 7: 3.2 mi
   Day 4: 3.9 mi

   Has Doris run at least 24 mi this week?

2. Alan Leibowitz is ordering the following equipment.

Parallel bars:	$119.99
12 mats:	$175.50
1 Vaulting horse:	$151.50

   Mr. Leibowitz has exactly $600. Does he have enough money to pay for his order?

Estimate to solve.

LINDEN SCHOOL—FIRST ROUND GYMNASTICS SCORES

	Floor exercise	Vaulting horse	Balance beam	Uneven bars
Nia:	6.275	7.95	7.9	6.5
Jenny:	6.875	7.25	5.875	6.88
Lisa:	6.75	8.0	5.5	5.5
Lois:	6.95	7.0	6.875	6.0
Karen:	5.75	7.0	6.0	6.0
Alexandra:	7.25	6.875	5.0	5.5

The top three scorers of the first round compete in a second round. The winner is the girl with the highest total for both rounds.

3. Which event is probably the most difficult? Which is probably the easiest?

4. Lois wanted to score at least 50 points for both rounds. Her second round scores are: 6.55, 7.2, 6.5, and 6.75. Did she reach her goal?

5. The Linden School competes against the Rickert School. Use Round 1 scores to find how many points Rickert must score in a round to be competitive in the floor-exercise event? (To figure the winner, the team members' scores are added together.)

6. The team from the Rickert School scores 45.65 points on the balance beam during the first round. They score 46.85 points during the second round. How many points must the Linden team score during Round 2 to win that event?

Addition of Decimals

A. Ingemar Stenmark, a Swedish skier, won a gold medal in the 1980 men's slalom. His first run was clocked at 53.89 seconds. His time for the second run was 50.37 seconds. What was his combined time?

Add 53.89 + 50.37.

Line up the decimal points. Add the hundredths. Regroup if necessary.	Add the tenths. Regroup if necessary.	Add the ones. Regroup if necessary.	Add the tens. Write the decimal point.
1	1 1	1 1	
5 3.8 9	5 3.8 9	5 3.8 9	5 3.8 9
+ 5 0.3 7	+ 5 0.3 7	+ 5 0.3 7	+ 5 0.3 7
6	2 6	4 2 6	1 0 4.2 6

His combined time was 104.26 seconds.

B. You may need to write equivalent decimals before adding.

Add 5.287 + 9.33 + 7.

Line up the decimal points. Add the thousandths.	Add the hundredths.	Add the tenths.	Add the ones. Write the decimal point.
	1	1	
5.2 8 7	5.2 8 7	5.2 8 7	5.2 8 7
9.3 3 0	9.3 3 0	9.3 3 0	9.3 3 0
+ 7.0 0 0	+ 7.0 0 0	+ 7.0 0 0	+ 7.0 0 0
7	1 7	6 1 7	2 1.6 1 7

9.33 = 9.330
7 = 7.000

Checkpoint Write the letter of the correct answer.

Add.

1. 34.8 + 71.23

a. 105.03
b. 105.21
c. 105.31
d. 106.03

2. $56.72 + $13.65

a. $60.37
b. $69.37
c. $70.37
d. $71.37

3. 4.19 + 0.782 + 47.8

a. 0.052172
b. 1.679
c. 52.772
d. 53.772

Add.

1. 10.12
+ 32.26

2. 4.30
+ 0.473

3. 25.231
+ 50.136

4. 21.138
+ 34.451

5. $6.22
+ 2.35

6. 3.19
+ 9.9

7. 2.474
+ 75.84

8. 9.1
+ 4.634

9. 6
+ 8.23

10. $9.74
+ 8.53

11. 3.072
8.653
+ 9.048

12. 7.651
1.492
+ 8.648

13. 7.00
25.67
+ 0.10

14. 0.1
0.2
+ 9.8

15. $0.08
2.47
+ 0.52

16. 2.34 + 12.1

17. 15.46 + 8.932

18. 756.9 + 0.78

19. 53.85 + 0.397

20. 1.374 + 85.066

21. $7.07 + $6.42

22. 19.765 + 9.53

23. 2.34 + 0.765 + 3.876

★24. (7.4 + 0.920) + 61

★25. 0.48 + (58.675 + 9.33)

Solve.

26. In 1980, Eric Heiden won the men's speed-skating event with a time of 38.03 seconds. Karin Enke won the women's event with a time of 41.78 seconds. Estimate to the nearest second how much faster Heiden was.

27. At the 1984 Olympics, Valerie Brisco-Hooks won the women's 200-meter run with a time of 21.81 seconds. She also won the women's 400-meter run in 48.83 seconds. What was her total time for the two events?

28. In a slalom, a skier had a time for her first run of 55.78 seconds and a time for her second run of 56.97 seconds. What was her combined time?

★29. In a 100-meter swimming relay, four swimmers had these times: 70.4 seconds, 62.4 seconds, 66.8 seconds, 59.6 seconds. Was their combined time faster or slower than 4 minutes?

CHALLENGE

Find the missing digits.

1. 5.▧
+ ▧.4
——
8.2

2. 4.▧3
+ ▧.5▧
——
8.02

3. 8.▧▧4
+ ▧.76▧
——
9.021

4. 5.▧▧3
+ ▧.65▧
——
13.642

Subtracting Decimals

A. At the 1976 Winter Olympics, Sheila Young of the United States set an Olympic record in the 500-meter speed-skating competition. Her time was 42.76 seconds. In a later competition, she finished the same race in 40.68 seconds. How much faster was her time in the later competition?

Find the difference: 42.76 − 40.68.

Line up the decimal points. Subtract the hundredths. Regroup if necessary.	Subtract the tenths. Regroup if necessary.	Subtract the ones and the tens. Write the decimal point.
$\begin{array}{r} \overset{6\ 16}{4\,2.7\,\cancel{6}} \\ -\,4\,0.6\,8 \\ \hline 8 \end{array}$	$\begin{array}{r} \overset{6\ 16}{4\,2.7\,\cancel{6}} \\ -\,4\,0.6\,8 \\ \hline 0\,8 \end{array}$	$\begin{array}{r} \overset{6\ 16}{4\,2.7\,\cancel{6}} \\ -\,4\,0.6\,8 \\ \hline 2.0\,8 \end{array}$

Her time was 2.08 seconds faster.

B. You may need to write equivalent decimals before subtracting.

Find 3.7 − 3.25.

Line up the decimal points.	Subtract the hundredths.	Subtract the tenths.	Subtract the ones. Write the decimal point.
$\begin{array}{r} 3.7\,0 \\ -\,3.2\,5 \\ \hline \end{array}$ $\boxed{3.7 = 3.70}$	$\begin{array}{r} \overset{6\ 10}{3.7\,\cancel{0}} \\ -\,3.2\,5 \\ \hline 5 \end{array}$	$\begin{array}{r} \overset{6\ 10}{3.7\,\cancel{0}} \\ -\,3.2\,5 \\ \hline 4\,5 \end{array}$	$\begin{array}{r} \overset{6\ 10}{3.7\,\cancel{0}} \\ -\,3.2\,5 \\ \hline 0.4\,5 \end{array}$ $\boxed{\text{This 0 must be written.}}$

Checkpoint Write the letter of the correct answer.

Subtract.

1. 3.4 − 2.61

2. $2.35 − $0.98

3. 5.345 − 0.468

1.	2.	3.
a. 0.79	**a.** $1.37	**a.** 2.987
b. 0.81	**b.** $2.47	**b.** 4.877
c. 1.21	**c.** $2.63	**c.** 5.813
d. 6.01	**d.** $3.33	**d.** 5.987

Subtract.

1. $\begin{array}{r} 9.78 \\ -\ 2.68 \end{array}$ 2. $\begin{array}{r} 7.79 \\ -\ 5.14 \end{array}$ 3. $\begin{array}{r} 6.89 \\ -\ 5.24 \end{array}$ 4. $\begin{array}{r} 92.34 \\ -\ 1.14 \end{array}$ 5. $\begin{array}{r} \$29.99 \\ -\ 1.92 \end{array}$

6. $\begin{array}{r} 7.98 \\ -\ 5.59 \end{array}$ 7. $\begin{array}{r} 7.34 \\ -\ 6.26 \end{array}$ 8. $\begin{array}{r} 67.31 \\ -\ 52.17 \end{array}$ 9. $\begin{array}{r} \$70.21 \\ -\ 36.42 \end{array}$ 10. $\begin{array}{r} \$572.92 \\ -\ 383.94 \end{array}$

11. $\begin{array}{r} 25.04 \\ -\ 7.16 \end{array}$ 12. $\begin{array}{r} 54.08 \\ -\ 5.04 \end{array}$ 13. $\begin{array}{r} 62.42 \\ -\ 5.90 \end{array}$ 14. $\begin{array}{r} \$36.00 \\ -\ 5.65 \end{array}$ 15. $\begin{array}{r} \$644.90 \\ -\ 4.95 \end{array}$

16. $\begin{array}{r} 80.6 \\ -\ 58.94 \end{array}$ 17. $\begin{array}{r} 60.34 \\ -\ 13.8 \end{array}$ 18. $\begin{array}{r} 7.73 \\ -\ 5.65 \end{array}$ 19. $\begin{array}{r} 7.31 \\ -\ 0.92 \end{array}$ 20. $\begin{array}{r} \$766.61 \\ -\ 397.84 \end{array}$

21. $6 - 0.08$ 22. $67.9 - 0.39$ 23. $18.8 - 8.27$ 24. $58.3 - 0.03$

Solve.

25. Thomas Burke of the United States was the winner of the 100-meter dash in the first modern Olympics held in 1896. Burke's time was 12.0 seconds. In 1980, Allen Wells won the same race in 10.15 seconds. Whose time was faster? How much faster?

★26. In the 400-meter relay race, four runners race legs of 100 meters each. If the first two runners took a total of 23.36 seconds, the third took 10.33 seconds, and the total time raced was 43.86 seconds, how much time did the last runner take?

FOCUS: REASONING

Some of Sally's friends are John's friends. All of Ron's friends are Sally's friends. From these statements, what can we say about John's friends?

More Practice, page 424

PROBLEM SOLVING
Writing a Number Sentence

When you have trouble figuring out how to solve a problem, try writing a number sentence. A number sentence can help you decide how to use the numbers you know to find the number you need.

> In 1980, Ludmila Kondratyeva of the Soviet Union won the women's 100-meter run. Her time was 11.6 seconds. In 1984, Evelyn Ashford, of the United States, won the 100-meter run in 10.97 seconds. How much faster than Kondratyeva did Ashford run the race?

1. List what you know and what you want to find.

Kondratyeva's time was 11.6 seconds. Ashford's time was 10.97 seconds.

How much faster was Ashford's time?

2. Think about how to use this information to solve the problem.

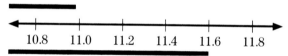

You know how many there are in two groups. You want to find how much larger one group is than another. You can subtract.

3. Write a number sentence. Use n to stand for the number you want to find.

Kondratyeva's − Ashford's = How
time   time   much
                         faster
                         Ashford
                         ran

11.6   −   10.97   = n

4. Solve. Write the answer. Ashford ran 0.63 seconds faster than Kondratyeva.

$$11.6 - 10.97 = 0.63$$
$$n = 0.63$$

Write the letter of the correct number sentence.

1. A diver scored 421.62 points in a platform-diving competition. Another diver scored 8.48 fewer points. What was the second diver's score?

   a. $421.62 + 6.48 = n$
   b. $421.62 - 8.48 = n$
   c. $421.62 + n = 8.48$

2. A skier made 2 slalom runs. His time for the first was 54.67 seconds. His time for the second was 52.48 seconds. What was his combined time?

   a. $54.67 + 52.48 = n$
   b. $54.67 - 52.48 = n$
   c. $52.48 + n = 54.67$

Write a number sentence. Solve.

3. In 1984, Valerie Brisco-Hooks ran the 400-meter race in 21.81 seconds. At the 1980 Olympics, the winning time was 0.22 seconds slower. What was the winning time in 1980?

4. The winning men's high jump in 1984 was 7.71 feet. In 1980, the winning jump was 7.73 feet. In which year was the winning jump higher? How much higher was the winning jump?

5. In 1984, Mary Meagher beat the old 100-meter butterfly swimming record by 1.16 seconds. The old record was 60.42 seconds. What was Mary's time?

6. At the twenty-third Summer Olympics, the United States won 83 gold medals, 61 silver medals, and 30 bronze medals. How many medals did the United States win?

7. One skier scored 225.8 points in ski jumping. Another skier scored 3.6 points higher. What was the second skier's score?

8. One springboard diver scored 503.98 points. A second diver scored 26.09 fewer points. What was the second diver's score?

★9. In running the hurdles, the first runner had a time of 47.79 seconds. The next runner's time was 0.08 seconds slower. The third runner was 1.3 seconds slower than the second runner. What was the third runner's time?

★10. A four-man team entered the 400-meter freestyle relay. The first three swimmers' individual times were 57.31 seconds, 54.83 seconds, and 58.46 seconds. The team's combined score was 222.67 seconds. What was the fourth swimmer's time?

LOGICAL REASONING

Sometimes you can solve a problem by eliminating every possible answer except one.

Mr. and Mrs. Jeffrey had a tennis tournament with their friends, Mr. and Mrs. Kaplan, Mr. and Mrs. Lamson, and Mr. and Mrs. Mossetto. Each player had one partner. Who was Mrs. Jeffrey's partner?

Clues: **a.** Each pair had a man and a woman.
b. Husbands and wives were not partners
c. Mr. Lamson played with Mrs. Kaplan.
d. Mr. Jeffrey's sister is Mrs. Mossetto, but they did not play together.

Copy the table. Write *no* for each pair that is not possible. For each correct pair that you find, write *no* for the remaining pairs that are no longer possible.

	Mrs. Jeffrey	Mrs. Kaplan	Mrs. Lamson	Mrs. Mossetto
Mr. Jeffrey	no	no		no
Mr. Kaplan		no		
Mr. Lamson	no	yes	no	no
Mr. Mossetto		no		no

1. Who was Mr. Jeffrey's partner?

2. Who was Mrs. Mossetto's partner?

3. Who was Mrs. Jeffrey's partner?

Mr. Andrews, Ms. Baker, Mrs. Carson, Mr. Drew, and Ms. Early each have a different occupation. They work as an electrician, a plumber, a teacher, a pilot, and a doctor. Who is the pilot?

Use the clues and make your own chart to solve.

Clues: **a.** Mr. Drew is on a bowling team with the teacher, the pilot, and Ms. Early.
b. Mrs. Carson's brother is the plumber.
c. Ms. Baker is the electrician.
d. Mr. Andrews is older than the teacher.

GROUP PROJECT

Making a Time Line

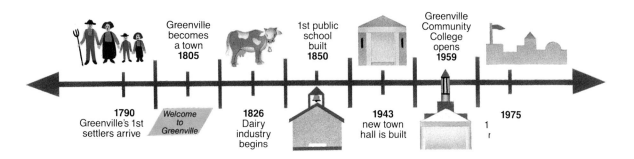

The problem: You and your classmates want to make a time line that shows the history of your town. A time line is a number line that shows dates of important events. Decide which events you want to include. Use the time line on this page as a model to make a time line of your town's history.

You might want to include events such as these:

- the date of the founding of your town
- the dates that famous people lived in your town
- when each public building was constructed
- when industries moved into town
- when your town was named
- when the first settlers moved to your area

CHAPTER TEST

Write as a decimal. (pages 44, 46, and 48)

1. two and three tenths

2. six hundredths

3. three thousandths

Write the word name for each decimal. (pages 44, 46, and 48)

4. 0.23

5. 6.357

Write the value of the blue digit. (pages 44, 46, and 48)

6. 0.7

7. 0.56

8. 2.031

Compare. Write >, <, or = for ●. (page 52)

9. 2.35 ● 2.350

10. 72.378 ● 72.37

Order from the least to the greatest. (page 52)

11. 25.03; 23.05; 25.35

Round to the nearest whole number, tenth, and hundredth. (page 54)

12. 13.258

13. 12.178

Estimate. Write > or < for ●. (page 56)

14. 5.23 + 3.76 ● 8

15. 25.323 + 17.282 ● 44

16. 12.65 − 7.87 ● 5

17. 35.36 − 5.93 ● 29

Add. (page 60)

18.   2.23
        + 4.2

19.   53.6
        + 1.539

20.   2.52
        + 0.475

21. $5.25
        + 0.83

22. 4.35 + 0.4 + 3.7

23. 3.251 + 4.73 + 8.9

Subtract. (page 62)

24.   2.37
        − 1.29

25.   50.53
        − 47.61

26. $73.83
        − 59.92

27.   47.7
        − 9.05

28. 6 − 0.08

29. 18.8 − 8.27

Write a number sentence and solve. (pages 64 and 65)

30. Joe runs the 440-yard dash in 73.56 seconds. Sam's time is 1.72 seconds slower than Joe's time. What is Sam's time?

31. Shirley runs the 220-yard dash in 37.47 seconds. Pat's time is 2.3 seconds faster than Shirley's time. What is Pat's time?

Use the information from the bar graph to solve. (pages 50 and 51)

32. In which year was the 440-yard dash run the fastest?

33. Which years had the closest times in the 440-yard dash?

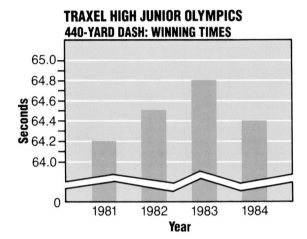

TRAXEL HIGH JUNIOR OLYMPICS
440-YARD DASH: WINNING TIMES

BONUS

This place-value chart shows numbers to ten-thousandths.

Ones	Tenths	Hundredths	Thousandths	Ten-thousandths
1	4	6	9	8

Decimal point

Read: one and 4,698 ten-thousandths.
Write 1.4698.

Add or subtract. Then order the answers from the least to the greatest.

1. 3.7126
     + 4.2352

2. 2.0967
     + 5.6384

3. 7.2635
     − 3.8036

4. 6.5938
     − 4.8709

5. 2.5352
     + 4.8658

RETEACHING

A. Sometimes you may need to write an equivalent decimal before you compare.

Compare 4.92 and 4.9.

Line up the decimal points. Think: 4.9 = 4.90.	Begin by comparing digits at the left.	Continue comparing.	Continue comparing.
4.92 4.90	4 = 4 4.92 4.90	9 = 9 4.92 4.90	2 > 0 4.92 4.90

2 > 0   So, 4.92 > 4.9.

You can see this on a number line.

B. You can order decimals by comparing them.
Compare and order 0.7, 2.08, and 2.069.

Line up the decimal points. Write equivalent decimals.	Begin to compare at the left.	Continue comparing.
0.700 2.080 2.069	0.700 Think: no ones. 2.080 So, 0.700 is the 2.069 smallest number.	8 > 6 2.080 > 2.069

From the least to the greatest, the numbers are 0.7, 2.069, and 2.08.
From the greatest to the least, the numbers are 2.08, 2.069, and 0.7.

Compare. Write >, <, or = for ●.

1. 2.762 ● 2.672   **2.** 53.62 ● 5.362   **3.** 0.372 ● 0.299

4. 9.251 ● 91.25   **5.** 0.040 ● 0.400   **6.** 271.1 ● 72.11

Write in order from the greatest to the least.

7. 2.32, 1.745, 9.33   **8.** 0.07, 0.70, 0.71   **9.** 3.14, 1.579, 4.01

Write in order from the least to the greatest.

10. 7.505, 57.7, 5.07   **11.** 0.002, 0.012, 0.2   **12.** 79.184, 74.189, 79.9

13. 6.112, 6.2, 6.02   **14.** 23.06, 23.07, 23.09   **15.** 4.5, 5.04, 0.54

ENRICHMENT

Checking Accounts

Fred keeps money in a checking account at the bank. He uses a **check register** to record the amounts of his checks and deposits. He sends a check for $21.98 to the Olympic Committee to buy an Olympic coin. Below is a page from Fred's check register.

ITEM NO.	DATE	DESCRIPTION OF TRANSACTION	(−) PAYMENT OR WITHDRAWAL	✓ T	( ) FEE	(+) DEPOSIT OR INTEREST	BALANCE	
							231	57
096	6/13	U.S. Olympic Comm	21 98				21	98
							209	59
	6/15	Birthday Check				25 00	25	00
							234	59
097	6/21	Ron's Hardware	12 97				12	97
							221	62

The *balance column* shows how much money Fred has in his account after he subtracts the amount of a check or adds the amount of a deposit.

Find the amounts of the missing checks, deposits, or balances.

	(−) PAYMENT OR WITHDRAWAL	✓ T	(−) FEE	(+) DEPOSIT OR INTEREST	BALANCE	
					313	25
	27 52				27	52
1.					■	■
				13 11	13	11
2.					■	■
				79 58	79	58
3.					■	■

	(−) PAYMENT OR WITHDRAWAL	✓ T	(−) FEE	(+) DEPOSIT OR INTEREST	BALANCE	
					215	50
				21 35	21	35
4.					■	■
	150 98				150	98
5.					■	■
	37 59				37	59
6.					■	■

On July 1, Fred's balance was $215.27. On July 2, he wrote a check to Harvey's Sock Shop for $13.22. On July 5, he deposited $55.63 in his account. Using this information, copy and complete the register below.

	ITEM NO.	DATE	DESCRIPTION OF TRANSACTION	(−) PAYMENT OR WITHDRAWAL	✓ T	(−) FEE	(+) DEPOSIT OR INTEREST	BALANCE	
								215	27
7.	031	■	■	■	■			■	■
8.								■	■
9.		■	■			■	■	■	■
10.								■	■

TECHNOLOGY

Here are some LOGO commands.

FD   This moves the turtle forward the number of steps shown.

BK   This moves the turtle backward.

RT   This makes the turtle turn to the right.

LT   This makes the turtle turn to the left.

HOME   This takes the turtle to the center of the screen.

PU   This moves the turtle without drawing a line.

PD   This tells the turtle to begin drawing lines again.

PE   This tells the turtle to erase a line.

A LOGO program is called a **procedure.** When you type a procedure's name, the turtle follows all the commands in the procedure. Two or more commands can be on a line. The last command in a procedure must be END.

If you draw a line that you wish to erase, use the command PE. Suppose you type FD 100, but meant to type FD 75. Type the command PE BK 25 PD. The turtle will move backward 25 steps and will erase that part of the line. Always remember to type PD after using PE. PD tells the turtle to stop erasing and to begin drawing again.

This line was drawn by the command FD 105.

_____

1. Write a command to make the line 130 steps long.

2. Write a command to change your new line to a length of 85 steps.

72

3. Write a procedure for drawing two lines, each one 30 steps long and 15 steps apart.

4. This square was drawn by the following procedure. Rewrite the procedure to turn the square into a rectangle.

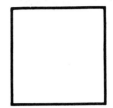

TO SQUARE
FD 40 RT 90 FD 40 RT 90
FD 40 RT 90 FD 40 RT 90
END

★5. This octagon was drawn by the following procedure. Write a procedure that would divide the octagon in half.

TO OCTAGON
FD 40 RT 45 FD 40 RT 45 FD 40 RT 45
FD 40 RT 45 FD 40 RT 45 FD 40 RT 45
FD 40 RT 45 FD 40 RT 45
END

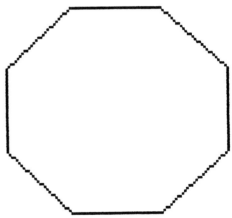

CUMULATIVE REVIEW

Write the letter of the correct answer.

1. Write in standard form:
   $500,000 + 60,000 + 4,000 + 300 + 9$.

   a. 564,300,009   **b.** 564,309
   c. 564,390   **d.** not given

2. Compare. Use $>$, $<$, or $=$ for ●.
   597,603 ● 597,630

   a. $=$ **b.** $<$
   c. $>$ **d.** not given

3. $597.07 + 789.24$

   a. $1,356.21   **b.** $1,386.31
   c. $1,376.31   **d.** not given

4. Round to the nearest ten thousand:
   56,000.

   a. 50,000   **b.** 60,000
   c. 58,000   **d.** not given

5. Estimate: $4,496 - 2,377$.

   a. 1,000   **b.** 3,000
   c. 4,000   **d.** not given

6. What is the value of the blue digit?
   9 58,789

   a. 5,000   **b.** 50,000
   c. 500,000   **d.** not given

7. Write in expanded form: 30,579.

   a. $3,000 + 500 + 70 + 9$
   b. $30,000 + 500 + 70 + 9$
   c. $300,000 + 500 + 70 + 9$
   d. not given

8. Write the short word name:
   507,313,896.

   a. 507 billion, 313 million, 896 thousand
   b. 507 million, 313 thousand, 896
   c. 507 billion, 313 million, 896
   d. not given

9. $7,807 - 6,968$

   a. 839   **b.** 965
   c. 1,949   **d.** not given

10. $5,798 + 6,055 + 940 + 11$

    a. 11,694   **b.** 12,704
    c. 12,804   **d.** not given

11. A commercial jet is flying at an altitude of 34,059 feet. An air-force jet is flying 7,670 feet higher than that. How high is the air-force jet flying?

    a. 26,389   **b.** 41,729
    c. 42,629   **d.** not given

12. An airport runway may be as long as 4,783 meters. The clear zone, over which the plane flies before touching ground, covers 823 meters of the total runway. Estimate how many meters are not clear zone.

    a. 4,000   **b.** 4,400
    c. 5,000   **d.** not given

Plan a white-water rafting trip down the Colorado River through the Grand Canyon. Use a map to decide where your stopping points will be. What kinds of food, supplies, and equipment will you take? How much hiking and rafting will you do? How will you organize your trip?

3 MULTIPLYING WHOLE NUMBERS

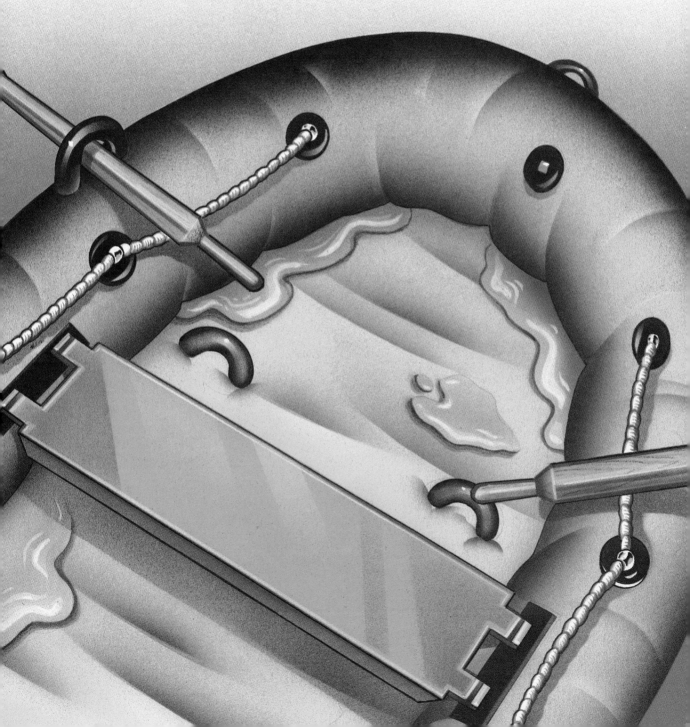

Multiplication Facts

A. Find 6 + 6 + 6.

You can add to find how many.

6 + 6 + 6

or

When the addends are the same, you can multiply.

3 × 6

B. Multiplication can be shown in two ways.

$$3 \times 6 = 18$$

factor   factor   product

or

$$
\begin{array}{r}
6 \leftarrow \text{factor} \\
\times\, 3 \leftarrow \text{factor} \\
\hline
18 \leftarrow \text{product}
\end{array}
$$

C. These properties of multiplication can help you find the products.

Commutative Property If the order of the factors is changed, the product remains the same.	2 × 8 = 16 8 × 2 = 16
Associative Property If the grouping of the factors is changed, the product remains the same.	(3 × 3) × 2 = 3 × (3 × 2) 9 × 2 = 3 × 6 18 = 18
Identity Property If one factor is 1, the product is always the other factor.	9 × 1 = 9 1 × 7 = 7
Property of Zero If one factor is 0, the product is always 0.	7 × 0 = 0 0 × 8 = 0
Distributive Property To find the product of a number times the sum of two addends, you can multiply each addend by the number and then add the products.	3 × (5 + 2) = (3 × 5) + (3 × 2) 3 × 7 = 15 + 6 21 = 21

Multiply.

1. 5 ×2	**2.** 9 ×3	**3.** 9 ×8	**4.** 6 ×8	**5.** 7 ×0
6. 9 ×6	**7.** 0 ×2	**8.** 1 ×4	**9.** 8 ×1	**10.** 4 ×3

11. 6×5 **12.** 5×9 **13.** 3×0 **14.** 7×1

15. $(3 \times 3) \times 6$ **16.** $6 \times (1 \times 2)$ **17.** $(4 \times 0) \times 3$ **18.** $4 \times (3 \times 2)$

Copy and complete. Write the name of the property.

19. $5 \times (3 + 8) = (5 \times 3) + (\blacksquare \times \blacksquare)$ **20.** $4 \times \blacksquare = 5 \times 4$

21. $9 \times \blacksquare = 9$ **22.** $\blacksquare \times 1 = 7$ **23.** $3 \times \blacksquare = 4 \times 3$

24. $(3 \times 4) \times 2 = 3 \times (\blacksquare \times 2)$ **25.** $6 \times (4 + 7) = (\blacksquare \times 4) + (6 \times \blacksquare)$

26. $5 \times \blacksquare = 0$ **27.** $(4 \times 8) + (4 \times 6) = \blacksquare \times (8 + 6)$

★29. $54 \times (65 \times 85) = (54 \times \blacksquare) \times 85$ **★29.** $27 \times 0 = 0 \times \blacksquare$

Solve.

30. Marge reads 3 articles about Columbus's voyage. Each article has 8 pages. How many pages does she read?

31. One article stated that Columbus explored the Caribbean for 9 weeks. For how many days did he explore the Caribbean?
(HINT: There are 7 days in a week.)

32. Marge reads that it took Columbus 10 weeks to sail to America, but only 8 weeks to sail back to Spain. How much shorter was the return trip?

★33. If there were 2 lookout teams on each of Columbus's 3 ships, and 5 sailors on each lookout team, how many sailors were lookouts?

ANOTHER LOOK

Subtract.

1. 6,000 − 461	**2.** 7,010 − 284	**3.** 4,300 − 602	**4.** 2,040 − 159	**5.** 1,306 − 278

6. $2,000 - 188$ **7.** $4,002 - 255$ **8.** $8,070 - 206$ **9.** $7,040 - 728$

More Practice, page 425

Multiples of 10

A. Early explorers of the American wilderness marked trails by cutting the bark of trees. This was called *blazing a trail*. If 70 miles of trail are blazed in 1 month, how many miles can be blazed in 10 months?

You can multiply to find how many miles can be blazed in 10 months.

Find 10×70.

Think: $10 \times 70 = (10 \times 10) \times 7 = 100 \times 7$.
So, $10 \times 70 = 700$.

In 10 months, 700 miles of trail can be blazed.

B. Look for a pattern.

$$5 \times 9 = 45 \qquad 50 \times 9 = 450$$
$$5 \times 90 = 450 \qquad 50 \times 90 = 4,500$$
$$5 \times 900 = 4,500 \qquad 50 \times 900 = 45,000$$
$$5 \times 9,000 = 45,000 \qquad 50 \times 9,000 = 450,000$$

Other examples:

$$\begin{array}{r} 50 \\ \times\ 60 \\ \hline 3,000 \end{array} \qquad \begin{array}{r} 7,000 \\ \times\ \ \ 200 \\ \hline 1,400,000 \end{array}$$

Checkpoint Write the letter of the correct answer.

Multiply.

1. $\begin{array}{r}80\\ \times\,10\\\hline\end{array}$	**2.** $\begin{array}{r}500\\ \times\ 10\\\hline\end{array}$	**3.** $\begin{array}{r}700\\ \times\ 50\\\hline\end{array}$	**4.** $\begin{array}{r}6,000\\ \times\ \ \ 200\\\hline\end{array}$
a. 80	**a.** 50	**a.** 350	**a.** 1,200
b. 800	**b.** 500	**b.** 3,500	**b.** 12,000
c. 8,000	**c.** 5,000	**c.** 35,000	**c.** 120,000
d. 80,000	**d.** 50,000	**d.** 350,000	**d.** 1,200,000

Multiply.

1. $\begin{array}{r} 10 \\ \times\ 4 \\ \hline \end{array}$

2. $\begin{array}{r} 20 \\ \times\ 3 \\ \hline \end{array}$

3. $\begin{array}{r} 50 \\ \times\ 5 \\ \hline \end{array}$

4. $\begin{array}{r} 100 \\ \times\ 2 \\ \hline \end{array}$

5. $\begin{array}{r} 800 \\ \times\ 7 \\ \hline \end{array}$

6. $\begin{array}{r} 200 \\ \times\ 8 \\ \hline \end{array}$

7. $\begin{array}{r} 5,000 \\ \times\ 4 \\ \hline \end{array}$

8. $\begin{array}{r} 3,000 \\ \times\ 3 \\ \hline \end{array}$

9. $\begin{array}{r} 70 \\ \times\ 60 \\ \hline \end{array}$

10. $\begin{array}{r} 10 \\ \times\ 30 \\ \hline \end{array}$

11. $\begin{array}{r} 40 \\ \times\ 30 \\ \hline \end{array}$

12. $\begin{array}{r} 300 \\ \times\ 60 \\ \hline \end{array}$

13. $\begin{array}{r} 600 \\ \times\ 50 \\ \hline \end{array}$

14. $\begin{array}{r} 1,000 \\ \times\ 60 \\ \hline \end{array}$

15. $\begin{array}{r} 9,000 \\ \times\ 300 \\ \hline \end{array}$

16. $80 \times 2,000$

17. $700 \times 6,000$

18. 40×50

19. 200×800

20. $3,000 \times 70$

21. $4,000 \times 300$

Find n.

★22. $6 \times n = 6,000$

★23. $100 \times n = 4,000$

★24. $70 \times n = 56,000$

Solve.

25. Jean-Pierre buys beads to trade during his journey. He buys 30 bags of beads. Each bag contains 600 beads. How many beads does he have to trade?

26. Simple items are good to trade. Jean-Pierre buys 40 cards of sewing needles. Each card holds 60 needles. How many needles does Jean-Pierre buy?

FOCUS: ESTIMATION

You can estimate the product of two numbers by multiplying their lead digits and writing in zeros.

Find the lead digit. Multiply.	Count the places after the lead digit.	Write zeros in the product.

$\begin{array}{r} 218 \\ \times\ 6 \\ \hline 12 \end{array}$

$\begin{array}{r} 218 \\ \times\ 6 \\ \hline 12 \end{array}$ Two places mean 2 zeros.

1,200
$6 \times 200 = 1,200$

Estimate.

1. 429×3
2. 815×6
3. 911×2
4. 641×5

5. $4,152 \times 3$
6. $7,341 \times 5$
7. $2,453 \times 4$
8. $6,321 \times 9$

More Practice, page 425

PROBLEM SOLVING
Checking for a Reasonable Answer

When you complete a problem, think about your answer. Is it a reasonable answer to the problem? You can often spot an error by thinking about what the value of the answer should be.

A group of students plan a one-day trip to Denver. They decide they will need to make 2 sandwiches for each person going on the trip. If 9 people go on the trip, how many sandwiches do they need?

a. 11 sandwiches **b.** 18 sandwiches **c.** 180 sandwiches

Without finding the exact answer, you can see that choice **a** is too small and choice **c** is too great. Choice **b** is the most reasonable. You can estimate to check if your choice is a reasonable one. About 5 people could eat 11 sandwiches, and 180 sandwiches would be enough for 90 people. For 9 people, 18 sandwiches would be about right.

Read each problem. Without computing the exact answer, write the letter of the correct answer.

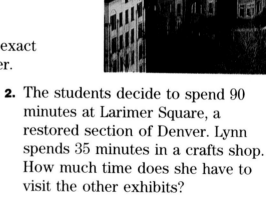

1. It will cost each student $8 for a bus ticket to Denver. If 9 students go, how much will their total bus fare be?

   a. $810
   b. $17
   c. $72

2. The students decide to spend 90 minutes at Larimer Square, a restored section of Denver. Lynn spends 35 minutes in a crafts shop. How much time does she have to visit the other exhibits?

   a. 3 minutes   **b.** 55 minutes
   c. 125 minutes

3. Julio bought a souvenir gold nugget. He gave the clerk $10. His change was $5.03. How much did he pay for the gold nugget?

   a. $4.97   **b.** $0.03   **c.** $50.30

4. Jill spent $3.00 for postcards and $0.88 for stamps. How much did Jill spend?

   a. $2.12   **b.** $30.88   **c.** $3.88

Read each problem. Without computing the exact answer, write the letter of the correct answer.

5. John can run 1 mile in 10 minutes. The 16 Street Mall is 1 mile long. How long will it take John to run to the end of the mall and back?

a. 2 hours
b. 2 minutes
c. 20 minutes

6. There are 1,500 animals in the Denver Zoo. On their visit, the students see 357 of these animals. How many of the animals at the Denver Zoo did they not see?

a. 11,430 of the animals
b. 11,857 of the animals
c. 1,143 of the animals

7. The Ballroom in the Children's Museum is filled with thousands of plastic balls. If there are 65 children in the ballroom, and 27 more arrive, how many children are there in the Ballroom?

a. 38 children
b. 92 children
c. 80,092 children

8. The students learn that Denver is called the Mile-High City because its altitude is 5,280 feet. Pikes Peak, also in Colorado, is 14,110 feet high. How much higher than Denver is Pikes Peak?

a. 8,830 feet
b. 18,830 feet
c. 19,650 feet

9. The tour guide explains that the Denver Mint produces about 14 million coins per day. About how many coins are produced there in 10 days?

a. about 14 million coins
b. about 140 million coins
c. about 14 billion coins

10. The mountain road drops in altitude at a rate of 8 meters per minute as it descends into Denver. At that rate, how many meters of altitude would the road drop in 8 minutes?

a. 24 meters
b. 64 meters
c. 180 meters

★11. The bus is caught in a traffic jam 78 miles from home. Traffic moves at only 7 miles per hour. At that rate, how far would the bus be from home after 6 hours?

a. 36 miles
b. 98 miles
c. 143 miles

★12. The group rides on the historic Silverton train. A ticket costs $28 for adults and $14 for children. What is the total fare for 7 children and 2 adults?

a. $1,540
b. $154
c. $252

Estimating Products

A. The Stillmans plan to explore the northwestern part of the United States in their camper. Their trip will last 21 days. They plan to travel 165 miles each day. About how many miles do they plan to travel?

You can estimate a product by rounding each factor to its largest place, and then multiplying.

Estimate 21 × 165.

Round each factor.

$$165 \longrightarrow 200$$
$$\times\ \ 21 \longrightarrow \ \ \ 20$$

Multiply the rounded factors.

$$\begin{array}{r} 200 \\ \times\ \ \ 20 \\ \hline 4{,}000 \end{array}$$

They plan to travel about 4,000 miles.

B. If one of the factors is a 1-digit number, you can estimate by rounding the other factor and multiplying.

Estimate 5 × 686.

$$\begin{array}{r} 686 \longrightarrow\ \ \ 700 \\ \times\ \ \ 5 \longrightarrow \times\ \ \ \ 5 \\ \hline 3{,}500 \end{array}$$

C. You can estimate money the same way you estimate whole numbers.

Estimate 9 × $6.89.

9 × $6.89 is about 9 × $7
9 × $7 = $63

The estimated product of 9 × $6.89 is about $63.00.

Estimate.

1. 43 × 18	**2.** 86 × 24	**3.** 79 × 36	**4.** 16 × 48	**5.** 87 × 59	**6.** 32 × 48
7. 324 × 8	**8.** 782 × 6	**9.** 535 × 79	**10.** 424 × 23	**11.** 376 × 34	**12.** 538 × 67
13. $3.67 × 3	**14.** $2.81 × 4	**15.** $6.98 × 8	**16.** $8.45 × 7	**17.** $8.75 × 3	**18.** $7.82 × 9

19. 17×453 **20.** 23×32 **21.** 34×764 **22.** 16×927

23. 88×32 **24.** 76×487 ★**25.** 98×62 ★**26.** 97×99

Solve. For Problem 29, use the Infobank.

27. Larry Stillman takes his camera and 18 rolls of film. Each roll can produce 36 pictures. About how many pictures can Larry take?

28. The family visits Old Faithful, a geyser at Yellowstone Park. The geyser erupts about 22 times a day. If there are 365 days in a year, about how many times will the geyser erupt per year?

29. Before the Stillmans began their trip, they bought camping supplies. Use the information on page 416 to solve. Estimate the total cost of camping supplies for the 5 members of the family.

★**30.** The Stillman camper travels 23 miles on 1 gallon of gasoline. If the Stillmans drive 225 miles per day, and gasoline costs $1.13 a gallon, about how much do the Stillmans spend on gasoline each day?

MIDCHAPTER REVIEW

Write the property.

1. $6 \times 0 = 0$ **2.** $7 \times 5 = 5 \times 7$ **3.** $6 \times (4 + 3) = (6 \times 4) + (6 \times 3)$

Multiply.

4. 80×20 **5.** 5×900 **6.** $400 \times 7,000$ **7.** 30×600

Estimate.

8. 82×12 **9.** 72×944 **10.** 6×543 **11.** $8 \times \$7.55$

PROBLEM SOLVING
Estimation

Many questions can be answered by using estimated amounts. Sometimes your estimate is too close to the amount you are comparing it to. In this case, you need to find an exact amount.

> The 21 members of the North Fork Bike Club are going to the Mashomack Preserve on Shelter Island, New York. They have to ride a ferryboat to reach the island. The director of the club will pay the ferryboat fare, which is $2.95 per rider. She has $80. Will that be enough?

Although the director can answer this question by finding the exact amount of the fare, she would prefer to round the numbers and estimate.

She rounds. $2.95 $\longrightarrow$ $3.00   21 members $\longrightarrow$ 20 members
Then she multiplies. $3.00 × 20 = $60.00 $60.00 < $80.00

She has enough money.

Alfred has $19.06 to spend on souvenirs. He wants 3 maps that cost $3.00 each, a $8.49 Shelter Island sweatshirt, and a bag of shells for $1.76. Does he have enough money?

Alfred estimates.

Maps	(3 × $3.00)	$9.00
Sweatshirt	($8.49)	8.00
Bag of shells	($1.76)	+ 2.00
		$19.00

Since $19.00 is so close to the spending limit—$19.06—Alfred decides to find an exact amount.

Maps	(3 × $3.00)	$9.00
Sweatshirt		8.49
Bag of shells		+ 1.59
		$19.08

Alfred does not have enough money.

Decide whether you need to find an estimate or an exact answer. Explain your decision.

1. As she enters the preserve, the club director sees a sign that suggests that each visitor contribute $1.85. How much should she put in the contribution box if she pays the suggested amount for each of the 21 members?

2. One member of the club wants to photograph the salt marshes, freshwater ponds, fields, and upland forest that make up the preserve. She has $56.31 with which to buy film. Does she have enough to buy ten $4.99 rolls of film?

Solve. Find an estimate or an exact answer as needed.

3. The operators hope to raise $150.00 per day through visitor contributions to maintain the preserve. The manager counts 101 names in that day's visitor book. If each visitor paid the $1.85 suggested contribution, would they reach their daily goal?

4. The photographer wants to enlarge 3 pictures of otters that were taken at the preserve. To enlarge them, she will have to pay $10 per photo. For how much should she write her check to the photo lab?

5. The operators of the preserve want to find out if they have reached their goal of collecting $450.00 for 3 days.

The amounts collected are:

Monday $134
Tuesday $128
Wednesday $187

Did they reach their goal?

6. The whole of Shelter Island is almost 3 times as large as the preserve. The preserve is exactly 2,039 acres. About how large is Shelter Island?

★7. One bike-club member decides to buy a seashell wind chime for $4.95, 3 souvenir plates for $2.00 each, and 4 T-shirts at $6.00 each. He has $34. Does he have enough?

★8. At the restaurant, 10 of the cyclists order the $4.95 broiled-flounder special. Another 5 order the spaghetti dinner at $6.00. The other 6 members order the $9.95 lobster special. These prices do not include tax or tip. The club director says that club dues can pay for $125.00 of the meal's cost. Will the dues cover all of the costs?

Multiplying by 1-Digit Factors

A. New Amsterdam was founded by Dutch traders in 1625. Many ships sailed into its harbor. If 87 ships docked each year, how many ships docked in 9 years?

Multiply 9 × 87.

Multiply the ones.
Regroup the 63 ones.

$$\begin{array}{r} {\scriptstyle 6} \\ 8\,7 \\ \times\quad 9 \\ \hline 3 \end{array} \qquad \begin{array}{r} 7 \\ \times\,9 \\ \hline 63 \end{array}$$

Multiply the tens.
Then add the 6 tens.

$$\begin{array}{r} {\scriptstyle 6} \\ 8\,7 \\ \times\quad 9 \\ \hline 7\,8\,3 \end{array} \qquad \begin{array}{r} 8 \\ \times\,9 \\ \hline 72 \\ +\ 6 \\ \hline 78 \end{array}$$

In 9 years, 783 ships docked.

B. Sometimes you must regroup several times. Find the product of 5 × 219.

Multiply the ones.
Regroup the 45 ones.

$$\begin{array}{r} {\scriptstyle 4} \\ 2\,1\,9 \\ \times\quad 5 \\ \hline 5 \end{array} \qquad \begin{array}{r} 9 \\ \times\,5 \\ \hline 45 \end{array}$$

Multiply the tens.
Then add the 4 tens.

$$\begin{array}{r} {\scriptstyle 4} \\ 2\,1\,9 \\ \times\quad 5 \\ \hline 9\,5 \end{array} \qquad \begin{array}{r} 1 \\ \times\,5 \\ \hline 5 \\ +\,4 \\ \hline 9 \end{array}$$

Multiply the hundreds.

$$\begin{array}{r} {\scriptstyle 4} \\ 2\,1\,9 \\ \times\quad 5 \\ \hline 1,0\,9\,5 \end{array} \qquad \begin{array}{r} 2 \\ \times\,5 \\ \hline 10 \end{array}$$

Multiply with money the same way you multiply with whole numbers.

Multiply 4 × $3.65.

$$\begin{array}{r} \$3.65 \\ \times\quad 4 \\ \hline \$14.60 \end{array}$$

Remember to write the dollar sign and the cents point.

Checkpoint Write the letter of the correct answer.

Multiply.

1. $7.26	2. 23	3. 902	4. 52
× 7	× 4	× 6	× 9

1.
a. $7.19
b. $15.82
c. $50.42
d. $50.82

2.
a. 86
b. 92
c. 122
d. 920

3.
a. 5,412
b. 5,462
c. 5,472
d. 54,012

4.
a. 468
b. 477
c. 4,518
d. 4,618

Multiply.

1.	42 × 4	2.	61 × 6	3.	73 × 3	4.	84 × 2	5.	50 × 4

6.	67 × 7	7.	86 × 5	8.	54 × 3	9.	$0.97 × 4	10.	$0.79 × 2

11.	146 × 3	12.	207 × 9	13.	807 × 8	14.	$4.37 × 2	15.	$9.16 × 4

16.	680 × 9	17.	234 × 4	18.	$7.50 × 2	19.	$3.54 × 3	20.	$1.84 × 7

21. 4×31 **22.** 6×16 **23.** $4 \times \$3.95$ **24.** 4×23

25. 8×312 **26.** 2×22 **27.** 6×408 **28.** $8 \times \$3.12$

Solve.

29. Jonas Bronck came from Denmark with 36 dairy cows. Within a few years, the number of cows increased 6 times. How many cows did Bronck have then?

30. New Yorkers owned 99 ships in 1747. In the next 20 years, the number of ships they owned increased by 5 times. How many ships did New Yorkers own in 1767?

CHALLENGE

Copy the number triangle.

What is the pattern of the numbers in the blue circles?

What is the pattern of the numbers in each row?

Fill in each ○ in the bottom row of the triangle.

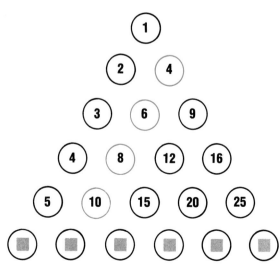

Multiplying Larger Numbers

People from all over the world come to America to explore its wonders. The Giroux family travels from Paris on the Concorde, a supersonic jet plane. It can fly at 1,424 miles per hour. How far can the Concorde fly in 3 hours?

You can multiply as if the same number of miles were flown each hour.

Multiply 3 × 1,424.

First estimate the product.

$$
\begin{array}{r}
1,424 \longrightarrow 1,000 \\
\times \quad 3 \longrightarrow \times \quad 3 \\
\hline
3,000
\end{array}
$$

Multiply the ones. Regroup if necessary.

$$
\begin{array}{r}
\overset{1}{1,42\,4} \\
\times \quad\quad 3 \\
\hline
2
\end{array}
\qquad
\begin{array}{r}
4 \\
\times\ 3 \\
\hline
12
\end{array}
$$

Multiply the tens. Regroup if necessary.

$$
\begin{array}{r}
\overset{1}{1,42\,4} \\
\times \quad\quad 3 \\
\hline
7\,2
\end{array}
\qquad
\begin{array}{r}
2 \\
\times\ 3 \\
\hline
6 \\
+\ 1 \\
\hline
7
\end{array}
$$

Multiply the hundreds. Regroup if necessary.

$$
\begin{array}{r}
\overset{1}{1,}\overset{1}{4}2\,4 \\
\times \quad\quad 3 \\
\hline
2\,7\,2
\end{array}
\qquad
\begin{array}{r}
4 \\
\times\ 3 \\
\hline
12
\end{array}
$$

Multiply the thousands. Regroup if necessary.

$$
\begin{array}{r}
\overset{1}{1,}\overset{1}{4}2\,4 \\
\times \quad\quad 3 \\
\hline
4,2\,7\,2
\end{array}
\qquad
\begin{array}{r}
1 \\
\times\ 3 \\
\hline
3 \\
+\ 1 \\
\hline
4
\end{array}
$$

The Concorde can fly 4,272 miles in 3 hours.
The product is reasonably close to the estimate.

Other examples:

$$
\begin{array}{r}
7,941 \\
\times \quad 4 \\
\hline
31,764
\end{array}
\qquad
\begin{array}{r}
63,402 \\
\times \quad 7 \\
\hline
443,814
\end{array}
\qquad
\begin{array}{r}
\$35.27 \\
\times \quad\quad 5 \\
\hline
\$176.35
\end{array}
$$

Checkpoint   Write the letter of the correct answer.

Multiply.

1.
$$
\begin{array}{r}
2,436 \\
\times \quad 7 \\
\end{array}
$$

2.
$$
\begin{array}{r}
\$22.46 \\
\times \quad\quad 4 \\
\end{array}
$$

3.
$$
\begin{array}{r}
18,076 \\
\times \quad\quad 7 \\
\end{array}
$$

1.
a. 14,812
b. 15,922
c. 17,052
d. 48,692

2.
a. $22.50
b. $85.34
c. $88.64
d. $89.84

3.
a. 76,092
b. 86,702
c. 126,032
d. 126,532

Multiply.

1. 3,440 × 2	**2.** 1,312 × 3	**3.** 2,101 × 4	**4.** 2,315 × 3	**5.** $34.23 × 2

6. 2,325 × 3	**7.** 3,207 × 3	**8.** 2,016 × 4	**9.** 1,605 × 8	**10.** $32.15 × 3

11. 12,169 × 4	**12.** 23,064 × 3	**13.** 36,495 × 2	**14.** 33,347 × 3	**15.** $11,248 × 4

16. 3 × 23,055 **17.** 2 × 44,675 **18.** 4 × 2,216 **19.** 8 × $652.43

20. 4 × 12,386 **21.** 2 × 9,909 **22.** 6 × 40,948 **23.** 7 × $732.18

Solve.

24. The Giroux family explores the United States by car. They drive about 1,947 miles per week for 4 weeks. How many miles do they drive?

25. At the Kennedy Space Center in Florida, the Giroux's are amazed to learn that a rocket that travels at 8,426 miles per hour is not fast enough to escape Earth's gravity. The rocket must travel 3 times faster. How fast must the rocket travel?

★**26.** When the Giroux's wrote the budget for their trip, they had two lists of choices. What is the least cost for 2 people for both hotel and food?

	Cost for 1 person	Cost for 1 person
Hotel	$75.00 per night	$45.00 per night
Food	$82.00 per day	$37.50 per day

FOCUS: ESTIMATION

To help you estimate products, look for factors that are near 10; 100; or 1,000.

Estimate 57 × 98. **Think:** 98 is close to 100.
$$57 × 100 = 5,700$$

Estimate.

1. 48 × 97 **2.** 73 × 978 **3.** 455 × 103 **4.** 288 × 87

5. 65 × 96 **6.** 387 × 102 **7.** 34 × 989 **8.** 23 × 992

PROBLEM SOLVING
Practice

Write the letter of the operation you would use to solve the problem.

1. The Santa Fe Trail once linked Independence, Missouri, with Santa Fe, New Mexico. The first stop along the trail was Council Grove. A stage coach traveled 66 miles the first day. It had 54 miles to go before it reached Council Grove. How far from Independence was the first stop?

   a. addition   **b.** subtraction

2. Because the trail was long and dangerous, traders charged more for goods sold at the end of the trip than they charged for goods sold at its beginning. Suppose a trader bought a knife for $1.72 in Independence and sold it for $8.45 in Santa Fe. How much profit did the trader make?

   a. addition   **b.** subtraction

Write a number sentence and solve.

3. The Council Grove sheriff bought 21 cents' worth of eggs and 16 cents' worth of potatoes. How much money did he spend?

4. A wagon train covered 270 miles of the 800-mile-long Santa Fe Trail. How much farther did it have to go?

Write the letter of the answer that seems most reasonable.

5. The Council Grove register showed that 5,819 mules, 478 horses, and 22,738 oxen passed through the town in 1860. How many oxen and horses passed through Council Grove that year?

   a. 316   **b.** 23,216
   c. 29,035   **d.** 83,216

6. The overland mail delivered mail to California in 25 days. The pony express, which replaced the overland mail in 1860, took about 8 days. How much time did the pony express save?

   a. 17 days   **b.** 33 days
   c. 173 days   **d.** 200 days

Solve.

7. In 1803, the Louisiana Purchase increased the size of the United States from about 890,000 square miles to 1,720,000 square miles. Estimate the number of square miles of land gained in the Louisiana Purchase.

★8. Fritz read about the Louisiana Purchase. He read 24 pages on Saturday and twice as many on Sunday. How many pages did Fritz read during the weekend?

PROBLEM SOLVING
Making Change

Sometimes you don't have the exact amount of money to buy an item. You can give the cashier more than the cost of an item, you will receive change.

Carl buys a map of the United States. It costs $7.49. He gives the salesclerk a $10 bill. How much change will he receive? Name the least number of coins and bills he will receive.

You can solve the problem by subtracting the price of the map from the amount given to the salesclerk.

$10.00 ⟵ amount given to salesclerk
− 7.49 ⟵ price of map
$2.51 ⟵ Carl's change

To find the number of coins and bills, begin with the total cost of the item, and count up to the amount given.

Count from:
$7.49 + 1¢ = $7.50; $7.50 + 0.50 = $8.00; $8.00 + 2.00 = $10.00

| 1 penny | 2 quarters or half-dollar | 2 dollars | amount given |

Solve.

1. Craig buys a coonskin cap. It costs $3.57. He gives the salesclerk a $5 bill. How much change will he receive?

2. Betsy bought moccasins for $12.34. She gave the salesclerk a $20 bill. How much change did she receive?

3. Arthur buys a pioneer powder horn. It costs $8.75. He gives the salesclerk a $10 bill. He receives the smallest number of coins and bills as change. Name them.

4. Jane buys a postcard at the giftshop for $0.20. She gives the salesclerk a $1 bill, and she receives the smallest number of coins or bills as change. Which coins or bills does she receive?

★5. Ted buys 6 commemorative pens at $0.15 each. He gives the salesclerk a $5 bill. How much change will he receive?

★6. Mona buys 4 tickets to the pioneer play at $2.75 each. She gives the salesclerk a $20 bill. She receives the smallest number of coins or bills as change. Name the coins or bills.

Multiplying by 2-Digit Factors

A. The Mason family traveled 129 miles to the start of the Oregon Trail. A guide told them that the trail was 20 times longer than the distance they had already traveled. How long was the Oregon Trail?

To multiply by a multiple of 10, write 0 in the ones place. Then multiply by the tens.

Multiply by ones.	Multiply by tens.
1 2 9	1 2 9
× 2 0	× 2 0
0	2,5 8 0

The Oregon Trail was 2,580 miles long.

B. Find 26 × 99.

Multiply by ones.	Multiply by tens.	Add.
9 9	9 9	9 9
× 2 6	× 2 6	× 2 6
5 9 4	5 9 4	5 9 4
	1 9 8 0	1 9 8 0
		2,5 7 4

Other examples:

79	206	$3.54
× 48	× 71	× 42
632	206	7 08
3 160	14 420	141 60
3,792	14,626	$148.68

Checkpoint Write the letter of the correct answer.

Multiply.

1. 96	**2.** 810	**3.** 95	**4.** 1,673
× 40	× 23	× 41	× 25
a. 384	**a.** 1,620	**a.** 475	**a.** 3,355
b. 3,640	**b.** 4,050	**b.** 3,695	**b.** 61,711
c. 3,740	**c.** 18,630	**c.** 3,895	**c.** 31,335
d. 3,840	**d.** 19,730	**d.** 3,945	**d.** 41,825

Multiply.

1. 33
× 20

2. 42
× 30

3. 63
× 50

4. 28
× 70

5. 121
× 90

6. 56
× 12

7. 64
× 16

8. 72
× 35

9. 82
× 28

10. 74
× 23

11. 367
× 70

12. 293
× 28

13. 105
× 52

14. 388
× 26

15. $5.75
× 83

16. $12.59
× 54

17. $10.45
× 65

18. $31.20
× 27

19. $14.56
× 22

20. $11.43
× 45

21. 68 × 7,847

22. 48 × 3,033

23. 31 × 9,114

24. $48 × 18.77

★25. (43 × 5) × 32

★26. (15 × 27) × 52

★27. (54 × 28) × 65

Solve.

28. Most of the wagons left from Independence in the spring. Usually, 187 wagons left each week. How many wagons left in 14 weeks?

29. The wagon train traveled about 13 miles each day along the 2,379-mile trail. If the wagons traveled for 63 days, how much farther did they have to travel?

30. The cooks bought food supplies for the wagon train. They estimated that they would use about 27 pounds of flour each day. If the trip took 165 days, how many pounds of flour did they need?

31. Suppose you are going west with a wagon train. Plan the supplies you would take. Determine the quantities that you would need of tools, nails, food staples, cooking utensils, and fabric.

CALCULATOR

Use your calculator to solve each problem. Look for a pattern in the products.

1. 15,873 × 7

2. 15,873 × 14

3. 15,873 × 21

4. 15,873 × 28

5. 15,873 × 35

6. 15,873 × 42

Use the pattern of the products to write the next two problems in this group. What are the products?

More Practice, page 426

Multiplying 3-Digit Numbers

The discovery of gold at Sutter's Mill in California started the famous gold rush of 1849. At one point, 239 gold-seekers, known as "forty-niners," arrived in California each day. How many arrived in that year (365 days)?

First estimate the product.

$$
\begin{array}{r}
365 \longrightarrow 400 \\
\times\,239 \longrightarrow \times\,200 \\
\hline
80{,}000
\end{array}
$$

Multiply by ones.	Multiply by tens.	Multiply by hundreds.	Add.
365	365	365	365
×239	×239	×239	×239
3285	3285	3285	3285
	10950	10950	10950
		73000	73000
			87,235

In that year, 87,235 forty-niners arrived.
The product is reasonably close to the estimate.

Other examples:

$$
\begin{array}{r}
396 \\
\times\,207 \\
\hline
2\,772 \\
0\,000 \\
79\,2 \\
\hline
81{,}972
\end{array}
$$
You can leave out these zeros.

$$
\begin{array}{r}
396 \\
\times\,207 \\
\hline
2\,772 \\
79\,200 \\
\hline
81{,}972
\end{array}
$$

$$
\begin{array}{r}
\$7.89 \\
\times\,\ \ 143 \\
\hline
23\,67 \\
315\,60 \\
789\,00 \\
\hline
\$1{,}128.27
\end{array}
$$

Checkpoint Write the letter of the correct answer.

Multiply.

1. 211×543

2. 89×617

3. $\$7.39 \times 862$

1.	2.	3.
a. 2,172	**a.** 10,489	**a.** $118.24
b. 103,573	**b.** 54,353	**b.** $5,074.08
c. 104,573	**c.** 54,913	**c.** $6,085.08
d. 114,573	**d.** 549,130	**d.** $6,370.18

Solve.

1.	434 × 500	2.	142 × 300	3.	322 × 600	4.	303 × 500	5.	$3.21 × 700

6.	144 × 827	7.	293 × 406	8.	328 × 516	9.	427 × 404	10.	$6.02 × 548

11.	701 × 210	12.	243 × 129	13.	834 × 201	14.	429 × 212	15.	$3.40 × 192

16.	632 × 378	17.	441 × 146	18.	701 × 219	19.	938 × 493	20.	$4.93 × 677

21. 312 × 104   22. 523 × 928   23. 214 × 310   24. $1.01 × 516

25. 169 × 805   26. 419 × 301   27. 789 × 938   28. $4.34 × 904

Solve.

29. In 1849, one mining company estimated that it would have to pay each of its miners $295 per year. If the company employed 687 miners, what is its yearly payroll?

30. During 1849, a total of 80,000 miners went to California. Of these, 39,000 took the sea route around Cape Horn. How many miners traveled overland?

31. One rich find was the Calaveras Nugget. It weighed 162 pounds and sold for $269 a pound. How much money was the Calaveras Nugget

★**32.** A gold miner pays his miners well, for 1849. Each miner must choose whether to be paid $2.17 per day for a 237-day work season or $0.37 per hour for the same number of days. If miners work for 12 hours per day, which rate would the miners probably choose?

CHALLENGE

Find each missing digit.

1.	326 × 1▓ —— 1▓04 + 32▓0 —— 4,564	2.	▓56 × 27 —— 3▓92 + ▓120 —— 12,312	3.	749 × ▓3 —— 2▓47 + 5▓920 —— ▓▓,▓▓▓

PROBLEM SOLVING
Solving Two-Step Problems/Making a Plan

Sometimes you have to use more than one step to solve a problem. Before you can answer the question in the problem, you have to find needed data. Then you use this data to answer the question that was asked. Making a plan can help you solve this kind of problem.

Visitors to California explore its natural wonders and visit its exciting cities. On Monday, a tour company took 148 people on a tour of Hollywood and its movie studios. Each person paid $19. On Tuesday, the tour company received $2,375 from ticket sales. How much more money did the tour company receive on Monday than on Tuesday?

Needed data: How much money was received on Monday?

Plan

Step 1: Find out how much money the tour company received on Monday.
Step 2: Find the difference between Monday's and Tuesday's amounts.

Step 1: Multiply to find out how much the tour company received on Monday.

$$
\begin{array}{rl}
148 & \text{(number of people)} \\
\times\ \$19 & \text{(amount paid for tour)} \\
\hline
\$2,812 & \text{(total amount received on} \\
& \text{Monday)}
\end{array}
$$

Step 2: Find the difference by subtracting.

$$
\begin{array}{rl}
\$2,812 & \text{(total received on Monday)} \\
-\ 2,375 & \text{(total received on Tuesday)} \\
\hline
\$\ \ 437 & \text{(more received on Monday} \\
& \text{than Tuesday)}
\end{array}
$$

The tour company received $437 more on Monday than on Tuesday.

Complete the plan by writing the missing step.

1. The Tuolumne River in California's Yosemite National Park is a favorite spot for white-water rafting. One group spent $1,853 for 5 days of rafting. In a second group, each of 7 people paid $280 for 5 days. Which group paid more money?

Step 1: Find how much money the second group paid.

Step 2:

2. A group of rafters traveled 168 miles in their first 5 days. In the next 5 days, they traveled 28 miles per day. How many miles did they travel in 10 days?

Step 1: Find how many miles the group traveled in the next 5 days.

Step 2:

Make a plan for each problem. Solve.

3. Sara visited Fisherman's Wharf in San Francisco. She had $25.00 in her purse. She paid $8.95 for a lobster lunch. She paid $6.00 for a T-shirt. She bought film for $4.50 and lemonade for $1.25. How much money did Sara have left?

4. The trip from San Francisco to San Diego along the Pacific Coast Highway is 514 miles. Marilyn and her family leave San Francisco. If they drive at 45 miles per hour for 11 hours, how far from San Diego will they be?

5. Marilyn and her family stopped at Carmel. Altogether, they bought 145 shell beads to make necklaces. The beads cost $0.39 each. Marilyn gave the clerk a $100 bill. How much change did she receive?

6. A group of scouts plans to hike 100 miles along the trails of Yosemite National Park. If they hiked 16 miles per day for 6 days, how many miles must they hike the next day?

7. In 1934, 18 farmers set up stands on the outskirts of Los Angeles to start the Farmers Market. Today, 9 times that number of farmers have stands at this popular market. How many more farmers have stands at the market today than did the farmers in 1934?

★8. A souvenir buyer for the San Diego Zoo ordered 175 adult-size T-shirts for $3.50 each. He also ordered 175 child-size T-shirts for $2.95 each. How much more did he spend on the adult-size T-shirts than on the child-size T-shirts?

CALCULATOR

Use your calculator to find the missing digits in the following multiplication problems. Study the example.

Example:

```
    3 6 7
×     2 3
  1 ▪ 0 1
  7 3 ▪
  ▪,4 4 1
```

Multiply:

$3 \times 367 = 1{,}1\,0\,1$

$2 \times 367 = 7\,3\,4$

$23 \times 367 = 8{,}4\,4\,1$

The missing digits are 1, 4, and 8

1.
```
      6 7 8
×       4 5
    3 3 ▪ 0
    2 ▪ 1 2
    3 0,▪ 1 0
```

2.
```
      2 2 9
×       5 7
    1 ▪ ▪ 3
    1 ▪ 4 5
    1 3,▪ 5 3
```

3.
```
      7 8 2
×       8 5
    ▪ 9 ▪ 0
    6 2 ▪ 6
    ▪ 6,▪ 7 0
```

4.
```
      7 0 9
×       6 3
    ▪ 1 2 7
    ▪ 2 ▪ 4
    ▪ 4,6 ▪ 7
```

5.
```
    3 6 2 1
×       8 2
    ▪ 2 ▪ 2
  2 ▪ 9 ▪ 8
  2 ▪ 6,9 2 ▪
```

6.
```
    4 7 9 6
×       7 6
  2 ▪ 7 ▪ 6
  3 ▪ 5 ▪ 2
  ▪ 6 ▪,4 ▪ 6
```

7.
```
      5 8 2
×     6 9 1
      5 ▪ 2
    5 2 ▪ 8
  ▪ 4 ▪ 2
  ▪ 0 ▪,1 ▪ 2
```

8.
```
    7 9 1 4
×     3 8 2
  ▪ 5 8 ▪ 8
  ▪ 3 3 ▪ 2
  ▪ 3 ▪ ▪ 2
  ▪,0 2 ▪,1 ▪ 8
```

9.
```
    4,6 8 2
×     5 4 ▪
    ▪ 6 ▪ 2
  1 ▪ 7 ▪ 8
  ▪ 3 4 ▪ 0
  ▪,5 3 ▪,9 ▪ 2
```

GROUP PROJECT

Banking on the Alphabet

The problem: Which letter of the alphabet is worth the most? When you write, *a* and *e* are among the most useful letters. You use them often, while you hardly ever use *x* or *z*. Here is a game you can play in which *x* and *z* are among the most valuable letters in the alphabet.

Give each letter a cent value. Begin with *a*, worth $0.01, and go on to *b* ($0.02), *c* ($0.03), and on up to *z*, which is worth $0.26. Use these letter values to answer each question.

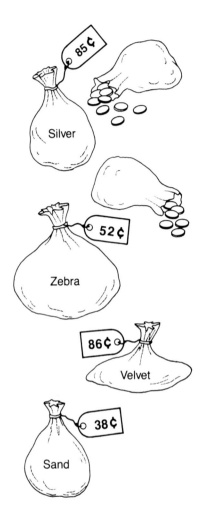

1. Which is more valuable, *gold* or *rust?*

2. How much more is a *diamond* worth than *wood?*

3. Which is worth more, *oysters* or *pearls?*

4. Which is worth the most, a *dime*, a *dollar*, a *penny*, a *nickel*, or a *quarter?* Which is worth exactly $1.00?

5. Find out how much this sentence is worth: *The treasure map is hidden under the floor.*

6. Can you think of a jewel that is worth more than a *ruby?* Don't stop at the ordinary jewels. Try unusual ones from the encyclopedia.

7. Can you come up with any words that are worth more than $1.00? What is the most valuable word you can find? What is the shortest word you can find that is worth more than $1.00?

8. What is the value of your name?

★9. These words are worth exactly $1.00: *chimpanzee*, *fountain*, and *whistled*. Can you come up with any $1.00 words or names? You might have a contest to see how many $1.00 words you and your classmates can come up with. Then put all your words worth $1.00 or more into a "bank." How much money do you have in the bank?

CHAPTER TEST

Copy and complete. Write the name of the property.
(page 76)

1. $4 \times \blacksquare = 4$

2. $3 \times (7 + 2) = (3 \times 7) + (\blacksquare \times 2)$

3. $(2 \times 6) \times 3 = 2 \times (\blacksquare \times 3)$

4. $9 \times \blacksquare = 8 \times 9$

Multiply.

5. $(4 \times 2) \times 6$

6. $(5 \times 8) \times 0$

Multiply. (pages 76, 78, 86, 88, 92, and 95)

7.
$$\begin{array}{r} 4 \\ \times\, 9 \\ \hline \end{array}$$

8.
$$\begin{array}{r} 10 \\ \times\, 3 \\ \hline \end{array}$$

9.
$$\begin{array}{r} 400 \\ \times\ \ 2 \\ \hline \end{array}$$

10.
$$\begin{array}{r} 600 \\ \times\ 50 \\ \hline \end{array}$$

11. 300×700

12. $2{,}000 \times 60$

13. $7{,}000 \times 400$

Estimate. (page 83)

14.
$$\begin{array}{r} 94 \\ \times\ 61 \\ \hline \end{array}$$

15.
$$\begin{array}{r} \$3.79 \\ \times\ \ \ \ 8 \\ \hline \end{array}$$

16.
$$\begin{array}{r} 342 \\ \times\ 53 \\ \hline \end{array}$$

17.
$$\begin{array}{r} \$4.65 \\ \times\ \ \ \ 8 \\ \hline \end{array}$$

Multiply. (pages 76, 78, 86, 88, 92, and 95)

18.
$$\begin{array}{r} 53 \\ \times\ 2 \\ \hline \end{array}$$

19.
$$\begin{array}{r} \$5.89 \\ \times\ \ \ \ 3 \\ \hline \end{array}$$

20.
$$\begin{array}{r} \$23.95 \\ \times\ \ \ \ \ \ 5 \\ \hline \end{array}$$

21.
$$\begin{array}{r} 38{,}244 \\ \times\ \ \ \ \ \ \ 7 \\ \hline \end{array}$$

22. 36×42

23. 73×19

24. 368×37

25. $\$4.57 \times 62$

26. $2{,}305 \times 84$

27. $\$15.82 \times 51$

28. 422×505

29. $\$7.39 \times 134$

Solve. (pages 80–81, 84–85, and 96–97)

30. Each year, the North Shore Hiking Club makes plans to hike through one of America's many national parks. As a souvenir for the members of the club who participate in the hike, park officials design T-shirts and caps with the park's name and the date of the hike. This year, the club budgets $200.00 for the purchase of the shirt/cap sets. If 9 members buy the sets for $15.25 each, how much is left of the $200.00 budget?

31. The club chooses to hike the Assateague National Seashore off the coast of Maryland and Virginia. The members plan the distance they want to cover well in advance. This way, if there is extra time left after the planned hike, they can do some exploring. They planned to hike 21 miles per day for 7 days. If they hike an extra 7 miles on the fifth day, how many miles will they have hiked after 5 days?

32. The North Shore Hiking Club has a total of 32 members. Each member pays $15.00 dues each year. The club has budgeted $400.00 for hikes and other activities. Estimate to find whether the membership dues will be enough to cover the budgeted items.

33. One club outing included a group lunch for the 12 members who went on the hike. If each hiker was expected to eat a sandwich, an apple, and juice, how many items needed to be packed. Write the letter of the reasonable answer.

a. 12 **b.** 36 **c.** 72

BONUS

Solve.

1. 3,267
   × 1,043

2. 6,500
   × 2,368

3. 8,492
   × 5,321

4. 7,378
   × 8,765

5. 5,480
   × 4,325

6. 4,073
   × 2,120

7. 8,245
   × 1,004

8. 1,239
   × 1,183

9. 7,132
   × 1,946

10. 3,747
   × 2,139

RETEACHING

When you multiply by more than 1 digit, be careful to line up your products in the correct place. You can use zeros to help you.

Multiply 329×487.

Multiply by ones.	Multiply by tens.	Multiply by hundreds.	Add.
487	487	487	487
$\times$ 329	$\times$ 329	$\times$ 329	$\times$ 329
4383	4383	4383	4 383
	9740	9740	9 740
		146100	146 100
			160,223

Use zeros to line up the products correctly.

Multiply.

1. 753
$\times$ 480

2. 420
$\times$ 141

3. 530
$\times$ 274

4. 603
$\times$ 578

5. 810
$\times$ 780

6. 825
$\times$ 104

7. 750
$\times$ 410

8. 885
$\times$ 480

9. 927
$\times$ 210

10. 390
$\times$ 168

11. 810
$\times$ 178

12. 630
$\times$ 265

13. 447
$\times$ 177

14. 396
$\times$ 279

15. 384
$\times$ 263

16. 565
$\times$ 454

17. 609
$\times$ 548

18. 761
$\times$ 648

19. 976
$\times$ 794

20. 650
$\times$ 190

21. 664×647

22. 465×118

23. 944×710

24. 329×138

25. 320×268

26. 811×440

27. 250×623

28. 638×684

29. 902×129

30. 312×390

31. 308×121

32. 917×375

ENRICHMENT

Exponents

Steve has an album of stamps that commemorate the exploration of America. Each page of his album has 10 rows of stamps, with space for 10 stamps in each row. How many stamps are on each page?

You can multiply 10×10. When you multiply a number by itself, you can write the factors in **exponent form.**

10×10 can be written as 10^2.

The 2 is an exponent. It shows how many times that 10 is used as a factor.

$10^2 = 10 \times 10 = 100$

There will be 100 stamps on each page.

Other examples:

$7^1 = 7$
$2^3 = 2 \times 2 \times 2 = 8$
$3^4 = 3 \times 3 \times 3 \times 3 = 81$

Write the number and the exponent.

1. $4 \times 4 \times 4$ 2. $6 \times 6 \times 6 \times 6 \times 6$

3. $8 \times 8 \times 8 \times 8$ 4. $10 \times 10 \times 10 \times 10 \times 10 \times 10$

5. $7 \times 7 \times 7$ 6. $2 \times 2 \times 2 \times 2 \times 2 \times 2 \times 2 \times 2$

7. 5×5 8. $3 \times 3 \times 3$

Write the product. You may use a calculator to help you.

9. 2^5 10. 4^4 11. 7^2 12. 10^4 13. 3^5

14. 8^1 15. 9^3 16. 5^2 17. 5^4 18. 8^2

19. 10^3 20. 3^2 21. 6^4 22. 9^2 23. 4^5

24. 5^5 25. 2^4 26. 10^1 27. 3^3 28. 1^7

CUMULATIVE REVIEW

Write the letter of the correct answer.

1. $17.91 - 9.17$

   a. 8.74   **b.** 8.84
   c. 8.86   **d.** not given

2. $83.97 + 17.48$

   a. 101.35   **b.** 101.45
   c. 101.55   **d.** not given

3. $59.73 - 0.07$

   a. 59.66   **b.** 59.76
   c. 59.80   **d.** not given

4. $6.39 + 14.8 + 0.013$

   a. 20.6   **b.** 21.203
   c. 21.23   **d.** not given

5. Order from the least to the greatest:
4.851, 0.485, 48.7, 4.80.

   a. 4.80, 4.851, 0.485, 48.7
   b. 48.7, 4.80, 4.851, 0.485
   c. 0.485, 4.80, 4.851, 48.7
   d. not given

6. Write as a decimal: six hundred
thirty-eight thousandths.

   a. 0.638   **b.** 6.380
   c. 600.38   **d.** not given

7. $497,603 - 19,937$

   a. 366,777   **b.** 387,666
   c. 377,666   **d.** not given

8. $759 + 623 + 1,114$

   a. 1,496   **b.** 2,486
   c. 2,496   **d.** not given

9. $615,101 - 317,011$

   a. 298,090   **b.** 302,111
   c. 308,191   **d.** not given

10. Compare. Choose $>$, $<$, or $=$ for ●.
340,597 ● 340,579

   a. $>$ **b.** $<$
   c. $=$ **d.** not given

11. $\$623.00 + \$319.00$

   a. $932.00   **b.** 942
   c. $942.00   **d.** not given

12. Simon's time in the 100-yard dash is
11.53 seconds. After practicing, he
improves his time to 11.29 seconds.
Let n = the amount of
improvement in his time. Choose
the correct number sentence.

   a. $11.53 + 11.29 = n$
   b. $11.53 - 11.29 = n$
   c. $11.53 \times 11.29 = n$
   d. not given

13. Deborah swims the 100-meter
freestyle in 58.67 seconds. This is
5.32 seconds faster than her time
last year. Choose the operation to
find Deborah's time last year.

   a. add   **b.** subtract
   c. compare   **d.** not given

Flick a switch. A light goes on. Push a button. The dryer starts. Does the electricity used in your home and school seem to come from nowhere? How far away is the source of electricity in your community? How is electricity usage measured in your community? How does the electric company decide how much to charge?

4 MULTIPLYING DECIMALS

Estimating Decimal Products

A. Wally burns coal in his fireplace for heat. He burns 1.75 kg of coal each hour. If he uses the fireplace for 4.5 hours, about how much coal does he use?

Estimate 4.5×1.75.

You can estimate decimal products the same way you estimate whole-number products.

Round both factors to their greatest place.

$$1.75 \rightarrow 2$$
$$\times\ \ 4.5 \rightarrow 5$$

Multiply the factors.

$$
\begin{array}{r}
2 \\
\times\ 5 \\
\hline
10
\end{array}
$$

Wally burns about 10 kg of coal in 4.5 hours.

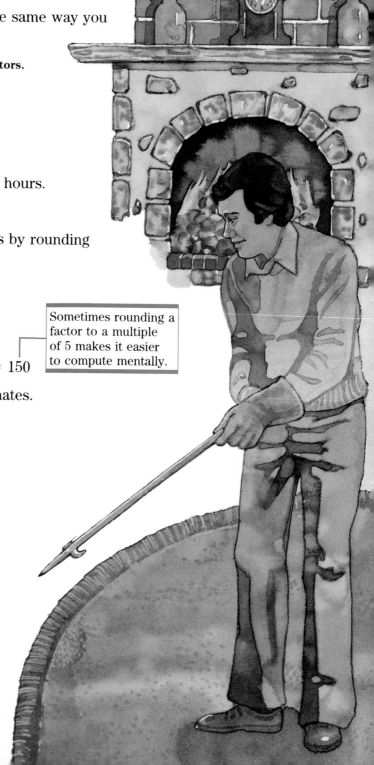

B. You can estimate decimal products by rounding several different ways.

Estimate 6.2×26.483.

$$6.2 \times 26.483$$
$$\downarrow \qquad \downarrow$$
$$6\ \times\ \ 30 = 180$$

$$6.2 \times 26.483$$
$$\downarrow \qquad \downarrow$$
$$6\ \times\ \ 25 = 150$$

> Sometimes rounding a factor to a multiple of 5 makes it easier to compute mentally.

Both 180 and 150 are reasonable estimates.

Other examples:

$$
\begin{array}{r}
24.6 \rightarrow\ \ 20 \qquad 25 \\
\times\ \ 4.7 \rightarrow \times\ \ 5 \ \text{or} \times\ \ 5 \\
\hline
100 \qquad 125
\end{array}
$$

$$
\begin{array}{r}
76.2 \rightarrow\ \ 80 \qquad 75 \\
\times\ 3.75 \rightarrow \times\ \ 4 \ \text{or} \times\ \ 4 \\
\hline
320 \qquad 300
\end{array}
$$

$$
\begin{array}{r}
\$46.75 \rightarrow\ \ \$50 \qquad \$45 \\
\times\ \ 2.03 \rightarrow \times\ \ \ 2 \ \text{or} \times\ \ 2 \\
\hline
\$100 \qquad \$90
\end{array}
$$

Estimate. Write > or < for ⬤.

1. 2.7×8.3 ⬤ 15
2. 3.4×7.4 ⬤ 32
3. 3.7×5.1 ⬤ 23
4. 5.4×23.7 ⬤ 126
5. 3.2×61.2 ⬤ 171
6. 6.3×47.7 ⬤ 312
7. 2.1×8.64 ⬤ 19
8. 3.72×8.5 ⬤ 33
9. 4.4×32.43 ⬤ 102
10. $2.1 \times \$4.68$ ⬤ $9
11. $3.5 \times \$15.76$ ⬤ $75
12. $5.4 \times \$36.25$ ⬤ $149
13. 2.46×9.15 ⬤ 20
14. 6.8×87.49 ⬤ 573
15. 6.34×75.9 ⬤ 485
16. 7.6×26.3 ⬤ 250
17. $7.92 \times \$7.58$ ⬤ $64
18. 3.89×56.47 ⬤ 242

Estimate.

19.
$$\begin{array}{r} 9.8 \\ \times\ 4.2 \\ \hline \end{array}$$

20.
$$\begin{array}{r} 15.8 \\ \times\ 8.23 \\ \hline \end{array}$$

21.
$$\begin{array}{r} 22.19 \\ \times\ 3.27 \\ \hline \end{array}$$

22.
$$\begin{array}{r} \$9.39 \\ \times\ 6.8 \\ \hline \end{array}$$

23.
$$\begin{array}{r} \$7.52 \\ \times\ 4.4 \\ \hline \end{array}$$

Solve.

24. The hydroelectric plant where Sam works produces 38,300 kilowatts of electricity. After new generators are installed, the power capacity of the plant will rise 1.5 times. About how many kilowatts of electricity will the plant produce?

★25. A steam-turbine plant produced 11.67 million kilowatts of electricity. New turbines were added to raise the plant's energy capacity 2.3 times, but it only rose 1.8 times. About how many kilowatts does the plant produce? About how much more is this than the plant's previous capacity?

CALCULATOR

Use your calculator to solve these multiplication exercises. Look for a pattern.

$0.1089 \times 9 = $ ▢
$0.10989 \times 9 = $ ▢
$0.109989 \times 9 = $ ▢

Use the pattern to compute these exercises mentally.

1. 0.10999989×9
2. 0.1099989×9
3. 0.109999989×9

PROBLEM SOLVING
Help File Review

Remember: If you can't solve a problem right away, keep trying. There are places where you can go for help. One of these places is the Help File on pages 411–414 of this book.

Each part of the file will help you answer a different question about solving a problem.

1. **Questions:** "What is the problem asking me to find?"
2. **Tools:** "What must I do to solve this problem?"
3. **Solutions:** "How can I do the math needed to solve this problem?"
4. **Checks:** "How can I check my answer to be sure that it is correct?"

Read the problem. Then choose the part of the Help File each student should use. Write the letter of the correct answer.

Solar energy is measured in units called quads. Solar cells can produce 4.8 quads of energy. Solar collectors can produce about 4 times that amount. Estimate the amount of energy that a solar collector can produce.

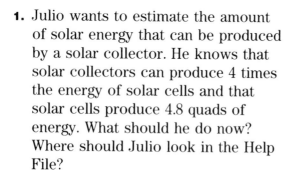

1. Julio wants to estimate the amount of solar energy that can be produced by a solar collector. He knows that solar collectors can produce 4 times the energy of solar cells and that solar cells produce 4.8 quads of energy. What should he do now? Where should Julio look in the Help File?

   a. Tools   b. Solutions
   c. Checks   d. Questions

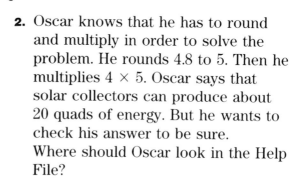

2. Oscar knows that he has to round and multiply in order to solve the problem. He rounds 4.8 to 5. Then he multiplies 4×5. Oscar says that solar collectors can produce about 20 quads of energy. But he wants to check his answer to be sure. Where should Oscar look in the Help File?

   a. Solutions   b. Checks
   c. Tools   d. Questions

Solve each problem. If you have trouble, go to the Help File.

3. In 1774, a solar furnace produced a temperature of more than 3,092°F. Usually, the temperature rose to only about 2,942°F. What is the difference between these two temperatures?

4. In the late 1800's, sunlight was used to produce electrical energy. For every 100 square feet of sunlight, 746 watts of power were produced. How much energy would have been produced by twice as many square feet of sunlight?

5. Waterpower used to be measured in horsepower. Today, it is measured in kilowatts. If 1.34 kilowatts equals 1 horsepower, estimate how many kilowatts would be produced by a 1,575-horsepower waterwheel.

6. In the United States in 1984, waterpower production measured 65.8 million kilowatts. The total world production was 6 times that amount. Estimate how many kilowatts the world produced.

7. In 1980, the world level of coal use was 2.3 tons of coal per person. The United States level was 5 times that number. Estimate the 1980 level of coal use per person for the United States.

8. In 1968, the world used enough oil to equal 2,892 million tons of coal. By 1978, the world was using 1.59 times that amount. Estimate how many million tons of oil the world was consuming by 1978.

★9. In 1984, the United States produced more than 2,300 million kilowatt-hours of electricity. In the same year, Canada produced more than 370 million kilowatt-hours. How much more electricity did the two countries produce together than the 1,300 million kilowatt-hours produced by the Soviet Union?

Multiplying Decimals and Whole Numbers

Most of the world's power plants use fossil fuels such as coal, oil, and natural gas to make electricity. If a small power plant burns 0.8 of its supply of 23 tons of coal in one day, how many tons of coal has it burned?

Multiply 0.8×23.

First, you can estimate the product.

$$\begin{array}{r} 23 \longrightarrow \ \ 20 \\ \times\,0.8 \longrightarrow \times\ \ 1 \\ \hline 20 \end{array}$$ The product is about 20.

Multiply as you would with whole numbers.

$$\begin{array}{r} 2\,3 \\ \times\,0.8 \\ \hline 1\,8\,4 \end{array}$$

Use your estimate to help you place the decimal point in the product.

$$\begin{array}{r} 2\,3 \\ \times\,0.8 \\ \hline 1\,8.4 \end{array}$$

The power plant has burned 18.4 tons of its coal supply in one day.

Other examples:

$$\begin{array}{r} 3.7 \\ \times\ \ 23 \\ \hline 11\,1 \\ 74\,0 \\ \hline 85.1 \end{array}$$

$$\begin{array}{r} \$16.73 \\ \times\ \ \ \ 4 \\ \hline \$66.92 \end{array}$$

$$\begin{array}{r} 0.742 \\ \times\ \ \ \ 26 \\ \hline 4\,452 \\ 14\,840 \\ \hline 19.292 \end{array}$$

Checkpoint Write the letter of the correct answer.

Multiply.

1.
$$\begin{array}{r} 12 \\ \times\,2.9 \end{array}$$

a. 3.48
b. 24.108
c. 34.8
d. 348

2. 9.3×46

a. 93.0
b. 416.8
c. 427.8
d. 4278

3.
$$\begin{array}{r} \$24.32 \\ \times\ \ \ \ 8 \end{array}$$

a. $162.46
b. 19.46
c. $194.56
d. 19,456

Multiply.

1. 35 × 0.3	**2.** 45 × 0.7	**3.** 36 × 0.9	**4.** 68 × 0.7	**5.** 49 × 0.8
6. 9 × 0.8	**7.** 21 × 0.6	**8.** 34 × 0.7	**9.** 48 × 0.9	**10.** 87 × 0.3
11. 67 × 5.9	**12.** 8.515 × 6.2	**13.** 6.346 × 4.5	**14.** 73 × 7.9	**15.** 11.734 × 9.5
16. 3.97 × 4	**17.** 2.9 × 15	**18.** 36.531 × 28	**19.** $4.38 × 37	**20.** $17.26 × 29

21. 15 × 2.9   **22.** 67 × 5.9   **23.** 45 × $3.27   **24.** 17 × $12.28

25. 97 × 0.115   **26.** 37 × 4.459   **27.** 128 × $9.18   **28.** 51 × $19.43

29. 78 × 12.9   **30.** 144 × 0.483   **31.** 67 × $36.04   **32.** 33 × $28.91

33. 543 × 8.54   **34.** 83 × 0.226   **35.** 43 × $7.92   **36.** 31 × $2.87

Solve. For Problem 39, use the Infobank.

37. Uranium 235 is used to produce nuclear power. A kilogram of uranium 235 provides the same amount of power as 2,720 metric tons of coal. How much power would 2.5 kilograms provide?

38. Wind power is used in many places to pump water and make electricity. If a windmill can pump 25 gallons of water in 1 hour, how many gallons can it pump in 9.5 hours?

39. Use the information on page 416 to write and solve your own problem.

MIDCHAPTER REVIEW

Estimate.

1. 6.2 × 7.8   **2.** 4.29 × 25.18   **3.** 18.8 × 7.32   **4.** 6.92 × $6.49

Multiply.

5. 0.9 × 8   **6.** 56 × 3.14   **7.** 0.16 × 17   **8.** 49 × 8.707

9. 23 × 7.28   **10.** 19 × 0.657   **11.** 29 × 65.33   **12.** 84 × 18.9

PROBLEM SOLVING
Making an Organized List

Sometimes you can solve a problem by organizing the given information in the form of a list.

The Wayne School is planning a "Sources of Energy" fair. Students will make exhibits that will be displayed on tables set up in the schoolyard. Each table can hold 2 exhibits.

Allen, Beth, Carl, Dina, Elliot, and Fay will each enter exhibits. How many pairs can be formed from these 6 exhibits?

Make a list to show all the possible pairs, or combinations. Let the first letters of the students' names stand for the 6 exhibits.

Allen's exhibit can be displayed alongside any of the 5 other exhibits.

A and B
A and C
A and D
A and E
A and F

Beth's exhibit can be displayed alongside 4 other exhibits. You don't need to include B and A, because that is the same as A and B.

B and C
B and D
B and E
B and F

Carl's exhibit can be paired with 3 other exhibits.

C and D
C and E
C and F

Dina's exhibit can be displayed alongside 2 other exhibits.

D and E
D and F

Elliot's exhibit can be paired with the 1 remaining exhibit.

E and F

Count the number of combinations.

$5 + 4 + 3 + 2 + 1 = 15$

So, 15 combinations can be formed from the 6 exhibits.

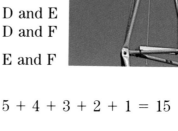

Solve. Use a list if needed.

1. George decides to add 1 more energy exhibit. Now there are 7 energy exhibits. How many different pairs of energy exhibits are possible?

2. Allen, Joan, Kim, and Larry want to prepare exhibits about nuclear power. Their teacher asks them to work together in pairs. How many possible pairs are there?

3. Working in pairs, 4 boys and 4 girls design exhibits about coal. Each pair is made up of 1 boy and 1 girl. How many possible pairs are there?

4. Ronnie, Dale, and Gerry are the finalists in the "Best Exhibit" contest. Prizes will be given for first, second, and third place. In how many different orders could the 3 finalists finish?

5. There will be 4 "Wind Power" exhibits on display at the fair. Dean, Emma, and Frank will take turns demonstrating the exhibits. Each of the 3 students will demonstrate each exhibit twice. How many demonstrations on wind power will there be?

6. Dean has 4 coins in his pocket: a quarter, a dime, a nickel, and a penny. If Dean uses only 2 coins, how many possible combinations that amount to more than 25 cents can he make?

7. Emma has 5 coins in her pocket: a half-dollar, a quarter, a dime, a nickel, and a penny. How many combinations of 4 coins can she make?

8. If each of the books on a display rack covers 2 of the following power-source topics—electricity, natural gas, coal, oil, and nuclear power—how many books are there on the display rack? If each book covered 3 of the topics, would there be more or fewer books on the rack?

9. Nora's demonstration about the use of energy is a prizewinner. After each demonstration, Nora and her 4 assistants each distribute 8 pamphlets about solar energy. Nora gave 6 demonstrations. How many pamphlets did Nora and her team distribute?

Multiplying Decimals

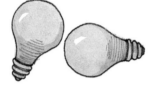
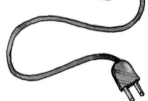

A. The average home in the United States uses about 8.4 kilowatt-hours of electricity every day. How much electricity would such a home use in 2.5 days?

Multiply 2.5 × 8.4.

Multiply as you would with whole numbers.

```
    8.4
 ×  2.5
    420
  1680
  2100
```

Place the decimal point so that the product has as many places as the sum of the decimal places in the factors.

```
    8.4      1 place
 ×  2.5    + 1 place
    420
  1680
  21.00     2 places
```

The average home would use 21 kilowatt-hours of electricity in 2.5 days.

Other examples:

```
   0.25              0.78
 ×  0.6            ×  3.9
  0.150              702
                    2340
                    3.042
```

Write this 0.

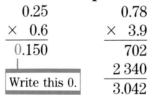

B. When multiplying money, round your answer to the nearest whole cent.

```
  $2.59     2 places
 ×  1.3   + 1 place
    777
  2590
  3.367     3 places
```

$3.367 rounds to $3.37.

Multiply.

1. $\begin{array}{r} 0.35 \\ \times\ \ 0.3 \\ \hline \end{array}$
2. $\begin{array}{r} 0.57 \\ \times\ \ 0.5 \\ \hline \end{array}$
3. $\begin{array}{r} 0.9 \\ \times\ 0.25 \\ \hline \end{array}$
4. $\begin{array}{r} 0.52 \\ \times\ \ 0.7 \\ \hline \end{array}$
5. $\begin{array}{r} 0.65 \\ \times\ \ 0.3 \\ \hline \end{array}$

6. $\begin{array}{r} 0.54 \\ \times\ \ 0.7 \\ \hline \end{array}$
7. $\begin{array}{r} 0.92 \\ \times\ \ 1.4 \\ \hline \end{array}$
8. $\begin{array}{r} 0.24 \\ \times\ \ 4.5 \\ \hline \end{array}$
9. $\begin{array}{r} 0.62 \\ \times\ \ 1.4 \\ \hline \end{array}$
10. $\begin{array}{r} 0.94 \\ \times\ \ 3.8 \\ \hline \end{array}$

11. $\begin{array}{r} 25.87 \\ \times\ \ \ 0.3 \\ \hline \end{array}$
12. $\begin{array}{r} 14.01 \\ \times\ \ \ 3.9 \\ \hline \end{array}$
13. $\begin{array}{r} 0.66 \\ \times\ \ 7.4 \\ \hline \end{array}$
14. $\begin{array}{r} 0.4 \\ \times\ 0.3 \\ \hline \end{array}$
15. $\begin{array}{r} 13.67 \\ \times\ \ \ 0.4 \\ \hline \end{array}$

16. 0.6×0.78
17. 0.52×13.4
18. 0.91×5.8
19. 0.42×6.3

20. 0.2×27.48
21. 0.8×4.84
22. 0.11×1.6
23. 0.6×3.27

Multiply. Round the product to the nearest cent.

24. $\begin{array}{r} \$0.31 \\ \times\ \ 0.6 \\ \hline \end{array}$
25. $\begin{array}{r} \$26.85 \\ \times\ \ \ \ 0.5 \\ \hline \end{array}$
26. $\begin{array}{r} \$7.37 \\ \times\ \ \ 0.4 \\ \hline \end{array}$
27. $\begin{array}{r} \$1.38 \\ \times\ \ \ 0.9 \\ \hline \end{array}$
28. $\begin{array}{r} \$2.75 \\ \times\ \ \ 0.7 \\ \hline \end{array}$

29. $\$0.50 \times 0.37$
30. $\$0.30 \times 28.38$
31. $\$3.67 \times 5.98$
32. $\$0.50 \times 22.09$

Solve.

33. The Wu's use an average of 391.4 kilowatt-hours of electricity each month. They pay $0.18 per kilowatt hour. What is their average monthly bill?

34. The Wu's install storm windows. They cost $212.49, but there is a rebate of 0.2 of the price. How much do the Wu's receive as their rebate?

35. To save on their electric bill, the Wu's stopped using a space heater. The heater had cost $0.43 per hour and had been used 3.5 hours daily. How much money was saved per day?

36. It costs $0.03 per hour to use a 100-watt lightbulb and $0.02 per hour for a 75-watt bulb. If you replace the 100-watt bulbs in your house with 75 watt bulbs, how much would you save in 5 hours?

ANOTHER LOOK

Subtract.

1. $\begin{array}{r} 2,006 \\ -\ 1,987 \\ \hline \end{array}$
2. $\begin{array}{r} 529,878 \\ -\ 432,090 \\ \hline \end{array}$
3. $\begin{array}{r} 500 \\ -\ 456 \\ \hline \end{array}$
4. $\begin{array}{r} 27,037 \\ -\ 15,953 \\ \hline \end{array}$
5. $\begin{array}{r} 3,010 \\ -\ 3,002 \\ \hline \end{array}$

6. $39,480 - 25,869$
7. $202 - 199$
8. $40,170 - 28,198$

More Multiplying Decimals

Some solar-energy cells supply only 0.005 kilowatts of energy per minute. Tom Pearson's small electric generator supplies 3.5 times this amount of energy. How much energy does the generator supply?

Multiply 3.5 × 0.005.

Sometimes you need to write zeros in the product to place the decimal point correctly.

Multiply as you would with whole numbers.	Add zeros to show the correct number of decimal places. Place the decimal point in the product.

$$\begin{array}{r} 3.5 \\ \times\, 0.0\,0\,5 \\ \hline 1\,7\,5 \end{array}$$

$$\begin{array}{rl} 3.5 & \text{1 place} \\ \times\, 0.0\,0\,5 & +\text{3 places} \\ \hline 0.0\,1\,7\,5 & \text{4 places} \end{array}$$

The generator supplies 0.0175 kilowatts per minute.

Other examples:

$$\begin{array}{rl} 0.1 & \text{1 place} \\ \times\, 0.08 & +\text{2 places} \\ \hline 0.008 & \text{3 places} \end{array}$$

$$\begin{array}{rl} 1.3 & \text{1 place} \\ \times\, 0.002 & +\text{3 places} \\ \hline 0.0026 & \text{4 places} \end{array}$$

$$\begin{array}{rl} \$1.01 & \text{2 places} \\ \times\, 0.05 & +\text{2 places} \\ \hline \$0.0505 & \text{4 places} \end{array}$$

> **$0.0505 rounded to the nearest cent is $0.05.**

Checkpoint Write the letter of the correct answer.

Multiply.

1. 0.08 × 1.2

2. $\begin{array}{r} 0.3 \\ \times\, 0.12 \end{array}$

3. 0.04 × 0.05

a. 0.86	**a.** 0.036	**a.** 0.0002
b. 0.096	**b.** 0.36	**b.** 0.002
c. 0.96	**c.** .360	**c.** 0.0200
d. 0.960	**d.** 360	**d.** 0.2000

Multiply.

1. $\begin{array}{r} 0.07 \\ \times\ 0.3 \\ \hline \end{array}$	**2.** $\begin{array}{r} 0.31 \\ \times\ 0.3 \\ \hline \end{array}$	**3.** $\begin{array}{r} 0.16 \\ \times\ 0.4 \\ \hline \end{array}$	**4.** $\begin{array}{r} 0.25 \\ \times\ 0.3 \\ \hline \end{array}$	**5.** $\begin{array}{r} 0.08 \\ \times\ 0.3 \\ \hline \end{array}$
6. $\begin{array}{r} 0.03 \\ \times\ 1.6 \\ \hline \end{array}$	**7.** $\begin{array}{r} 0.02 \\ \times\ 2.5 \\ \hline \end{array}$	**8.** $\begin{array}{r} 0.04 \\ \times\ 1.7 \\ \hline \end{array}$	**9.** $\begin{array}{r} 0.03 \\ \times\ 2.2 \\ \hline \end{array}$	**10.** $\begin{array}{r} 0.02 \\ \times\ 2.4 \\ \hline \end{array}$
11. $\begin{array}{r} 0.27 \\ \times\ 0.3 \\ \hline \end{array}$	**12.** $\begin{array}{r} 0.02 \\ \times\ 1.9 \\ \hline \end{array}$	**13.** $\begin{array}{r} 0.42 \\ \times\ 0.2 \\ \hline \end{array}$	**14.** $\begin{array}{r} 0.04 \\ \times\ 1.4 \\ \hline \end{array}$	**15.** $\begin{array}{r} 0.51 \\ \times\ 0.1 \\ \hline \end{array}$

16. 1.1×0.07 **17.** 0.08×0.8 **18.** 3.6×0.02 **19.** 0.03×0.5

20. 0.06×0.5 **21.** 3.1×0.03 **22.** 4.3×0.02 **23.** 0.05×0.9

Multiply. Round the product to the nearest cent.

24. $\begin{array}{r} \$4.17 \\ \times\ 0.02 \\ \hline \end{array}$	**25.** $\begin{array}{r} \$3.92 \\ \times\ 0.02 \\ \hline \end{array}$	**26.** $\begin{array}{r} \$3.17 \\ \times\ 0.03 \\ \hline \end{array}$	**27.** $\begin{array}{r} \$1.03 \\ \times\ 0.07 \\ \hline \end{array}$	**28.** $\begin{array}{r} \$2.43 \\ \times\ 0.04 \\ \hline \end{array}$

Solve. For Problem 30, use the Infobank.

29. A power plant produces 0.048 kilowatt-hours of electricity per person. A second plant produces twice as much. How much electricity is produced by the second plant?

30. Use the information on page 416 to find the cost of using 8 hours of electricity in New York, El Paso, and Chicago. Then write and solve your own word problem.

CALCULATOR

Try to multiply $\boxed{0.00004}$ $\boxed{\times}$ $\boxed{0.00007}$. Most calculators cannot display the answer because there are not enough spaces to show the complete product. To find the correct answer, multiply $\boxed{4}$ $\boxed{\times}$ $\boxed{7}$. Then place the decimal point in the correct place. Your product should be 0.0000000028.

Use this technique with your calculator to find the product.

1. 0.00006×0.00003 **2.** 0.00008×0.00007 **3.** 0.00005×0.00003

4. 0.000004×0.000002 **5.** 0.00009×0.00009 **6.** 0.00006×0.000021

More Practice, page 427

PROBLEM SOLVING
Identifying Needed Information

Some problems do not contain all the information needed to solve them. Sometimes you can find this information. Sometimes you cannot. Use the checklist to help you solve the problem.

> If you burned a 100-watt light bulb day and night for a whole year, you would use about 880 kilowatt-hours of electricity. How much would this cost?

Use the checklist to solve the problem.

What information is needed? the price of a kilowatt hour of electricity

Checklist	Yes	No
Do I already know the information?		✓
Can I find the information		
in a magazine article?		✓
in a reference book?		✓
from another person/the electric company?	✓	
Am I really stuck?		✓

You can contact your local electric company to find the cost per kilowatt-hour in your area. You can then multiply that amount by 880 to find the total cost.

Choose the information you would need to solve each problem. Write the letter of the correct answer.

1. The Lees spend $67.40 yearly to run their air conditioner. How much money could they save by using a large fan instead?

    a. the number of summer months
    b. the cost of running a large fan
    c. the size of the fan

2. The price of heating oil tripled between 1978 and 1983. How much did the price go up?

    a. the price of oil in 1978
    b. the amount of oil produced in 1983
    c. the amount of oil used in 1978

Solve. If there is not enough information, write what information you would need.

3. From a 42-gallon barrel of crude oil, 19.5 gallons of gasoline can be made. How many gallons of gasoline can be made from 50 barrels of crude oil?

4. In 1984, Hawaiians paid 11.29¢ per kilowatt hour for electricity. To the nearest cent, how much did 8 kilowatt-hours cost in Hawaii in 1984?

5. The Grand Coulee Dam and the John Day Plant are the first- and the second-largest hydroelectric plants in this country. The Grand Coulee can produce 6,494,000 kilowatts of power. How much less power does the John Day produce?

6. The Steger family's color TV used 320 kilowatt-hours of electricity last year. The cost per kilowatt-hour was 7.52¢. To the nearest cent, how much did it cost the Stegers to use their TV for one year?

7. In 1984, the normal yearly cost of operating a clothes dryer was $74.67. How much higher was the cost in 1984 than in 1983?

8. If the price of gasoline was $1.36 per gallon, how much would it cost to drive a car a distance of 5,000 miles?

9. In 1983, coal produced 0.25 of our electric power, gas produced 0.1, and nuclear power produced 0.125. What part of our electrical power was not produced by coal, gas, nuclear power, or hydroelectric power?

10. About 5.7 barrels of crude oil provide the same amount of energy as 2,000 pounds of coal. How many barrels of oil would provide the same amount of energy as 4,000 pounds of coal?

READING MATH

Read this poem.

The moon shining in the water
Looks like a silver coin
Dropped at my feet.

What is the second word you read? What is the sixth word? What are the last four words?

How do you read a poem? Your eyes follow a specific path.

- You start with the first word at the left in the top line.

- You read across the line, from left to right to the end.

- Then you return to the left and move down a line.

- Then you read from left to right again.

You don't read across, from left to right, in math.
How would you solve this problem?
What number would you write first? second? third?

$$
\begin{array}{r}
48 \\
\times\ 7 \\
\hline
\end{array}
\qquad
\begin{array}{r}
48 \\
\times\ 7 \\
\hline
6
\end{array}
\qquad
\begin{array}{r}
{}^{5}\ \\
4\,8 \\
\times\ 7 \\
\hline
6
\end{array}
\qquad
\begin{array}{r}
{}^{5}\ \\
4\,8 \\
\times\ 7 \\
\hline
3\,3\,6
\end{array}
$$

When you multiply those numbers, you work from top to bottom and move from right to left.

Read each problem below. Write down the order in which you read and write numbers.

1. $\begin{array}{r} 59 \\ +\ 43 \\ \hline \end{array}$
2. $\begin{array}{r} 625 \\ -\ 86 \\ \hline \end{array}$
3. $\begin{array}{r} 63 \\ \times\ 9 \\ \hline \end{array}$

GROUP PROJECT

Fuel Costs—A Burning Question

The problem: Your family wants to cut down its heating bill. They are considering buying a fireplace or a wood stove to replace the oil burner. Read the facts below. Decide whether or not your family should replace the oil burner with a fireplace or a wood stove. Discuss the reasons with your classmates.

Key Facts

- There will be a cost to convert from oil to wood-burning methods of heating.
- It would cost about $220.00 per month to heat your home with oil during the winter.
- It would cost about $130.00 per month during the winter to heat your home with a fireplace or a wood stove.
- A fireplace will not be affected by a power shortage.
- A fireplace needs to be cleaned daily.
- Oil is delivered to your house.
- You must cut and transport wood yourself.
- You must obey any regulations about fireplaces and wood stoves.

CHAPTER TEST

Estimate. Write > or < for ●. (page 106)

1. 1.3×5.9 ● 10

2. 2.6×8.4 ● 20

3. 7.2×4.1 ● 30

4. $3.7 \times \$54.19$ ● $200

5. $5.4 \times \$91.20$ ● $500

Estimate. (page 106)

6.   6.2
   $\times 7.4$

7. $12.75
   $\times \quad 5.2$

8.   9.4
   $\times 3.8$

9. $25.39
   $\times \quad 6.3$

Multiply. (pages 110, 114, and 116)

10.   27
   $\times 0.4$

11.   69
   $\times 0.3$

12.   31
   $\times 0.9$

13.   47
   $\times 0.2$

14.   55
   $\times 4.1$

15.   88
   $\times 6.4$

16.   11
   $\times 8.7$

17.   36
   $\times 9.2$

18. $27.11
   $\times \quad 0.5$

19.   32.56
   $\times \quad 0.4$

20.   0.73
   $\times 8.2$

21. $0.94
   $\times \quad 3.3$

22. 0.04×0.2

23. 0.004×3.1

24. $0.03 \times \$2.03$

25. 0.008×4.2

26. 0.3×41.84

27. 0.19×50.6

28. $4.2 \times \$12.37$

29. $0.06 \times \$31.94$

Solve. If there is not enough information, write what information is needed to solve. (pages 118 and 119)

30. Pat helps his family conserve gasoline by riding his bike to school. Pat figures out that his average speed while bicycling to school is 4.5 miles per hour. Pat's older brother, Phil, can ride 3 times as fast as that. What is the speed of Pat's younger brother?

31. Sally and her family are conserving energy by riding bicycles and walking whenever possible. One day, Sally rode her bicycle 27.53 miles. Her sister, Jane, rode her bicycle 4 times that distance. How many miles did Jane ride?

Solve. (pages 112 and 113)

32. The fifth grade class at Palston Elementary prepares a report on energy. The students are given a choice of 4 topics from which they must choose 2. The topics are solar energy, windmills, nuclear power, and water power. How many different reports can be written?

33. To go with the energy reports, 6 students will make drawings of 3 of the 4 topics. How many different combinations of drawings will there be?

BONUS

Solve.

1. $(2.35 \times 4.76) + (3.03 \times 4.1)$

2. $(7.21 \times 4.36) - (0.74 \times 5.63)$

3. $(0.98 \times 4.31) + (6.2 \times 4.07)$

4. $(5.32 \times 0.63) - (0.06 \times 0.14)$

5. $(3.75 \times 4.09) - (0.11 \times 0.23)$

6. $(0.57 \times 0.35) + (7.59 \times 3.16)$

7. $(0.12 \times 4.97) + (5.07 \times 3.1)$

8. $(4 \times 2.5986) - (0.1 \times 3.54)$

RETEACHING

You need to count the decimal places in the factors in order to place the decimal point correctly in the product.

Multiply 3.7×4.052.

Multiply as you would with whole numbers.

$$\begin{array}{r} 4.052 \\ \times\ \ \ 3.7 \\ \hline 149924 \end{array}$$

Place the decimal point so that the product has as many places as the sum of the decimal places in the factors.

$$\begin{array}{rl} 4.052 \rightarrow & 3 \text{ places} \\ \times\ \ \ 3.7 \rightarrow & +\ 1 \text{ place} \\ \hline 14.9924 & 4 \text{ places} \end{array}$$

Other examples:

$$\begin{array}{rl} 1.34 \rightarrow & 2 \text{ places} \\ \times\ \ 0.2 \rightarrow & +\ 1 \text{ place} \\ \hline 0.268 \rightarrow & 3 \text{ places} \end{array}$$

$$\begin{array}{rl} 1.23 \rightarrow & 2 \text{ places} \\ \times\ \ \ \ 3 \rightarrow & +\ 0 \text{ places} \\ \hline 3.69 \rightarrow & 2 \text{ places} \end{array}$$

$$\begin{array}{rl} 0.09 \rightarrow & 2 \text{ places} \\ \times\ \ 7.5 \rightarrow & +\ 1 \text{ place} \\ \hline 0.675 \rightarrow & 3 \text{ places} \end{array}$$

Multiply.

1. $\begin{array}{r} 0.445 \\ \times\ \ \ 0.6 \\ \hline \end{array}$

2. $\begin{array}{r} 3.35 \\ \times\ \ 0.1 \\ \hline \end{array}$

3. $\begin{array}{r} 0.769 \\ \times\ \ \ 0.8 \\ \hline \end{array}$

4. $\begin{array}{r} 0.448 \\ \times\ \ 16.9 \\ \hline \end{array}$

5. $\begin{array}{r} 41.2 \\ \times\ 0.06 \\ \hline \end{array}$

6. $\begin{array}{r} 0.012 \\ \times\ \ 10.6 \\ \hline \end{array}$

7. $\begin{array}{r} 7.12 \\ \times\ 0.04 \\ \hline \end{array}$

8. $\begin{array}{r} 3.42 \\ \times\ \ 6.9 \\ \hline \end{array}$

9. $\begin{array}{r} 9.8 \\ \times\ 7.4 \\ \hline \end{array}$

10. $\begin{array}{r} 0.06 \\ \times\ \ 8.9 \\ \hline \end{array}$

11. $\begin{array}{r} 21.02 \\ \times\ \ 0.03 \\ \hline \end{array}$

12. $\begin{array}{r} 0.073 \\ \times\ \ \ 3.8 \\ \hline \end{array}$

13. $\begin{array}{r} 6.108 \\ \times\ \ \ 0.6 \\ \hline \end{array}$

14. $\begin{array}{r} 3.81 \\ \times\ \ 0.3 \\ \hline \end{array}$

15. $\begin{array}{r} 8.12 \\ \times\ \ 0.6 \\ \hline \end{array}$

16. $\begin{array}{r} 7.644 \\ \times\ \ \ 0.7 \\ \hline \end{array}$

17. $\begin{array}{r} 0.699 \\ \times\ \ \ 6.5 \\ \hline \end{array}$

18. $\begin{array}{r} 2.397 \\ \times\ \ \ 0.7 \\ \hline \end{array}$

19. $\begin{array}{r} 3.981 \\ \times\ \ \ 0.9 \\ \hline \end{array}$

20. $\begin{array}{r} 1.07 \\ \times\ 0.33 \\ \hline \end{array}$

21. 2.7×0.35

22. 0.89×3.06

23. 0.37×0.64

24. 3.4×1.067

25. 0.4×2.39

26. 1.44×0.23

27. 0.374×6.3

28. 6.07×2.4

ENRICHMENT

Scientific Notation

The speed of light is about 300,000 kilometers per second. Large numbers like this are often written in **scientific notation.**

Scientific notation is based on powers of 10.

$$10 = 10^1 \qquad 100 = 10^2 \qquad 1,000 = 10^3 \qquad 10,000 = 10^4$$

Notice that the exponent equals the number of zeros in the standard numeral.

Scientific notation has two factors. One factor is a number between 1 and 10. The other factor is a power of 10.

Write 300,000 in scientific notation.

Move the decimal point to the right of the first digit.

$$3 \,.\, 0\ 0\ 0\ 0\ 0$$

Count the number of places you moved the decimal point to find the power of 10.

$$3 \,.\, 0\ 0\ 0\ 0\ 0 \qquad \boxed{\text{5 decimal places}}$$

$$300,000 = 3 \times 100,000 = 3 \times 10^5$$

Written in scientific notation, 300,000 is 3×10^5.

The speed of light is also expressed as 186,000 miles per second. Write 186,000 in scientific notation.

Move the decimal to the right of the first digit.

$$1 \,.\, 8\ 6\ 0\ 0\ 0$$

Count the number of places you moved the decimal point to find the power of 10.

$$186,000 = 1.86 \times 100,000 = 1.86 \times 10^5$$

Written in scientific notation, 186,000 is 1.86×10^5.

Write each number in scientific notation.

1. 80,000

2. 500,000

3. 2,000,000

4. 36,000

5. 783,000

6. 18,000,000

Write the number.

7. 4×10^4

8. 5.5×10^4

9. 9×10^5

10. 2.1×10^4

11. 3.04×10^5

12. 1.3×10^7

TECHNOLOGY

This procedure in LOGO draws the square that is shown in the picture.

```
TO SQUARE
FD 40 RT 90 FD 40 RT 90 FD 40 RT 90 FD 40 RT 90
END
```

Here is a shorter way to write the same procedure.

```
TO SQUARE
REPEAT 4 [FD 40 RT 90]
END
```

The REPEAT 4 command means that the two other commands are repeated 4 times.

Suppose you wanted to change the lengths of the square's sides. An easy way to do this is to use a **variable.**

A LOGO variable is part of a procedure that changes. When a procedure uses a variable, you must give the variable a value. This example draws a square.

```
TO SQUARE :SIDE
REPEAT 4 [FD :SIDE RT 90]
END
```

The variable is :SIDE. Always use the colon (:) in a variable. When you type the procedure name, you also type a number that tells the turtle the value to give the variable. So, to draw a square that has sides equal to the ones above, type this.

```
SQUARE 40
```

To change the lengths of the square's sides, substitute a different number for 40.

1. Identify the variable in this procedure name.

   TO DASHEDLINE :DASH

   REPEAT 3 [FD :DASH PU FD:DASH PD]

   END

2. How long is each dash if you type this command?

   DASHEDLINE 15

3. Write a command to tell the turtle to draw SQUARE with sides that are 75 steps long.

4. Identify the variable in the procedure below. Then copy the figure, and follow the procedure to finish the drawing. (The sides are 50 steps long in this drawing.)

   TO GUESS :SIZE
   REPEAT 5 [FD   RT 72]
   END

5. What is the figure?

6. Write a command to tell the turtle to use sides of 100 steps for this figure.

CUMULATIVE REVIEW

Write the letter of the correct answer.

1. $6 \times 13,459$

    a. 60,754   **b.** 68,404
    c. 80,754   **d.** not given

2. Estimate: 728×3.

    a. 2,000   **b.** 2,010
    c. 2,100   **d.** not given

3. 32×47

    a. 1,404   **b.** 1,504
    c. 2,305   **d.** not given

4. $(6 \times 0) \times 9$

    a. 0   **b.** 54
    c. 540   **d.** not given

5. $609 \times \$7.58$

    a. \$4,615.22   **b.** \$4,617.32
    c. \$4,719.00   **d.** not given

6. $40 \times 70,000$

    a. 280,000   **b.** 2,800,000
    c. 28,000,000   **d.** not given

7. Compare. Choose $>$, $<$, or $=$ for ●.
0.117 ● 0.12

    a. $<$ **b.** $>$
    c. $=$ **d.** not given

8. Write the standard form:
3 million, 62 thousand, thirty.

    a. 3,062,030   **b.** 3,620,030
    c. 3,620,300   **d.** not given

9. $53,906 - 29,749$

    a. 23,657   **b.** 24,007
    c. 24,157   **d.** not given

10. $0.11 + 7.971 + 6.5$

    a. 13.581   **b.** 14.581
    c. 145.81   **d.** not given

11. Write as a decimal:
sixty-eight thousandths.

    a. 0.68   **b.** 6.8
    c. 0.068   **d.** not given

Interpret the bar graph.

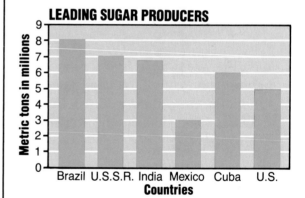

12. What country produces the most sugar?

    a. India   **b.** Brazil
    c. U.S.S.R.   **d.** not given

13. How much sugar is produced by both Cuba and the U.S.S.R.?

    a. 11,000,000 tons
    b. 12,000,000 tons
    c. 13,000,000 tons
    d. not given

Help your class organize a Sports Festival Day. Plan ten different games. Each person must play a minimum of 3 games. Most of the games should be group games, but you may include one or two individual games.

5 DIVIDING WHOLE NUMBERS

Division Facts

A. Jim's baseball team calls itself the Doubledays in honor of Abner Doubleday, the founder of the game. The team buys 36 uniform shirts with the team name on them. The shirts are packed 9 to a box. How many boxes do they buy?

You can subtract to find how many groups of 9 there are in 36.

$36 - 9 = 27$; $27 - 9 = 18$; $18 - 9 = 9$; $9 - 9 = 0$
There are four 9's in 36.

It is easier to divide to find the number of groups.

Find $36 \div 9$.

Think: ■ **× 9 = 36.**
    $4 \times 9 = 36$ **So, $36 \div 9 = 4$.**

The team buys 4 boxes of shirts.

B. The team also purchases 48 new baseball caps. They are packed equally into 6 boxes. How many hats are packed into each box?

You can divide to find the number in each group.

Divide $6\overline{)48}$.

Think: ■ **× 6 = 48.** 8
    $8 \times 6 = 48$ **So, $6\overline{)48}$.**

There are 8 hats packed into each box.

C. Division can be shown in three ways.

$$\begin{array}{r} 7 \\ 9\overline{)63} \end{array} \quad \longleftarrow \text{quotient}$$
$$\longleftarrow \text{dividend}$$
$$\uparrow$$
divisor

$$63 \div 9 = 7$$
$$\uparrow \quad \uparrow \quad \uparrow$$
dividend divisor quotient

$$\text{dividend} \longrightarrow \frac{63}{9} = 7$$
$$\text{divisor} \longrightarrow$$
$$\uparrow$$
quotient

Divide.

1. $6\overline{)48}$ 2. $8\overline{)32}$ 3. $7\overline{)56}$ 4. $4\overline{)28}$ 5. $3\overline{)27}$

6. $9\overline{)72}$ 7. $8\overline{)64}$ 8. $5\overline{)30}$ 9. $3\overline{)21}$ 10. $6\overline{)54}$

11. $5\overline{)15}$ 12. $9\overline{)18}$ 13. $4\overline{)36}$ 14. $6\overline{)36}$ 15. $7\overline{)35}$

16. $6\overline{)42}$ 17. $9\overline{)81}$ 18. $7\overline{)28}$ 19. $3\overline{)12}$ 20. $4\overline{)36}$

21. $5\overline{)45}$ 22. $8\overline{)72}$ 23. $3\overline{)18}$ 24. $7\overline{)42}$ 25. $9\overline{)27}$

26. $54 \div 6$ 27. $20 \div 4$ 28. $24 \div 3$ 29. $28 \div 4$ 30. $63 \div 9$

31. $14 \div 2$ 32. $49 \div 7$ 33. $63 \div 7$ 34. $40 \div 5$ 35. $18 \div 6$

36. $56 \div 8$ 37. $42 \div 6$ 38. $27 \div 9$ 39. $32 \div 4$ 40. $20 \div 5$

41. $63 \div 7$ 42. $30 \div 6$ 43. $45 \div 5$ 44. $81 \div 9$ 45. $24 \div 8$

46. $\frac{42}{7}$ 47. $\frac{25}{5}$ 48. $\frac{16}{4}$ 49. $\frac{35}{7}$ 50. $\frac{28}{4}$

Solve.

51. The team runs 3 laps around the field before every game. At the end of the 8-game season, how many laps has the team run?

52. The Doubledays' bleachers can seat 81 people. If there are 9 rows and each row has an equal number of seats, how many seats are there in each row?

53. When the Doubleday team won the championship, the coach divided a box of 18 brand-new baseballs equally among his 9 players. How many baseballs did each player receive?

★54. At the end of the season, 77 people were invited to the team's party. Of the people invited, 5 did not go. The rest of the guests were seated at 8 tables, with an equal number at each table. How many guests were there at each table?

CHALLENGE

1. The divisor is 6 and the dividend is 54. What is the quotient?

2. The quotient is 7 and the dividend is 49. What is the divisor?

3. The divisor is 9 and the quotient is 8. What is the dividend?

4. The dividend is 32 and the divisor is 4. What is the quotient?

Related Facts

A. The Explorer Club is planning a canoe trip to retrace French explorer Louis Joliet's 1673 journey up the Mississippi River. They plan to paddle 63 miles in all, traveling the same number of miles each day for 7 days. How many miles will they travel each day?

Find $7\overline{)63}$.

Think: $\blacksquare \times 7 = 63$.

$9 \times 7 = 63$ So, $63 \div 7 = 9$.

They will travel 9 miles each day.

B. You can write four number sentences using 7, 9, and 63. The number sentences make up a family of facts.

$9 \times 7 = 63$ $63 \div 9 = 7$
$7 \times 9 = 63$ $63 \div 7 = 9$

Use these facts to help you find missing factors.

Complete the number sentence $9 \times \blacksquare = 63$.

Think: $63 \div 9 = 7$.   So, $\blacksquare = 7$.

C. These rules will help you divide.

Any number divided by 1 is that number.	$8 \div 1 = 8$
0 divided by another number, except 0, is 0.	$0 \div 6 = 0$
Any number, except 0, divided by itself is 1.	$5 \div 5 = 1$
A number can never be divided by 0. Think: $\blacksquare \times 0 = 7$. There is no solution.	$7 \div 0 = \blacksquare$

Divide.

1. $5\overline{)35}$ **2.** $9\overline{)9}$ **3.** $4\overline{)32}$ **4.** $6\overline{)36}$ **5.** $7\overline{)49}$

6. $3\overline{)24}$ **7.** $9\overline{)45}$ **8.** $5\overline{)40}$ **9.** $7\overline{)63}$ **10.** $9\overline{)27}$

11. $7\overline{)0}$ **12.** $5\overline{)20}$ **13.** $6\overline{)54}$ **14.** $3\overline{)18}$ **15.** $8\overline{)64}$

16. $4\overline{)24}$ **17.** $3\overline{)21}$ **18.** $6\overline{)48}$ **19.** $5\overline{)45}$ **20.** $8\overline{)32}$

21. $36 \div 4$ **22.** $16 \div 4$ **23.** $48 \div 6$ **24.** $54 \div 9$ **25.** $9 \div 1$

26. $64 \div 8$ **27.** $14 \div 2$ **28.** $\frac{42}{6}$ **29.** $\frac{18}{2}$ **30.** $\frac{72}{9}$

Write a family of facts for each set of numbers.

31. 3, 4, 12   **32.** 30, 5, 6   **33.** 21, 7, 3   **34.** 7, 56, 8   **35.** 9, 7, 63

Complete each number sentence.

36. $6 \times \blacksquare = 48$ **37.** $4 \times \blacksquare = 36$ ★**38.** $20 \div \blacksquare = 4$ ★**39.** $45 \div \blacksquare = 5$

Solve.

40. The club plans a picnic at Stony Meadows Park for members and their families. They expect 72 people. The park's picnic tables can seat 8 people each. How many picnic tables will the club need?

★**41.** The club sponsors a day of sailboat races. In one race, 4 boats compete. Each boat has 7 people aboard. In a second race, twice the number of people take part, and they race 7 sailboats. If there is an equal number of people per boat, how many are aboard each boat in the second race?

FOCUS: MENTAL MATH

You can use division rules and related facts to mentally compute these exercises.

1. $2{,}350 \div \blacksquare = 1$ **2.** $\blacksquare \div 84{,}650 = 0$ **3.** $3{,}450 \div \blacksquare = 3{,}450$

4. $5 \times 145 = 725$ **5.** $38 \times 3{,}552 = 134{,}976$ **6.** $27 \times 6{,}854 = 185{,}058$

     $\blacksquare \div 5 = 145$ $\blacksquare \div 38 = 3{,}552$ $185{,}058 \div 6{,}854 = \blacksquare$

Quotients and Remainders

Tina is a photographer for her hometown newspaper. At a recent hang-glider meet, she took 36 photos of gliders in flight. After she developed the photos, she divided them equally among 5 friends. How many photos did each friend receive? How many photos did Tina keep for herself?

Find $36 \div 5$.

Think: ▢ × 5 is close to 36.

$6 \times 5 = 30$
$7 \times 5 = 35$
$8 \times 5 = 40$ Too great.   So, use 7.

$$\begin{array}{r} 7 \text{ R1} \\ 5\overline{)3\ 6} \\ 3\ 5 \\ \hline 1 \end{array}$$

> **Multiply: 7 × 5 = 35.**
> **Subtract: 36 − 35 = 1.**
> **Compare: 1 < 5.**
> **Write the remainder.**

The remainder tells you how many Tina kept for herself.

Each of Tina's friends received 7 photographs. She kept 1 for herself.

You can check division by multiplying and adding the remainder.

$$\begin{array}{r} 7 \longleftarrow \text{quotient} \\ \times\ 5 \longleftarrow \text{divisor} \\ \hline 35 \\ +\ 1 \longleftarrow \text{remainder} \\ \hline 36 \longleftarrow \text{dividend} \end{array}$$

The answer should be the same as the dividend.

Checkpoint   Write the letter of the correct answer.

Divide and check.

1. $55 \div 6$

a. 8 R7
b. 9
c. 9 R1
d. 10

2. $6\overline{)41}$

a. 5 R11
b. 6 R5
c. 7
d. 7 R1

3. $74 \div 9$

a. 7 R11
b. 8
c. 8 R2
d. 9

Divide.

1. $4\overline{)34}$ **2.** $3\overline{)26}$ **3.** $5\overline{)32}$ **4.** $7\overline{)45}$ **5.** $5\overline{)21}$

6. $3\overline{)20}$ **7.** $6\overline{)37}$ **8.** $6\overline{)41}$ **9.** $7\overline{)50}$ **10.** $8\overline{)44}$

11. $5\overline{)28}$ **12.** $9\overline{)74}$ **13.** $7\overline{)60}$ **14.** $8\overline{)52}$ **15.** $5\overline{)41}$

16. $6\overline{)39}$ **17.** $8\overline{)31}$ **18.** $9\overline{)84}$ **19.** $5\overline{)26}$ **20.** $4\overline{)13}$

21. $9\overline{)88}$ **22.** $5\overline{)43}$ **23.** $2\overline{)21}$ **24.** $9\overline{)47}$ **25.** $3\overline{)23}$

26. $40 \div 6$ **27.** $32 \div 4$ **28.** $80 \div 9$ **29.** $65 \div 8$ **30.** $25 \div 5$

31. $53 \div 7$ **32.** $62 \div 9$ **33.** $73 \div 8$ **34.** $38 \div 9$ **35.** $48 \div 9$

36. $31 \div 9$ **37.** $43 \div 5$ **38.** $19 \div 2$ **39.** $75 \div 7$ **40.** $60 \div 8$

41. $89 \div 9$ **42.** $44 \div 6$ **43.** $30 \div 4$ **44.** $46 \div 5$ **45.** $21 \div 2$

46. $\frac{62}{8}$ **47.** $\frac{50}{6}$ **48.** $\frac{28}{3}$ **49.** $\frac{75}{8}$ **50.** $\frac{58}{7}$

Solve.

51. Tina takes photos of each of the 8 teams. She promised each team an equal number of prints. If she has only 79 sheets of printing paper, how many prints can she make for each team?

★52. Tina's photographs of the hang gliding are so good that many people want to buy copies. Tina makes 18 prints and sells them in sets of 3. She sells another 45 prints in sets of 5. How many sets of prints does Tina sell?

CHALLENGE

Find the missing digits.

1. $6\overline{)}$ R1 The dividend is between 15 and 20.

2. $7\overline{)}$ R2 The dividend is between 30 and 40.

3. $9\overline{)}$ R2 The dividend is between 70 and 80.

More Practice, page 428

PROBLEM SOLVING
Choosing the Operation

You can use the hints in a problem to help you decide whether you should multiply or divide the numbers in the problem to find the answer.

Jan wears a weight belt when she goes scuba diving. She puts 6 weights on her belt. Each weight has a mass of 0.25 kg. What is the total mass of the weights on Jan's belt?

Hints:

If you know	and you want to find	you can
• how many groups there are • that the number in each group is the same • how many there are in each group	how many in all	multiply.
• how many there are in all • that the number in each group is the same • how many there are in each group	how many groups	divide.
• that the number in each group is the same • how many there are in all • how many groups there are	how many in each group	divide.

Once you have decided, you can solve the problem.

how many in each group		how many groups		how many in all
0.25	×	6	=	1.5

The weights have a total mass of 1.5 kg.

Write the letter of the operation you should use to solve the problem.

1. A group of 24 students wants to rent rowboats. Each boat will hold 4 students. How many boats should be rented?

   a. multiplication
   b. division

2. During an exciting softball game, 3 runs were scored each inning. To break a tie, 10 innings were played. How many runs were scored during the game?

   a. multiplication
   b. division

Solve.

3. When the Dixieland Debs played at the bandshell, 156 students went to hear them. Before the concert ended, 17 of the students left to go to swimming practice. How many students remained at the concert?

4. All 96 students visit the City Museum while they are at the park. The teachers separate them into 8 groups of equal size. How many students are there in each group?

5. Each student is given an apple or an orange for lunch. One teacher hands out 26 apples and twice that number of oranges. How many oranges does the teacher hand out?

6. After lunch, the students organize a volleyball tournament. There are 117 students who want to play, and each team has 9 players. How many teams are there in the tournament?

7. It costs $23.50 to rent a tennis court for an hour. Each court holds 4 players. Some students want to play, but they do not want to spend more than $5.50 each. If they split the cost of renting a court equally among 4 players, would each pay more or less than $5.50.

★8. Before leaving, 73 of the students bought Central Park T-shirts. The shirts were on sale. The sale price was $7 for 3 shirts. If each student received a shirt and the total spent was $172, how many sets of 3 did they buy? How many shirts were left?

Divisibility

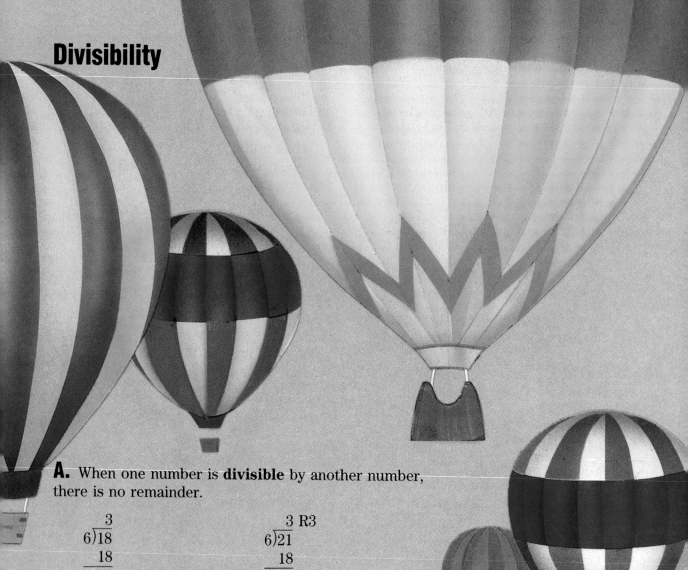

A. When one number is **divisible** by another number, there is no remainder.

$$\begin{array}{r} 3 \\ 6\overline{)18} \\ \underline{18} \\ 0 \end{array} \qquad \begin{array}{r} 3\ \text{R}3 \\ 6\overline{)21} \\ \underline{18} \\ 3 \end{array}$$

So, 18 is divisible by 6. So, 21 is not divisible by 6.

B. You can use the following rules to help you decide whether a number is divisible by 2, 5, or 10.

Even numbers have 0, 2, 4, 6, or 8 in the ones place. They are divisible by 2.

12; 14; 136; 4,468; 13,792

Odd numbers have a 1, 3, 5, 7, or 9 in the ones place. They are not divisible by 2.

21; 43; 255; 6,337; 48,289

Numbers that have a 0 or 5 in the ones place are divisible by 5.

75; 130; 985; 8,640; 94,325

Numbers that have a 0 in the ones place are divisible by 10.

60; 450; 790; 3,600; 87,560

Is the number divisible by 2? Write *yes* or *no*.

1. 17   **2.** 34   **3.** 55   **4.** 70   **5.** 29

6. 4,702   **7.** 6,847   **8.** 34,686

9. 96,755   **10.** 35,423   **11.** 46,758   **12.** 37,959

Is the number divisible by 5? Write *yes* or *no*.

13. 85   **14.** 60   **15.** 45   **16.** 66

17. 30   **18.** 525   **19.** 1,500   **20.** 5,630

21. 49,365   **22.** 26,583   **23.** 17,647   **24.** 45,559

Is the number divisible by 10? Write *yes* or *no*.

25. 62   **26.** 40   **27.** 35   **28.** 800

29. 755   **30.** 8,470   **31.** 6,283   **32.** 7,380

33. 75,050   **34.** 62,427   **35.** 17,850   **36.** 20,008

Copy the chart. Write a check in the box if the number is divisible by the divisor.

Dividend

		37.	**38.**	**39.**	**40.**	**41.**	**42.**	**43.**	**44.**	**45.**	**46.**
		27	72	90	318	775	870	1,150	4,025	8,006	57,330
	2										
Divisor	5										
	10										

CHALLENGE

If the sum of the digits in a number is divisible by 3, the number is divisible by 3.

861     $8 + 6 + 1 = 15$ $15 \div 3 = 5$

861 is divisible by 3.

Is the number divisible by 3? Write *yes* or *no*.

1. 672   **2.** 1,740   **3.** 3,675   **4.** 1,111   **5.** 3,235   **6.** 4,357

7. 1,260   **8.** 3,790   **9.** 2,181   **10.** 1,530   **11.** 938   **12.** 3,438

Estimating Quotients

A. In preparation for the Craft and Hobby Fair, 7 friends plan to make a patchwork quilt. They will need 1,974 squares of fabric. If each one brings an equal number of squares, about how many squares will each person bring?

Estimate 1,974 ÷ 7.

Decide on the number of digits in the quotient.

Divide the thousands: Think: 7)1. Not enough thousands.
Divide the hundreds: Think: 7)19.

So, the quotient begins in the hundreds place.

It will have 3 digits.

$$\overset{---}{7)1{,}974}$$

Think: 2 × 7 = 14.
   3 × 7 = 21   Too great. So, use 2.

$$\overset{2}{7)1{,}974}$$

Write zeros for the other digits. $7)\overline{1{,}974}$ ⟶ 200
You can say that each person will bring about 200 squares.

B. The next step is to decide how accurate your estimate is.

Think: 1,974 squares will be needed. If each person brings 200 squares, will that be enough?
7 × 200 = 1,400; 1,400 < 1,972.
So, 200 squares will not be enough.
It is an underestimate.

Each person will bring more than 200 squares.

Write how many digits the quotient will contain.

1. $6\overline{)358}$ 2. $8\overline{)927}$ 3. $5\overline{)236}$ 4. $4\overline{)647}$ 5. $9\overline{)493}$

6. $8\overline{)6,531}$ 7. $7\overline{)4,346}$ 8. $2\overline{)2,659}$ 9. $5\overline{)5,893}$ 10. $9\overline{)8,453}$

11. $3\overline{)24,876}$ 12. $9\overline{)82,543}$ 13. $4\overline{)45,208}$ 14. $6\overline{)72,349}$ 15. $7\overline{)49,876}$

Estimate. Write the letter of the correct answer.

16. $4\overline{)347}$ a. 8   b. 80   c. 800

17. $7\overline{)784}$ a. 1   b. 10   c. 100

18. $9\overline{)6,453}$ a. 7   b. 70   c. 700

19. $3\overline{)3,642}$ a. 10   b. 100   c. 1,000

20. $4\overline{)14,386}$ a. 30   b. 300   c. 3,000

Estimate.

21. $5\overline{)359}$ 22. $6\overline{)285}$ 23. $4\overline{)138}$ 24. $2\overline{)359}$ 25. $9\overline{)897}$

26. $4\overline{)6,398}$ 27. $3\overline{)2,497}$ 28. $6\overline{)3,621}$ 29. $5\overline{)4,672}$ 30. $8\overline{)5,789}$

31. $3\overline{)26,786}$ 32. $8\overline{)13,463}$ 33. $7\overline{)45,671}$ 34. $6\overline{)92,526}$ 35. $4\overline{)15,628}$

Solve.

36. Marge sews pillows at the crafts fair. She uses 2 yards of fabric for every pillow. If Marge used 386 yards of fabric, could she sew 200 pillows?

37. Jim makes ceramic coffee mugs for the crafts fair. He uses 3 pounds of clay for each mug. If Jim has 230 pounds of clay, about how many coffee mugs can he make?

MIDCHAPTER REVIEW

Divide.

1. $64 \div 8$ 2. $81 \div 9$ 3. $28 \div 4$ 4. $\frac{63}{7}$ 5. $\frac{45}{9}$

6. $7\overline{)59}$ 7. $9\overline{)34}$ 8. $6\overline{)41}$ 9. $\frac{70}{8}$ 10. $\frac{69}{9}$

Is the number divisible by 2? Write *yes* or *no*.

11. 37 12. 48 13. 56 14. 85 15. 74

PROBLEM SOLVING
Writing a Number Sentence

You can write a number sentence to help solve a word problem. Number sentences help you find the number you need by using the numbers you know.

> Tara belongs to the Roamers Bicycle Club. The club cycled 72 miles through the Napa Valley during an 8-day trip. They cycled the same number of miles each day. How many miles did the Roamers ride each day?

— **1.** List what you know and what you want to find.

In 8 days, the club biked 72 miles. They biked the same distance each day. How many miles did they bike each day?

— **2.** Think about how to use this information to solve the problem.

You know the number in each group is the same. You can divide to find how many in each group.

— **3.** Write a number sentence. Use n to stand for the number you want to find.

$$\begin{array}{ccccc} \text{miles biked} & \div & \text{hours} & = & \text{miles biked} \\ \text{in all} & & \text{biked} & & \text{each day} \\ 72 & \div & 8 & = & n \end{array}$$

— **4.** Solve. Write the answer. The club biked 9 miles each day.

$n = 9$

Read the problem. Write the letter of the correct number sentence.

1. One Saturday, Jamie rode his 10-speed bicycle 81 miles. That is 9 times the number of miles his little brother, Pete, rode that day. How many miles did Pete ride?

   a. $9 \div 81 = n$
   b. $81 \times 9 = n$
   c. $81 \div 9 = n$

2. Lynn will ride in a bike-a-thon to raise money for a community park. Lynn's pledges amount to $10.97 for every mile she rides. If Lynn rides 25 miles, how much will she raise for the park?

   a. $\$10.97 \div 25 = n$
   b. $1 \times \$10.97 = n$
   c. $25 \times \$10.97 = n$

Write a number sentence. Solve.

3. On one outing, the Hudson Valley Bike Club rode 45 miles from Newburgh to West Point. The bikers made the trip in 5 hours. How many miles did they travel each hour?

4. The Walla Walla Bike Club plans a three-week bike trip to San Diego. The club needs $1,550.00 for the hotel and food. There is $344.83 in the club's treasury. How much more money is needed?

5. There are 3 bikers who ride in a 45-mile relay race. If each biker rides an equal number of miles, how many miles does each biker ride?

6. The 8 members of a bike club decide to split equally the cost of bicycling caps. The total cost is $32.00. How much must each club member pay?

7. The bikers decide to ride 83 miles per day until they reach the beaches of northern California. If they stop after 15 days, how many miles have they ridden?

8. The fastest recorded speed ever ridden on a bicycle is 78 miles in one hour. At that speed, how many miles would be traveled in 7 hours of biking?

★9. Renee takes $85.00 on a bike trip. She spends $7.95 for a sweatshirt and $6.95 for film. If she spends $15.25 for a concert ticket, how much money will she have left?

★10. Tom and Sandy rode on a bicycle built for two. They biked 6 miles in 36 minutes. At that rate, how far would Tom and Sandy travel in 54 minutes?

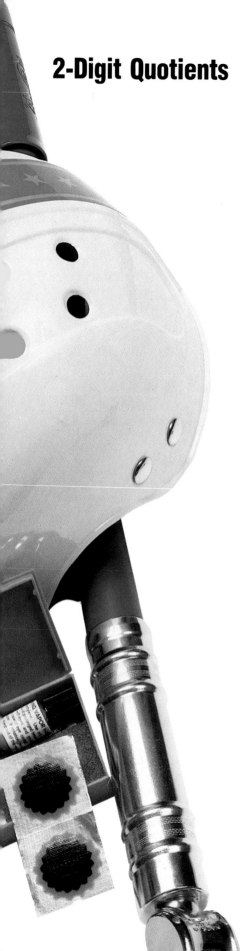

2-Digit Quotients

The *Tour d'Avalon* bicycle race attracts 283 riders. Before the race, all the cyclists ride by the reviewing stand in rows of 9. How many full rows of riders are there? How many riders are there in the last row?

Divide to find how many rows: $9\overline{)283}$.

Divide the hundreds. Think: $9\overline{)2}$. Not enough hundreds.

Divide the tens.

Think: $9\overline{)28}$.
Write 3.

$$\begin{array}{r} 3 \\ 9\overline{)2\,8\,3} \\ 2\,7 \\ \hline 1 \end{array}$$

Multiply.
Subtract.
Compare.

Divide the ones.
Bring down the 3.
Think: $9\overline{)13}$.
Write 1.

$$\begin{array}{r} 3\,1 \ \textbf{R}4 \\ 9\overline{)2\,8\,3} \\ 2\,7 \downarrow \\ \hline 1\,3 \\ 9 \\ \hline 4 \end{array}$$

Multiply.
Subtract.
Compare.
Write the remainder.

Check.

$$\begin{array}{r} 31 \\ \times \quad 9 \\ \hline 279 \\ + \quad 4 \\ \hline 283 \end{array}$$

There are 31 full rows of riders.
There are 4 riders in the last row.

Other examples:

$$\begin{array}{r} 15 \ \text{R}3 \\ 6\overline{)93} \\ 6 \\ \hline 33 \\ 30 \\ \hline 3 \end{array} \qquad \begin{array}{r} 97 \\ 5\overline{)485} \\ 45 \\ \hline 35 \\ 35 \\ \hline 0 \end{array} \qquad \begin{array}{r} 41 \ \text{R}3 \\ 4\overline{)167} \\ 16 \\ \hline 7 \\ 4 \\ \hline 3 \end{array}$$

Checkpoint Write the letter of the correct answer.

Divide.

1. $7\overline{)84}$

a. 10 R4
b. 11 R3
c. 11 R7
d. 12

2. $163 \div 6$

a. 20 R3
b. 27
c. 27 R1
d. 216

3. $5\overline{)127}$

a. 21 R2
b. 25
c. 25 R2
d. 26

Divide.

1. $2\overline{)64}$ **2.** $3\overline{)72}$ **3.** $2\overline{)94}$ **4.** $4\overline{)76}$ **5.** $6\overline{)90}$

6. $2\overline{)49}$ **7.** $4\overline{)78}$ **8.** $3\overline{)85}$ **9.** $4\overline{)97}$ **10.** $2\overline{)55}$

11. $2\overline{)164}$ **12.** $9\overline{)162}$ **13.** $7\overline{)182}$ **14.** $8\overline{)392}$ **15.** $9\overline{)315}$

16. $7\overline{)573}$ **17.** $3\overline{)229}$ **18.** $3\overline{)631}$ **19.** $9\overline{)650}$ **20.** $6\overline{)486}$

21. $6\overline{)427}$ **22.** $8\overline{)310}$ **23.** $7\overline{)217}$ **24.** $4\overline{)386}$ **25.** $9\overline{)709}$

26. $124 \div 8$ **27.** $212 \div 5$ **28.** $321 \div 3$ **29.** $654 \div 9$ **30.** $336 \div 8$

31. $417 \div 5$ **32.** $556 \div 8$ **33.** $301 \div 4$ **34.** $628 \div 8$ **35.** $374 \div 9$

36. $\frac{616}{7}$ **37.** $\frac{525}{5}$ **38.** $\frac{488}{6}$ **39.** $\frac{268}{4}$ **40.** $\frac{715}{8}$

Solve.

41. The race will last for 8 days and will cover a total of 528 miles. If the riders travel the same number of miles each day, how many miles do they cover the first day?

42. The Woodside Team practiced for a total of 144 hours in 8 weeks. They practiced for the same number of hours each week. For how many hours did they practice each week?

43. A bicycle team from Claremont will donate $0.25 to a local charity for each mile they race. They race a total of 528 miles. How much money will the charity receive?

★44. After the race, each of the 287 riders and 64 staff members received 2 free T-shirts. The shirts were packed 9 to a box. How many boxes of shirts were needed?

ANOTHER LOOK

Subtract.

1. $\begin{array}{r} 32,465 \\ -\ 3,523 \end{array}$ **2.** $\begin{array}{r} 654 \\ -571 \end{array}$ **3.** $\begin{array}{r} 1,786 \\ -\ 787 \end{array}$ **4.** $\begin{array}{r} 23,777 \\ -19,499 \end{array}$

5. $\begin{array}{r} 65,434 \\ -65,404 \end{array}$ **6.** $\begin{array}{r} 5,748 \\ -4,999 \end{array}$ **7.** $\begin{array}{r} 22,222 \\ -21,311 \end{array}$ **8.** $\begin{array}{r} 98,000 \\ -27,564 \end{array}$

9. $76,298 - 54,323$ **10.** $6,040 - 3,506$ **11.** $60,763 - 28,975$

3-Digit Quotients

Each year, volunteers help to clean Central Park. If 3,548 volunteers are separated into 4 groups of the same size, how many are in each group?

You can divide to find the number in each group.

$$4\overline{)3,548}$$

First estimate by finding the first number and writing zeros in the other places.

$$4\overline{)3,548} \longrightarrow 800$$
$$\quad 8$$

Divide the thousands. Think $4\overline{)3}$. Not enough thousands.

Divide the hundreds. Think: $4\overline{)35}$.	Divide the tens. Think: $4\overline{)34}$.	Divide the ones. Think: $4\overline{)28}$.
$\begin{array}{r} 8 \\ 4\overline{)3,548} \\ \underline{32} \\ 3 \end{array}$	$\begin{array}{r} 88 \\ 4\overline{)3,548} \\ \underline{32}\downarrow \\ 34 \\ \underline{32} \\ 2 \end{array}$	$\begin{array}{r} 887 \\ 4\overline{)3,548} \\ \underline{32} \\ 34\downarrow \\ \underline{32}\downarrow \\ 28 \\ \underline{28} \\ 0 \end{array}$

There are 887 volunteers in each group.
The answer is reasonably close to the estimate.

Other examples:

$$\begin{array}{r} 767 \text{ R2} \\ 6\overline{)4,604} \\ \underline{42} \\ 40 \\ \underline{36} \\ 44 \\ \underline{42} \\ 2 \end{array} \qquad \begin{array}{r} 212 \text{ R1} \\ 4\overline{)849} \\ \underline{8} \\ 4 \\ \underline{4} \\ 9 \\ \underline{8} \\ 1 \end{array}$$

Checkpoint   Write the letter of the correct answer.

Divide.

1. $4\overline{)1,172}$ **a.** 29 R3   **b.** 213   **c.** 290 R2   **d.** 293

2. $4,382 \div 7$ **a.** 620 R2   **b.** 625 R7   **c.** 626   **d.** 711 R5

3. $3,431 \div 5$ **a.** 610 R1   **b.** 680 R1   **c.** 686 R1   **d.** 687

Divide.

1. $2\overline{)430}$ **2.** $3\overline{)981}$ **3.** $2\overline{)650}$ **4.** $3\overline{)684}$ **5.** $4\overline{)864}$

6. $6\overline{)679}$ **7.** $6\overline{)710}$ **8.** $3\overline{)449}$ **9.** $8\overline{)973}$ **10.** $5\overline{)682}$

11. $5\overline{)4,560}$ **12.** $9\overline{)7,128}$ **13.** $8\overline{)4,472}$ **14.** $3\overline{)2,847}$ **15.** $6\overline{)2,813}$

16. $9\overline{)5,717}$ **17.** $2\overline{)1,463}$ **18.** $6\overline{)1,900}$ **19.** $7\overline{)2,485}$ **20.** $5\overline{)9,056}$

21. $423 \div 2$ **22.** $3,159 \div 4$ **23.** $1,984 \div 7$ **24.** $788 \div 4$ **25.** $840 \div 7$

26. $4,308 \div 6$ **27.** $5,672 \div 9$ **28.** $5,323 \div 6$ **29.** $1,629 \div 3$ **30.** $6,204 \div 7$

31. $\frac{832}{7}$ **32.** $\frac{2,067}{4}$ **33.** $\frac{4,405}{5}$ **34.** $\frac{729}{6}$ **35.** $\frac{5,887}{9}$

Solve.

36. A running club is holding a race in the park. There are 1,967 runners entered, and they are divided equally into 7 heats. How many runners are there in each heat?

37. Tickets for the park merry-go-round cost $1.25. One afternoon, 252 people buy tickets. How much money do they spend for merry-go-round tickets?

38. The 5 soccer leagues that play in the park were looking for new players. There were many more applicants than expected. Each league has at least 165 players. Refer to the chart to decide if there are enough players to have 6 leagues play each day.

SOCCER LEAGUES

Day	Number of applicants
Monday	1,040
Wednesday	935
Friday	1,105

CHALLENGE

If the sum of the digits of a number is 9, then the number is divisible by 9.

342     $3 + 4 + 2 = 9$ $9 \div 9 = 1$
342 is divisible by 9.

Is the number divisible by 9? Write *yes* or *no*.

1. 1,234 **2.** 3,105 **3.** 4,203 **4.** 7,104 **5.** 226 **6.** 4,112

Dividing Larger Numbers

Stamp collecting has been called the "king of hobbies and the hobby of kings." Michael has been collecting stamps since he was very young. He has divided his collection of 13,418 stamps equally among 7 albums. How many stamps are there in each album? How many additional stamps does he have?

Find 13,418 ÷ 7.

To place the first digit of the quotient

Think: 7)1.   Not enough ten thousands.

Think: 7)13.   Place the first digit of the quotient in the thousands place.

$$
\begin{array}{r}
1{,}9\,1\,6 \text{ R6} \\
7\overline{)1\,3{,}4\,1\,8} \\
7\downarrow \\
6\,4 \\
6\,3\downarrow \\
\overline{1\,1} \\
7 \\
\overline{4\,8} \\
4\,2 \\
\overline{6}
\end{array}
$$

Check.

$$
\begin{array}{r}
1{,}916 \\
\times\quad 7 \\
\hline
13{,}412 \\
+\qquad 6 \\
\hline
13{,}418
\end{array}
$$

There are 1,916 stamps in each album.
He has 6 additional stamps.

Other examples:

$$
\begin{array}{r}
5{,}484 \text{ R3} \\
8\overline{)43{,}875} \\
40 \\
\hline
3\,8 \\
3\,2 \\
\hline
6\,7 \\
6\,4 \\
\hline
3\,5 \\
3\,2 \\
\hline
3
\end{array}
\qquad
\begin{array}{r}
14{,}571 \text{ R3} \\
5\overline{)72{,}858} \\
5 \\
\hline
2\,2 \\
2\,0 \\
\hline
2\,8 \\
2\,5 \\
\hline
3\,5 \\
3\,5 \\
\hline
8 \\
5 \\
\hline
3
\end{array}
\qquad
\begin{array}{r}
2{,}283 \text{ R1} \\
3\overline{)6{,}850} \\
6 \\
\hline
8 \\
6 \\
\hline
2\,5 \\
2\,4 \\
\hline
1\,0 \\
9 \\
\hline
1
\end{array}
\qquad
\begin{array}{r}
24{,}234 \\
4\overline{)96{,}936} \\
8 \\
\hline
1\,6 \\
1\,6 \\
\hline
9 \\
8 \\
\hline
1\,3 \\
1\,2 \\
\hline
1\,6 \\
1\,6 \\
\hline
0
\end{array}
$$

Divide. Multiply to check your answer.

1. $4\overline{)4{,}484}$ 2. $7\overline{)9{,}485}$ 3. $6\overline{)7{,}686}$ 4. $3\overline{)9{,}399}$

5. $3\overline{)8{,}553}$ 6. $8\overline{)9{,}068}$ 7. $4\overline{)5{,}133}$ 8. $2\overline{)6{,}785}$

9. $3\overline{)28{,}143}$ 10. $6\overline{)35{,}076}$ 11. $7\overline{)22{,}225}$ 12. $5\overline{)47{,}358}$

13. $7\overline{)57{,}154}$ 14. $9\overline{)38{,}267}$ 15. $8\overline{)74{,}684}$ 16. $4\overline{)65{,}823}$

17. $6\overline{)36{,}726}$ 18. $7\overline{)63{,}798}$ 19. $8\overline{)63{,}456}$ 20. $4\overline{)35{,}105}$

21. $7\overline{)4{,}347}$ 22. $9\overline{)13{,}212}$ 23. $4\overline{)22{,}796}$ 24. $8\overline{)45{,}056}$

25. $5{,}143 \div 4$ 26. $7{,}389 \div 6$ 27. $35{,}169 \div 3$ 28. $83{,}330 \div 2$

29. $89{,}185 \div 3$ 30. $6{,}375 \div 5$ 31. $15{,}724 \div 8$ 32. $77{,}777 \div 6$

33. $6{,}408 \div 9$ 34. $1{,}244 \div 4$ 35. $9{,}786 \div 3$ 36. $3{,}084 \div 6$

37. $4{,}984 \div 7$ 38. $9{,}858 \div 3$ 39. $36{,}732 \div 6$ 40. $23{,}120 \div 5$

Solve.

41. Michael finds a box of old stamp albums in his attic. He takes a few stamps for his own collection and divides the remaining 35,682 into 6 equal groups to sell as starter kits. How many stamps are there in each starter kit?

★42. Michael files stamps of different countries in envelopes with 5 stamps in each. He has 750 Israeli stamps, 506 Spanish stamps, and 231 German stamps. How many full envelopes will Michael have?

FOCUS: REASONING

There are 16 students in a fourth-grade class. The diagram shows how many students belong to school clubs.

There are 2 students who only belong to the computer club.

There are 3 students who belong to both the chorus club and the art club only.

1. How many are in all three clubs?

2. How many students are in the computer club?

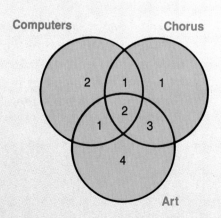

More Practice, page 429

PROBLEM SOLVING
Choosing/Writing a Sensible Question

Asking the right questions can help you organize information and make appropriate decisions.

The Calamity Falls Jug Band wants to go to the Bluegrass Festival at the state fair to compete in the band contest. The members decide to give a concert to raise money to cover their expenses. The band members meet to discuss their plans.

> First, the band members have to decide how much money they need to raise. Answering which of these questions will help them determine what their expenses will be?

- What is the cost of transportation? The cost of transportation is important information. It will cost the band more to travel farther.
- How many people are there in the band? The number of people in the band is important information. Each person's expenses will have to be covered.
- Which of the band members plays the banjo? It is not important to know which of the band members plays the banjo. This information will not affect the band's expenses.
- How long will the band stay at the fair? The length of time that the band will spend at the fair is important information. The cost of meals and hotel rooms will be an important part of the band's expenses.

In order to determine the band's expenses, it is important to know the cost of transportation, the number of people in the band, and the length of time that the band will stay at the fair. It will not help to know which of the band members plays the banjo.

Read each statement. Then write the letter of the question that the band does *not* need to answer before making a decision.

1. The band wants to know what its chances are of winning the contest.

   a. How long has the band played together?
   b. Has the band entered a contest before?
   c. Has the band ever played under another name?
   d. What other bands have entered the contest?

2. The band members have to decide whether they want to stay overnight at the fair.

   a. How much money does the band have?
   b. Who has organized the contest?
   c. Have many entertainers agreed to perform at the fair?
   d. At what time will the band play during the contest?

Read each statement. Then formulate two questions that the band should answer before making a decision.

3. The band has to decide where to hold the fund-raising concert.

4. A date has to be set for the concert.

5. The band has to decide which songs to play at the contest.

6. The band has to decide how many songs to play at the contest.

7. The band has to decide how much to charge for tickets to the fund-raising concert.

8. The band has to decide when to arrive at the state fair.

9. The band has to decide where to eat at the state fair.

10. The band members have to decide what to do with extra money they have left after their trip.

Zeros in the Quotient

Kite flying is a national pastime in China, Korea, Japan, and Malaysia. Some of the kites are so big that it takes 3 people to fly 1. If 624 people were flying these large kites in a contest, how many kites were in the air? Divide $3\overline{)624}$.

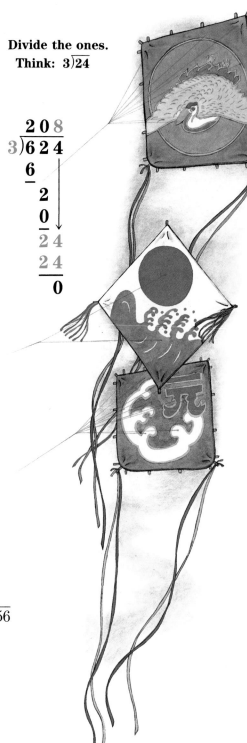

Divide the hundreds.
Think: $3\overline{)6}$

$$\begin{array}{r} 2 \\ 3\overline{)6\,2\,4} \\ \underline{6} \end{array}$$

Divide the tens.
Think: $3\overline{)2}$
Not enough tens.
Write 0

$$\begin{array}{r} 2\,0 \\ 3\overline{)6\,2\,4} \\ \underline{6}\downarrow \\ 2 \\ \underline{0} \\ 2 \end{array}$$

Remember to write the 0.

Divide the ones.
Think: $3\overline{)24}$

$$\begin{array}{r} 2\,0\,8 \\ 3\overline{)6\,2\,4} \\ \underline{6}\downarrow \\ 2 \\ \underline{0}\downarrow \\ 2\,4 \\ \underline{2\,4} \\ 0 \end{array}$$

There were 208 large kites in the air.

Other examples:

$$\begin{array}{r} 10\text{ R}3 \\ 5\overline{)53} \\ \underline{5} \\ 3 \end{array}$$

$$\begin{array}{r} 2{,}006 \\ 3\overline{)6{,}018} \\ \underline{6} \\ 0 \\ \underline{0} \\ 1 \\ \underline{0} \\ 18 \\ \underline{18} \\ 0 \end{array}$$

Checkpoint Write the letter of the correct answer.

Divide.

1. $4\overline{)8{,}052}$

a. 213
b. 2,010 R2
c. 2,013
d. 2,130

2. $18{,}540 \div 6$

a. 309
b. 3,009
c. 3,090
d. 3,900

3. $7\overline{)14{,}056}$

a. 208
b. 2,008
c. 2,080
d. 2,800

Divide.

1. $3\overline{)612}$ 2. $9\overline{)945}$ 3. $4\overline{)832}$ 4. $3\overline{)903}$ 5. $6\overline{)648}$

6. $4\overline{)83}$ 7. $7\overline{)72}$ 8. $3\overline{)62}$ 9. $2\overline{)611}$ 10. $4\overline{)810}$

11. $5\overline{)751}$ 12. $8\overline{)967}$ 13. $6\overline{)845}$ 14. $5\overline{)604}$ 15. $7\overline{)769}$

16. $3\overline{)1,521}$ 17. $5\overline{)3,540}$ 18. $3\overline{)1,590}$ 19. $6\overline{)5,760}$ 20. $4\overline{)2,428}$

21. $6\overline{)16,234}$ 22. $8\overline{)48,363}$ 23. $7\overline{)35,526}$ 24. $6\overline{)12,430}$ 25. $9\overline{)36,430}$

26. $5\overline{)27,753}$ 27. $3\overline{)62}$ 28. $3\overline{)27,029}$ 29. $6\overline{)3,644}$ 30. $2\overline{)34,001}$

31. $30,060 \div 6$ 32. $749 \div 7$ 33. $13,206 \div 8$ 34. $74 \div 9$

35. $54,200 \div 2$ 36. $67,451 \div 5$ 37. $6,328 \div 7$ 38. $66,003 \div 5$

39. $2,418 \div 3$ 40. $36,160 \div 6$ 41. $1,452 \div 7$ 42. $3,417 \div 2$

43. $45,445 \div 5$ 44. $12,031 \div 4$ 45. $18,907 \div 9$ 46. $42,498 \div 6$

Solve. For Problem 49, use the Infobank.

47. The box kite was invented in Australia in the 1890's by Lawrence Hargrave. He used 8 sticks for each kite. How many kites could he have built with 13,616 sticks?

48. The United States Weather Bureau flew long trains of box kites to collect weather data. A total of 408 kites in 4 separate trains were flown. How many kites were there in each train if each train had the same number of kites?

49. Use the information on page 415 to solve. Which kite requires the most sticks to build? Which kites need a tail longer than 3 feet? Make a list of the kites in order from the longest tail to the shortest tail.

50. Japanese families celebrate Children's Day by flying 1 fish kite for each son. If 6 families have 2 sons each and 3 families have 3 sons each, how many kites do these families fly in all?

FOCUS: MENTAL MATH

Compute mentally.

1. $(63 \div 9) \times 3$ 2. $(48 \div 6) \times 8$ 3. $(35 \div 7) \times 4$ 4. $(72 \div 8) \times 5$

5. $(14 \div 2) \times 6$ 6. $(24 \div 3) \times 4$ 7. $(27 \div 9) \times 6$ 8. $(32 \div 8) \times 6$

Dividing with Money

The Silver Star Amusement Park has a special weekday price for tickets: $14.25 for 3 tickets. Tim, Patty, and Carrie buy tickets together. What is the cost of each of their tickets at the weekday rate?

Find: $14.25 ÷ 3.

To divide amounts of money, think of the amounts as whole numbers. For 3)$14.25, Think: 3)1,425.

```
      $4.7 5
  3)$1 4.2 5
    1 2
      2 2
      2 1
        1 5
        1 5
          0
```

> Remember to write the dollar sign and the cents point.

Check.

```
   $4.75
 ×    3
 $14.25
```

The cost of each ticket is $4.75.

Other examples:

```
    $0.73
 8)$5.84
    5 6
      24
      24
       0
```

```
    $0.05
 6)$0.30
      30
       0
```

Checkpoint Write the letter of the correct answer.

Divide.

1. $75.42 ÷ 6

a. $12.07
b. $12.57
c. $14.23 R4
d. $125.70

2. 8)$876.48

a. $19.56
b. $109 R4
c. $109.06
d. $109.56

3. $185.80 ÷ 4

a. $41.45
b. $46.20
c. $46.45
d. $46.00 R1

Divide. Check by multiplying.

1. $6\overline{)\$0.54}$ 　　2. $3\overline{)\$0.81}$ 　　3. $7\overline{)\$0.84}$ 　　4. $5\overline{)\$0.65}$

5. $4\overline{)\$2.36}$ 　　6. $5\overline{)\$1.55}$ 　　7. $3\overline{)\$4.86}$ 　　8. $7\overline{)\$6.51}$

9. $8\overline{)\$70.56}$ 　　10. $5\overline{)\$14.85}$ 　　11. $4\overline{)\$22.52}$ 　　12. $3\overline{)\$65.73}$

13. $5\overline{)\$800.55}$ 　　14. $6\overline{)\$651.24}$ 　　15. $8\overline{)\$209.20}$ 　　16. $2\overline{)\$140.56}$

17. $\$19.80 \div 6$ 　　18. $\$29.37 \div 3$ 　　19. $\$69.65 \div 7$ 　　20. $\$305.84 \div 2$

21. $\$475.15 \div 5$ 　　22. $\$10.84 \div 4$ 　　23. $\$229.36 \div 2$ 　　24. $\$706.23 \div 9$

25. $\$26.88 \div 4$ 　　26. $\$318.51 \div 9$ 　　27. $\$728.49 \div 7$ 　　28. $\$521.04 \div 6$

29. $\$972.54 \div 3$ 　　30. $\$612.18 \div 6$ 　　31. $\$408.80 \div 2$ 　　32. $\$23.45 \div 7$

Solve.

33. The Upside-Down-Inside-Out is Tim's favorite ride. He spends $4.55 to ride it 7 times. How much does each ride cost?

34. Carrie likes the Looper-Dooper ride. She rides it 12 times. She spends $1.35 on each ride. How much money does she spend on the Looper-Dooper?

35. Tim, Carrie, and Patty eat lunch at the park. They each have a hamburger and a glass of milk. Patty buys an apple, Tim buys a bag of peanuts, and Carrie buys an orange. They decide to split the bill evenly. Look at the menu. How much does each pay?

SILVER STAR PARK MENU

Hamburger $1.35
Apple $0.50
Orange $0.40
Peanuts $0.60
Milk $0.75

ANOTHER LOOK

Multiply.

1. $\begin{array}{r} 727 \\ \times\ 11 \\ \hline \end{array}$ 　　2. $\begin{array}{r} 8{,}207 \\ \times\ 10 \\ \hline \end{array}$ 　　3. $\begin{array}{r} 23{,}654 \\ \times\ 21 \\ \hline \end{array}$ 　　4. $\begin{array}{r} 98{,}001 \\ \times\ 30 \\ \hline \end{array}$

5. $\begin{array}{r} 111 \\ \times\ 30 \\ \hline \end{array}$ 　　6. $\begin{array}{r} 4{,}004 \\ \times\ 17 \\ \hline \end{array}$ 　　7. $\begin{array}{r} 39{,}100 \\ \times\ 41 \\ \hline \end{array}$ 　　8. $\begin{array}{r} 75{,}320 \\ \times\ 50 \\ \hline \end{array}$

More Practice, page 429

PROBLEM SOLVING
Interpreting the Quotient and the Remainder

Sometimes when you divide to solve a problem, the quotient is not a whole number. If the answer is a quotient that has a remainder, read the question again. Be sure that the answer you write really answers the question. You may need to:

1. drop the remainder
2. round the quotient to the next-greater whole number, or
3. use only the remainder.

The Lenox family is planning a gathering at Lake Tiorati State Park. They want to reserve cabins for 27 family members. No more than 4 people are allowed to sleep in a cabin.

Divide:

$$\begin{array}{r} 6 \text{ R}3 \\ 4\overline{)27} \\ \underline{24} \\ 3 \end{array}$$

Read each question. Think about how the answers differ for each question.

Question	Action	Answer
1. How many cabins can be completely filled?	Drop the remainder.	6 cabins can be completely filled.
2. How many cabins are needed to house all the people?	Round the quotient to the next-greater whole number.	7 cabins are needed. (There will be 3 people left after the 6 cabins are filled.)
3. How many people will be in the cabin that is not completely full?	Use only the remainder.	The cabin that is not completely full will house 3 people.

Write the letter of the correct answer.

1. There are 17 members of the Lenox family who plan to drive to the lake together. If they can fit 5 people into each car, how many cars will they need for the trip?

$$\begin{array}{r} 3 \text{ R2} \\ 5\overline{)17} \end{array}$$

a. 2
b. 3
c. 4

2. The Lenox children want to rent a rowboat. It costs $9 per hour to rent a boat. The children have $22. For how many hours can they rent a rowboat?

$$\begin{array}{r} 2 \text{ R4} \\ 9\overline{)22} \\ \underline{18} \\ 4 \end{array}$$

a. 2 hours
b. 3 hours
c. 4 hours

Solve.

3. Joseph Lenox's family decides to camp in tents. There are 5 people in the family. If each tent holds 2 people, how many tents will they need?

4. Anna Lenox estimates that they will need at least 54 hot dogs for the cookout. If hot dogs are sold 8 to a package, how many packages must she buy?

5. Lou Lenox baked enough bran muffins so that each of the 27 family members could have one muffin. He packed the muffins 6 to a box. He brought only full boxes. How many boxes did he bring?

6. Maria Lenox makes 2 quarts (64 ounces) of lemonade. How many 6-ounce glasses of lemonade can she fill completely with the lemonade she has made?

7. The Lenox children decide to organize volleyball games on the beach. There are 17 family members who want to play. Each team will have 8 players. Extra players will be substitutes. How many substitutes will there be?

8. Some boys and girls at the lake want to form separate teams for relay races. There are 9 boys and 10 girls. Each relay team can have 4 members. Any extra people will be judges. How many teams are there?
How many judges will be girls?

How many judges will be boys?

CALCULATOR

You can use a calculator to practice estimating and finding quotients.

Use the numbers given in the boxes to complete each exercise.

Divisors		
2	3	4
5	6	7
8	9	

Dividends	
567	1,842
585	2,265
816	2,344
1,408	3,440
1,456	6,510
952	1,530
	2,608

Quotients	
107	247
176	293
189	453
208	652
238	

1. ▨)1,4 8 2

2. 9 3 0 ; ▨)▨,▨▨▨

3. 3)▨▨▨

4. 5)▨,▨▨▨

5. ▨)9 6 3

6. 4)▨▨▨

7. 2 0 4 ; ▨)▨▨▨

8. 8)2,▨▨▨

9. 9 2 1 ; ▨)▨,▨▨▨

10. 1 9 5 ; ▨)▨▨▨

11. 8)▨,▨▨ 8

12. 4)▨,▨▨▨

13. 3 0 6 ; ▨)▨,▨▨▨

14. 7)▨,▨▨▨

15. 4 3 0 ; ▨)▨,▨▨▨

GROUP PROJECT

Car Wash or Raffle

The problem: Your class needs to raise $500 in two months for a trip to the state fair. They're considering sponsoring a car wash or holding a raffle. Discuss these two ideas with your classmates. Choose one based on the information below.

Key Facts

Car Wash

- You can wash cars in the school parking lot.
- The school will provide soap and water, but you need to provide the buckets, rags, and brushes.

Raffle

- You think students will be willing to pay $0.50 for a raffle ticket.
- Local businesses or parents can contribute prizes to the raffle.
- The sixth-grade class held a successful raffle last month.

Key Questions

- How many cars do you need to wash?
- How many students will help?
- How would you advertise?
- What price will you charge?
- Do you need to hold more than one car wash?

- Where will you hold the raffle?
- How many items should you raffle?
- How many tickets do you need to sell?
- How would you advertise?

CHAPTER TEST

Divide. (pages 130, 136, 144, 146, and 148)

1. $5\overline{)45}$ **2.** $6\overline{)6}$ **3.** $\frac{0}{9}$ **4.** $\frac{30}{7}$

5. $9\overline{)166}$ **6.** $8\overline{)384}$ **7.** $\frac{474}{6}$ **8.** $\frac{623}{7}$

9. $3\overline{)635}$ **10.** $5\overline{)426}$ **11.** $\frac{727}{7}$ **12.** $\frac{27,755}{5}$

13. $\frac{2,421}{3}$ **14.** $\frac{12,424}{6}$ **15.** $9\overline{)36,439}$

16. $8\overline{)489,216}$ **17.** $5\overline{)\$765.45}$ **18.** $4\overline{)\$24.52}$

19. $17,521 \div 7$ **20.** $2,613 \div 4$

21. $\$6.56 \div 7$ **22.** $\$0.92 \div 3$

Copy the chart. Write a check in the box if the dividend is divisible by the divisor. (pages 132 and 138)

		Dividend		
		23.	**24.**	**25.**
		695	1,258	26,780
Divisor	2			
	5			
	10			

Estimate. (page 140)

26. $6\overline{)435}$ **27.** $3\overline{)714}$ **28.** $8\overline{)4,729}$ **29.** $6\overline{)55,289}$

Solve. (pages 134–135, 150–151, and 156–157)

30. There were 777 people entered in the Rocktown Easter-egg hunt last year. The entrants were divided into groups of 7 people each. How many groups took part in the hunt? Write the letter of the operation you would use to solve the problem.

a. multiplication
b. division

31. The Rocktown Racquet Rompers held a tennis tournament. Each day, all the tickets were sold out, for a total 1,928 tickets for the 4-day event. How many tickets were sold each day? Write a number sentence and solve.

32. There were 358 people at the Racquet Rompers Club dinner. If 4 people could sit at each table, how many tables were needed? How many people were seated at the last table?

33. The Racquet Rompers want to send 3 members to the New York State championship. Write the letter of the question that the band does *not* need to answer before deciding who should go.

a. Who is the best player?
b. Which tennis racquets should they buy?
c. How much will the trip cost?

BONUS

An input-output chart is a table that lists a series of operations to be performed on a number. The number you start with is the input. The number you end up with is the output. What happens if you input the number 11 to the chart at the right?

The output is 66.

Use the input-output chart to find the outputs of these numbers

1. 7   **2.** 10   **3.** 12   **4.** 9

5. 4   **6.** 5   **7.** 2   **8.** 3

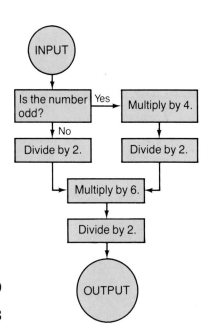

RETEACHING

Sometimes when you divide whole numbers, you need to write zeros in the quotient. Remember to multiply the divisor by the zero and write the product.

Divide 4)8,375.

Divide the thousands.
Think: 4)8.

$$\begin{array}{r} 2 \\ 4\overline{)8,375} \\ \underline{8} \end{array}$$

Divide the hundreds.
Think: 4)3.
Not enough hundreds. Write 0.

$$\begin{array}{r} 2\,0 \\ 4\overline{)8,375} \\ \underline{8}\downarrow \\ 3 \\ \underline{0} \\ 3 \end{array}$$
← Write 0 in the quotient.
← Remember to write the 0.

Divide the tens.
Think: 4)37.

$$\begin{array}{r} 2\,0\,9 \\ 4\overline{)8,375} \\ \underline{8} \\ 3 \\ \underline{0}\downarrow \\ 37 \\ \underline{36} \\ 15 \end{array}$$

Divide the ones.
Think: 4)15.

$$\begin{array}{r} 2\,0\,9\,3 \text{ R3} \\ 4\overline{)8,375} \\ \underline{8} \\ 3 \\ \underline{0} \\ 37 \\ \underline{36}\downarrow \\ 15 \\ \underline{12} \\ 3 \end{array}$$

Divide.

1. 6)1,254   2. 7)1,456   3. 5)535   4. 6)5,418   5. 4)3,216

6. 8)5,684   7. 3)992   8. 6)2,462   9. 9)2,797   10. 5)2,554

11. 7)35,342   12. 3)91,911   13. 8)56,723   14. 8)56,640   15. 4)28,152

16. 4,824 ÷ 6   17. 2,748 ÷ 9   18. 28,270 ÷ 7   19. 61,612 ÷ 4

20. 81,814 ÷ 8   21. 4,593 ÷ 9   22. 18,161 ÷ 6   23. 30,632 ÷ 3

ENRICHMENT

Short Division

You can use short division when you have a 1-digit divisor. Multiply and subtract mentally.

Find the quotient: $4\overline{)2{,}696}$

Divide the thousands. $4\overline{)2}$ Not enough thousands.

Divide the hundreds.

$4\overline{)26}$

Write the remainder next to the tens.

$$\frac{6}{4\overline{)2{,}6^2 96}}$$

Divide the tens.

$4\overline{)29}$

Write the remainder next to the ones.

$$\frac{6\ 7}{4\overline{)2{,}6^2 9^1 6}}$$

Divide the ones.

$4\overline{)16}$

Write the remainder if necessary.

$$\frac{6\ 7\ 4}{4\overline{)2{,}6^2 9^1 6}}$$

The Spoke-n-Four Bicycle Company tests its new 15-speed models with four long-distance relays. The cyclists pass through mountains, deserts, and unpaved sections of road. Within each relay, each cyclist rides an equal number of miles.

Below is a mileage chart that lists the four relays. Copy the chart, and use short division to find the number of miles ridden by each cyclist.

	Relay 1	Relay 2	Relay 3	Relay 4
Total number of miles	1,107	1,024	788	935
Number of cyclists	9	8	4	5
Number of miles ridden by each cyclist	▪	▪	▪	▪

CUMULATIVE REVIEW

Write the letter of the correct answer.

1. 0.079×53

    a. 4.177   **b.** 4.187

    c. 41.87   **d.** not given

2. 0.05×0.08

    a. 0.004   **b.** 0.040

    c. 0.40   **d.** not given

3. 72×36.519

    a. 2,529.368   **b.** 2,629.358

    c. 2,629.378   **d.** not given

4. 0.07×3.9

    a. 0.0273   **b.** 0.273

    c. 2.73   **d.** not given

5. $\$4.50 \times 7.3$

    a. $29.75   **b.** $32.75

    c. $32.85   **d.** not given

6. Estimate: 7.12×14.7.

    a. 78   **b.** 105

    c. 200   **d.** not given

7. 197×373

    a. 66,981   **b.** 72,481

    c. 73,481   **d.** not given

8. $4.081 + 9.764 + 10.159$

    a. 20.004   **b.** 21.040

    c. 24.004   **d.** not given

9. Compare. Write $>$, $<$, or $=$ for ●.
0.9 ● 0.899

    a. $<$ **b.** $>$

    c. $=$ **d.** not given

10. $938,620 - 419,598$

    a. 419,021   **b.** 519,022

    c. 519,022   **d.** not given

11. $370,004 - 298,557$

    a. 71,447   **b.** 72,557

    c. 82,557   **d.** not given

12. What is the value of the blue digit?
658,932,541

    a. 500   **b.** 5,000

    c. 50,000   **d.** not given

13. Jerry and Bob saw two baseball games last month. Tickets for the first game cost $4.50 each. Tickets for the second game cost $5.75 each. How much did Jerry and Bob spend on their tickets?

    a. $18.40   **b.** $19.50

    c. $20.50   **d.** not given

14. Attendance at Game 1 of the baseball season was 4,327 people. Total attendance for the season was about 8 times that amount. About how many people attended the games for the season?

    a. 3,461   **b.** 34,616

    c. 340,616   **d.** not given

People do amazing things to have their names entered into record books. Think of an activity that a group of friends could do. You might try to make the world's biggest sandwich or play the world's longest game of baseball. How many people and how much time would you need to break the record?

6 DIVISION: 2-DIGIT DIVISORS

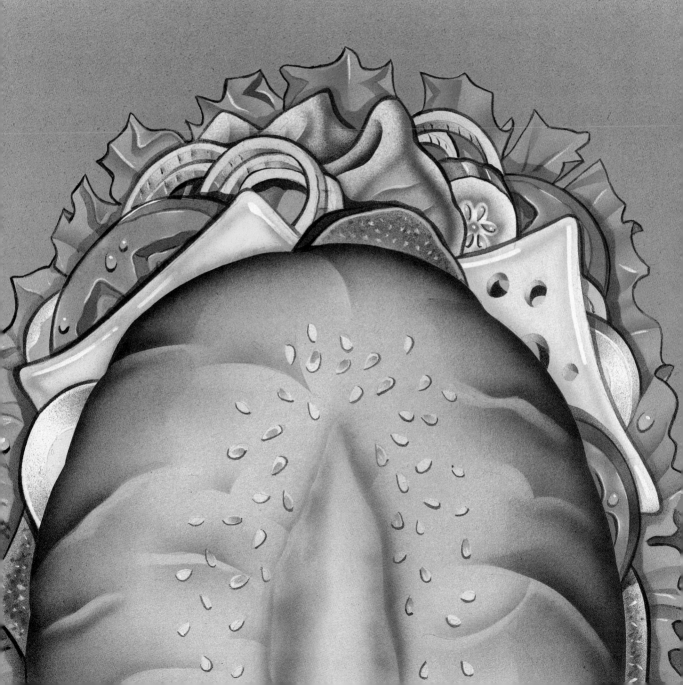

Dividing by Multiples of 10

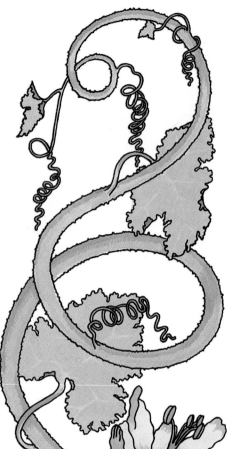

People try to become listed in record books in many ways. One way is to grow giant fruits and vegetables.

A. Marcy grows a 560-pound pumpkin. If she weighs 70 pounds, how many times as heavy as Marcy is her pumpkin?

Divide $560 \div 70$.

You can use division facts to help you divide greater numbers.

Think: $56 \div 7 = 8$.
$560 \div 70 = 8$.

The pumpkin is 8 times as heavy as Marcy.

Other examples:

$45 \div 9 = 5$ $30 \div 5 = 6$ $63 \div 7 = 9$
$450 \div 90 = 5$ $300 \div 50 = 6$ $630 \div 70 = 9$

B. Patterns can help you divide by multiples of 10.

$56 \div 8 = 7$ $36 \div 6 = 6$
$560 \div 80 = 7$ $360 \div 60 = 6$
$5,600 \div 80 = 70$ $3,600 \div 60 = 60$

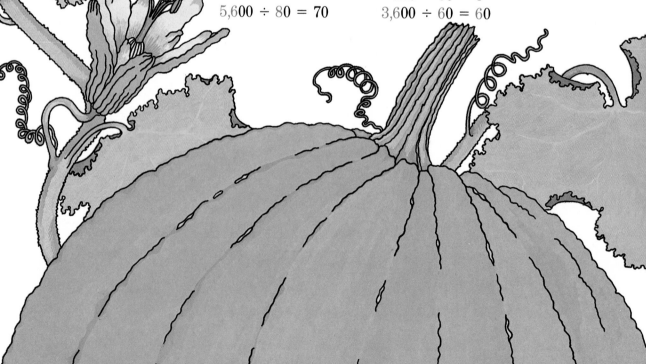

Divide.

1. $10\overline{)80}$ 2. $30\overline{)60}$ 3. $20\overline{)80}$ 4. $30\overline{)90}$ 5. $20\overline{)40}$

6. $60\overline{)60}$ 7. $30\overline{)60}$ 8. $40\overline{)80}$ 9. $10\overline{)80}$ 10. $10\overline{)50}$

11. $60\overline{)180}$ 12. $90\overline{)540}$ 13. $20\overline{)180}$ 14. $30\overline{)210}$ 15. $20\overline{)120}$

16. $30\overline{)180}$ 17. $20\overline{)100}$ 18. $40\overline{)360}$ 19. $40\overline{)320}$ 20. $30\overline{)210}$

21. $40\overline{)80}$ 22. $20\overline{)60}$ 23. $30\overline{)30}$ 24. $10\overline{)40}$ 25. $40\overline{)280}$

26. $20\overline{)180}$ 27. $30\overline{)180}$ 28. $40\overline{)80}$ 29. $10\overline{)10}$ 30. $50\overline{)50}$

31. $90 \div 30$ 32. $60 \div 20$ 33. $80 \div 20$ 34. $80 \div 40$ 35. $90 \div 90$

36. $\frac{70}{70}$ 37. $\frac{90}{10}$ 38. $\frac{360}{40}$ 39. $\frac{490}{70}$ 40. $\frac{450}{50}$

Solve.

41. Marcy reads in the *Guinness Book of World Records* that the world's heaviest orange weighed 90 ounces. If this orange had been sectioned into 10 equal slices, how much would each slice have weighed?

42. A farmer whom Marcy knows grows a giant 240-pound watermelon. The average watermelon weighs about 30 pounds. How many average watermelons would equal the weight of this giant watermelon?

CHALLENGE

Multiply or divide to find ▨.

1.

÷	7	▨	▨	▨
▨	28	24	▨	20
▨	21	18	12	15
8	▨	▨	▨	▨
▨	35	30	20	25

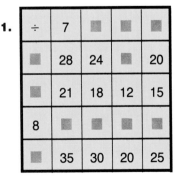

2.

÷	▨	▨	60	▨	70
5	200	50	▨	150	▨
▨	▨	70	▨	210	490
3	120	▨	▨	90	▨
▨	240	60	360	180	420

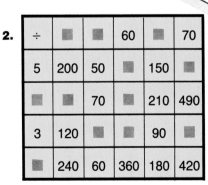

More Practice, page 430

Estimating Quotients

A. Macy's in New York is one of the largest department stores in the world. In one year, it uses almost 6,000 miles of twine and ribbon to wrap packages, 4,000 miles of tape, and about 7,000,000 folding boxes!

The department store in Black Hills is small, but it still uses many supplies. In 22 days, the store uses 6,975 folding boxes. On the average, how many boxes did it use every day?

Since the store probably did not use the same number of boxes each day, this is a good situation in which to estimate the quotient.

Decide on the number of digits in the quotient.

Divide the hundreds. Think: $22\overline{)69}$.
The quotient begins in the hundreds place.

It will have 3 digits.

$$\overset{\text{- - -}}{22\overline{)6,975}}$$

Think: $3 \times 22 = 66$. Write 3.

$$\overset{3}{22\overline{)6,975}}$$

Write zeros for the other digits. $22\overline{)6,975} \rightarrow 300$

You can say that the store uses about 300 boxes every day.

B. The next step is to decide how accurate your estimate is.

Think: 300 boxes are used every day. If the store uses 300 boxes each day, will it use 6,975 in 22 days?

$300 \times 22 = 6,600$; $6,600 < 6,975$.
So, more than 300 boxes are used.
300 is an underestimate.

The store uses more than 300 folding boxes every day.

Other examples:

$21\overline{)646} \longrightarrow 30$ $21\overline{)987} \longrightarrow 40$ $32\overline{)978} \longrightarrow 30$

$22\overline{)8,956} \longrightarrow 400$ $75\overline{)9,825} \longrightarrow 100$ $37\overline{)7,955} \longrightarrow 200$

Write how many digits the quotient will contain.

1. $69\overline{)828}$ **2.** $13\overline{)702}$ **3.** $57\overline{)855}$ **4.** $16\overline{)848}$ **5.** $42\overline{)630}$

6. $12\overline{)1,428}$ **7.** $79\overline{)4,898}$ **8.** $17\overline{)1,751}$ **9.** $62\overline{)4,526}$ **10.** $19\overline{)2,698}$

11. $36\overline{)4,356}$ **12.** $43\overline{)688}$ **13.** $77\overline{)9,702}$ **14.** $14\overline{)686}$ **15.** $81\overline{)8,667}$

Estimate. Write the letter of the correct answer.

16. $11\overline{)297}$ **a.** 2   **b.** 20   **c.** 200

17. $29\overline{)319}$ **a.** 1   **b.** 10   **c.** 100

18. $45\overline{)3,290}$ **a.** 80   **b.** 800   **c.** 8,000

19. $71\overline{)1,988}$ **a.** 20   **b.** 200   **c.** 2,000

20. $96\overline{)8,514}$ **a.** 9   **b.** 90   **c.** 900

Estimate.

21. $22\overline{)286}$ **22.** $24\overline{)744}$ **23.** $32\overline{)982}$ **24.** $22\overline{)857}$ **25.** $91\overline{)876}$

26. $81\overline{)7,452}$ **27.** $72\overline{)1,686}$ **28.** $64\overline{)3,508}$ **29.** $42\overline{)3,159}$ **30.** $22\overline{)1,848}$

31. $23\overline{)7,383}$ **32.** $93\overline{)5,247}$ **33.** $11\overline{)2,453}$ **34.** $61\overline{)1,037}$ **35.** $82\overline{)7,532}$

Solve.

36. Janice's hardware store sells 800 boxes of nails every year. About how many boxes of nails does the store sell each month?

37. The True-Built Lumber Store sold 9,750 board feet of lumber last summer. About how much lumber was sold each day? (HINT: the store was open 78 days during the summer.)

FOCUS: MENTAL MATH

Compute mentally. Write the division examples in each row that have

1. 6 as a quotient. $60\overline{)420}$ $20\overline{)100}$ $40\overline{)240}$ $30\overline{)276}$

2. a remainder less than 4. $60\overline{)365}$ $30\overline{)180}$ $80\overline{)644}$ $40\overline{)283}$

3. 10 as the remainder. $60\overline{)490}$ $90\overline{)450}$ $20\overline{)130}$ $50\overline{)370}$

PROBLEM SOLVING
Practice

Write the letter of the number sentence you would use
to solve the problem.

1. Bamboo is one of the fastest-
   growing plants. Some kinds grow 3
   feet per day. At that rate, how long
   would it take the bamboo to reach a
   height of 50 feet?

   a. $50 \times 3 = n$ **b.** $50 \div 3 = n$
   c. $50 + 3 = n$ **d.** $50 - 3 = n$

2. A human being is about 15 times
   bigger than a cat. Yet a cat has 14
   more bones in its body than a
   human has. A cat has 230 bones.
   How many bones does a human
   being have?

   a. $15 \times 230 = n$ **b.** $230 \div 14 = n$
   c. $230 + 14 = n$ **d.** $230 - 14 = n$

Write the letter of the information you need to solve
the problem.

3. It is estimated that by the age of 18,
   the average child in the United
   States has watched 17,000 hours of
   television. About how many years is
   that?

   a. number of hours in a year
   b. number of weeks in 18 years
   c. number of television sets

4. *Moving at a snail's pace* means a
   top speed of only 0.0313 miles per
   hour. At that rate, how many feet
   can a snail go in 3 hours?

   a. number of minutes in 1 hour
   b. number of feet in 1 mile
   c. number of minutes in 3 hours

Write a number sentence and solve.

5. Some snails move slowly even for
   snails. One kind of snail can't move
   faster than 0.00036 miles per hour.
   At that speed, how far would that
   kind of snail travel in 24 hours?

Write the letter of the question you
would ask to solve the problem.

6. Lulu visits Mount Waialeale in
   Hawaii. It is the rainiest place on
   Earth. Rain falls about 350 days per
   year. Lulu wants to buy postcards of
   the mountain. How many postcards
   should Lulu buy?

   a. Do her friends like rainy places?
   b. How many people does she want
      to send cards to?
   c. How tall is Mount Waialeale?
   d. How far is Hawaii from her
      home?

Use the line graph to solve.

8. What was the first year in which the winning time was less than 23 seconds?

9. How much faster was the winning time in 1980 than in 1960?

10. What trend does the graph show?

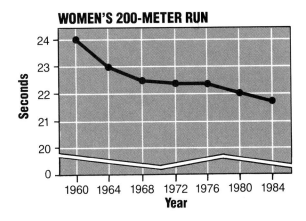

WOMEN'S 200-METER RUN

Solve.

11. In 1980, an English pilot flew around the world in slightly more than 44 hours. The trip covered 23,068 miles. About how many miles per hour did the pilot travel?

12. How fast can you write? The English novelist John Creasey wrote 564 books in 42 years. About how many books did he write per year?

13. Two of the best-selling videocassettes of all time are *Raiders of the Lost Ark* and *Star Wars*. *Raiders* is the leader. More than 600,000 cassettes have been sold. That's 4 times the number of *Star Wars* cassettes. About how many Star Wars cassettes have been sold?

14. The greatest recorded snowfall took place in Silver Lake, Colorado, in 1921. During a 24-hour period, an average of 3.17 inches of snow fell per hour. About how much snow fell in 24 hours.

15. The average price of gasoline in the United States shot up from about $0.65 per gallon in 1978 to $1.35 in 1981. How much more would it have cost Ms. Stroad to fill her tank in 1981 than in 1978?

16. In 1983, cars traveled an average of 16.33 miles per gallon. A car that had a 12-gallon tank was driven 500 miles. How many times did the driver have to fill the tank to complete the trip?

1-Digit Quotients

Deltiology, or collecting postcards, is the third most-popular hobby in the world. It probably began in 1869 when the first postcards were issued.

Jimmy decides to become an ace deltiologist. During the month of August, he collects 217 postcards. He collected about the same number of cards each day. On the average, how many postcards does he collect daily?

There are 31 days in August. Divide 217 ÷ 31.

Divide the hundreds. Think: $31\overline{)2}$. Not enough hundreds. Divide the tens. Think: $31\overline{)21}$. Not enough tens.

Divide the ones.
  Think: $31\overline{)217}$, or $3\overline{)21}$.
  Estimate 7.

$$
\begin{array}{r}
7 \\
31\overline{)217} \\
217 \\
\hline
0
\end{array}
$$
  Multiply.
  Subtract and compare.

Check.
$$
\begin{array}{r}
31 \\
\times\ 7 \\
\hline
217
\end{array}
$$

Jimmy collects an average of 7 postcards daily.

Other examples:

$$
\begin{array}{r}
4\ R1 \\
42\overline{)169} \\
168 \\
\hline
1
\end{array}
\qquad
\begin{array}{r}
3\ R4 \\
21\overline{)67} \\
63 \\
\hline
4
\end{array}
\qquad
\begin{array}{r}
9\ R18 \\
40\overline{)378} \\
360 \\
\hline
18
\end{array}
$$

Checkpoint   Write the letter of the correct answer.

Divide.

1. $53\overline{)424}$

   a. 7 R3
   b. 7 R73
   c. 8
   d. 8 R20

2. $365 \div 88$

   a. 3 R61
   b. 3 R101
   c. 4
   d. 4 R13

3. $\frac{196}{32}$

   a. 5 R36
   b. 6
   c. 6 R4
   d. 6 R14

Divide.

1. $23 \overline{)69}$ 2. $36 \overline{)72}$ 3. $12 \overline{)48}$ 4. $33 \overline{)99}$ 5. $20 \overline{)80}$

6. $31 \overline{)97}$ 7. $42 \overline{)94}$ 8. $56 \overline{)78}$ 9. $40 \overline{)85}$ 10. $24 \overline{)77}$

11. $34 \overline{)238}$ 12. $56 \overline{)224}$ 13. $21 \overline{)168}$ 14. $91 \overline{)728}$ 15. $30 \overline{)270}$

16. $73 \overline{)528}$ 17. $68 \overline{)359}$ 18. $40 \overline{)198}$ 19. $84 \overline{)428}$ 20. $93 \overline{)567}$

21. $32 \overline{)99}$ 22. $86 \overline{)602}$ 23. $54 \overline{)436}$ 24. $30 \overline{)90}$ 25. $72 \overline{)393}$

26. $67 \overline{)409}$ 27. $90 \overline{)368}$ 28. $43 \overline{)387}$ 29. $54 \overline{)498}$ 30. $91 \overline{)637}$

31. $341 \div 74$ 32. $69 \div 11$ 33. $326 \div 81$ 34. $83 \div 41$ 35. $277 \div 70$

36. $93 \div 31$ 37. $139 \div 23$ 38. $254 \div 63$ 39. $560 \div 80$ 40. $39 \div 13$

41. $\frac{320}{53}$ 42. $\frac{264}{44}$ 43. $\frac{76}{35}$ 44. $\frac{49}{22}$ 45. $\frac{495}{70}$

Solve.

46. At first, Jimmy keeps his postcards in a shoe box. When his collection grows to 500 cards, he decides to buy albums. How many albums does he need if he can fit 80 postcards in each one?

47. Jimmy buys postcards at flea markets and antique shops. He counts 105 listings for antique shops in the telephone book. How long will it take him to contact every shop if he calls 35 shops each day?

48. Jimmy buys several postcard collections. He sorts through them and finds 132 duplicate European travel cards. A friend offers him 3 United States cards in exchange for each one. How many United States cards can Jimmy get?

★49. Jimmy organizes a postcard exhibit at the local historical society. One afternoon he sorts through the museum's 182 golden-age postcards. He discards 42 cards and arranges the remaining cards in groups of 35. How many groups of cards are there?

ANOTHER LOOK

Multiply.

1. 254×27 2. 723×59 3. 174×92 4. 632×12 5. 774×41

2-Digit Quotients

The same number of calories is used in 8 hours of sleep as in running for 52 minutes.

A. Pete trained to run in the New York City Marathon. He ran for a total of 1,092 hours during a year's time. If he ran the same number of hours each week, how many hours did he run each week?

There are 52 weeks in a year. Divide $52\overline{)1,092}$.

Divide the thousands. Think: $52\overline{)1}$. Not enough thousands.
Divide the hundreds. Think: $52\overline{)10}$. Not enough hundreds.

Divide the tens.
Think: $52\overline{)109}$, or $5\overline{)10}$.
Estimate 2.

$$
\begin{array}{r}
2 \\
52\overline{)1,092} \\
\underline{104} \\
5
\end{array}
$$

Multiply.
Subtract and compare.

Divide the ones.
Think: $52\overline{)52}$, or $5\overline{)5}$.
Estimate 1.

$$
\begin{array}{r}
21 \\
52\overline{)1,092} \\
\underline{104}\downarrow \\
52 \\
\underline{52} \\
0
\end{array}
$$

Multiply.
Subtract and compare.

Pete ran 21 hours each week.

B. To divide amounts of money, think of the amounts as whole numbers. Remember to write the dollar sign and the cents point in the quotient.

$$
\begin{array}{r}
\$0.11 \\
43\overline{)\$4.73} \\
\underline{43}\downarrow \\
43 \\
\underline{43} \\
0
\end{array}
$$
Think: $43\overline{)473}$.

Checkpoint Write the letter of the correct answer.

1. $41\overline{)897}$

   a. 21 R36
   b. 21 R6
   c. 21
   d. 22

2. $62\overline{)5,273}$

   a. 13 R25
   b. 75 R23
   c. 85 R3
   d. 85

3. $71\overline{)6,668}$

   a. 93
   b. 93 R65
   c. 98
   d. 98 R61

Divide.

1. $39\overline{)819}$ 2. $55\overline{)605}$ 3. $22\overline{)880}$ 4. $67\overline{)\$9.38}$ 5. $56\overline{)\$8.40}$

6. $34\overline{)767}$ 7. $28\overline{)294}$ 8. $54\overline{)774}$ 9. $21\overline{)526}$ 10. $72\overline{)962}$

11. $83\overline{)6,806}$ 12. $79\overline{)4,819}$ 13. $21\overline{)1,701}$ 14. $62\overline{)\$35.96}$ 15. $71\overline{)\$60.35}$

16. $52\overline{)4,273}$ 17. $61\overline{)5,595}$ 18. $88\overline{)3,612}$ 19. $83\overline{)5,588}$ 20. $42\overline{)2,794}$

21. $36\overline{)468}$ 22. $61\overline{)5,663}$ 23. $80\overline{)5,079}$ 24. $52\overline{)1,924}$ 25. $74\overline{)969}$

26. $4,588 \div 62$ 27. $2,065 \div 64$ 28. $\$9.66 \div 23$ 29. $489 \div 34$ 30. $6,338 \div 80$

31. $\frac{2,785}{75}$ 32. $\frac{1,786}{21}$ 33. $\frac{\$32.68}{76}$ 34. $\frac{846}{35}$ 35. $\frac{332}{22}$

Solve.

36. As a warm-up for the New York City Marathon, Pete enters the Parker City Minimarathon. There are 1,035 runners that enter the race. They line up in 45 rows with an equal number of runners in each row. How many runners are there in each row?

37. The New York City Marathon begins in Staten Island. Some of the runners are brought to the starting line from Manhattan by bus. If 5,345 runners sign up for bus transportation, and each bus holds 65 people, how many buses will be needed?

★38. Greta sells marathon buttons along the route. The buttons come packed 7 dozen to a box. Each box costs $12.60. She sells the buttons for $0.75 each. What is her profit on each button?

MIDCHAPTER REVIEW

Divide.

1. $20\overline{)40}$ 2. $30\overline{)96}$ 3. $60\overline{)554}$ 4. $\frac{675}{50}$ 5. $70\overline{)772}$

6. $54\overline{)441}$ 7. $33\overline{)104}$ 8. $\frac{179}{26}$ 9. $\$6.46 \div 28$ 10. $23\overline{)184}$

11. $83\overline{)1,117}$ 12. $333 \div 17$ 13. $21\overline{)\$8.75}$ 14. $37\overline{)445}$ 15. $6,447 \div 70$

PROBLEM SOLVING
Estimation

Sometimes when you make plans, you have to provide estimated amounts.

Penny wrote a report about the Pentagon for her class. She explained that the Pentagon is located in Washington, D.C., and is the world's largest office building. It covers 29 acres of land. It also has the world's largest private phone system. Penny's class decides to visit the Pentagon on their class trip to Washington. The bus trip will be 240 miles long. The class wants to arrive at 2:00 P.M. What is the best time to leave?

To solve this problem, you need to provide an average speed of travel. There is no way to know the exact speed of the bus, but the class can provide an estimated figure. They base their estimate on past experiences.

- The speed limit is 55 mph on highways and 35 mph on side streets. Most of the trip is made on highways.
- There may be tie-ups due to construction or traffic.
- There may be delays due to bus problems.

The class looks at several possible averages they could use: 35 mph, 40 mph, 55 mph, and 60 mph.

They decide to use 40 mph as the most reasonable estimate of how quickly they will travel.

240 mi ÷ 40 mph = 6 h
The class decides to allow 6 hours for the bus ride. They agree to leave at 8:00 A.M.—6 hours before they want to reach the Pentagon.

Some of the students have decided to go to Washington by plane instead of by bus. The students have to decide the time to board the plane and the amount of money to allow for transportation. Decide whether each question needs to be answered. If it does, decide whether an exact answer needs to be found, or whether an estimate is enough. Write *need not answer*, *need exact answer*, or *estimate*.

1. How much is plane fare?

2. At what time is the plane scheduled to take off and land?

3. How long is the cab ride to and from the airport?

4. How much time will it take to register at the hotel and take a bus to the Pentagon?

5. How many pounds of luggage can be taken on board?

6. How much are souvenirs at the airport?

7. How much is cab fare to and from the airport? (Cabs in both cities use a set rate, not a meter.)

8. What is the cost of a hotel room?

Use the information to answer the questions.

9. The hotel bill and 2 lunches at the Pentagon will be paid for with funds raised by the class's yard sale. About how much money will they spend on 2 lunches and 6 hotel rooms for 2 nights? Will this year's yard-sale earnings cover the expenses?

- There are 12 students in the group.
- A luncheonette near school charges $2.75 for a tuna sandwich, $3.50 for a hamburger, and $0.70 for a glass of milk.
- Hotel rooms with double occupancy—2 people to a room—range from $45 to $65 per night.
- The last four-years' yard sales earned:
  Year 1—$892.46   Year 2—$936.12
  Year 3—$1045.10   Year 4—$898.26

Correcting Estimates

A. Sometimes you need to correct your estimate.

Divide $36\overline{)1{,}528}$.

Divide the thousands. Think: $36\overline{)1}$. Not enough thousands.
Divide the hundreds. Think: $36\overline{)15}$. Not enough hundreds.

Divide the tens.
  Think: $36\overline{)152}$, or $3\overline{)15}$.
  Estimate 5.

$$\begin{array}{r} 5 \\ 36\overline{)1{,}528} \\ 180 \\ \hline \end{array}$$ Multiply. Too great. You need to correct the estimate.

Try 4.
$$\begin{array}{r} 4 \\ 36\overline{)1{,}528} \\ 144 \\ \hline 8 \end{array}$$ Multiply. Subtract and compare.

Divide the ones.
  Think: $36\overline{)88}$, or $3\overline{)8}$.
  Estimate 2.

$$\begin{array}{r} 42 \text{ R16} \\ 36\overline{)1{,}528} \\ 144\downarrow \\ \hline 88 \\ 72 \\ \hline 16 \end{array}$$ Multiply. Subtract and compare. Write the remainder.

B. Divide $53\overline{)1{,}370}$.

Divide the thousands. Think: $53\overline{)1}$. Not enough thousands.
Divide the hundreds. Think: $53\overline{)13}$. Not enough hundreds.

Divide the tens.
  Think: $53\overline{)137}$, or $5\overline{)13}$.
  Estimate 2.

$$\begin{array}{r} 2 \\ 53\overline{)1{,}370} \\ 106 \\ \hline 31 \end{array}$$ Multiply. Subtract and compare.

Divide the ones.
  Think: $53\overline{)310}$, or $5\overline{)31}$.
  Estimate 6.

$$\begin{array}{r} 26 \\ 53\overline{)1{,}370} \\ 106\downarrow \\ \hline 310 \\ 318 \\ \hline \end{array}$$ Multiply. Too great. You need to correct the estimate.

Try 5.
$$\begin{array}{r} 25 \text{ R45} \\ 53\overline{)1{,}370} \\ 106 \\ \hline 310 \\ 265 \\ \hline 45 \end{array}$$ Multiply. Subtract and compare. Write the remainder.

Divide.

1. $13\overline{)677}$ 2. $23\overline{)828}$ 3. $62\overline{)3,048}$ 4. $99\overline{)\$65.34}$ 5. $84\overline{)7,351}$

6. $26\overline{)348}$ 7. $57\overline{)792}$ 8. $44\overline{)2,724}$ 9. $71\overline{)4,002}$ 10. $87\overline{)3,905}$

11. $39\overline{)891}$ 12. $11\overline{)\$6.27}$ 13. $21\overline{)1,019}$ 14. $96\overline{)7,421}$ 15. $95\overline{)1,592}$

16. $59\overline{)1,453}$ 17. $43\overline{)2,278}$ 18. $66\overline{)840}$ 19. $22\overline{)837}$ 20. $57\overline{)\$26.22}$

21. $912 \div 31$ 22. $390 \div 12$ 23. $859 \div 11$ 24. $658 \div 44$ 25. $4,048 \div 24$

26. $1,934 \div 57$ 27. $1,267 \div 23$ 28. $\$20.16 \div 32$ 29. $416 \div 30$ 30. $749 \div 12$

31. $\frac{407}{34}$ 32. $\frac{5,773}{79}$ 33. $\frac{749}{49}$ 34. $\frac{\$9.62}{74}$ 35. $\frac{5,303}{70}$

Solve. For Problem 38, use the Infobank.

36. The longest single unbroken-apple-peel record was set in 1976 by Kathy Waffler. Her apple peel measured 172 feet 4 inches. Would a 408-inch apple peel equal or break that record?
(HINT: 12 inches = 1 foot)

37. Barbara Jean Sonntag set a record in 1981 by crocheting 147 stitches per minute. At that speed, how many stitches would she be able to crochet in 1 hour?
(HINT: 60 minutes = 1 hour)

38. Use the information on page 416 to write and solve your own word problem.

39. The record for the longest "Rockathon" belongs to Maureen Weston. In 1972, she rocked in her rocking chair for 18 days. By how many additional days and hours would a 526-hour rockathon break the record?

CHALLENGE

The Ironweave Hair Tonic Company received an order from one of its distributors.

Purchase order:
For immediate delivery
42,432,832 bottles of hair tonic

Unfortunately, the people who wrote the order could not be reached to clarify the unreadable number. The people at Ironweave were sure the distributor knew the tonic was packed 36 bottles to a box. How many boxes of hair tonic had been ordered?

3-Digit Quotients

Driving is the most popular means of transportation for vacationing Americans.

The Grippo family drives 4,374 miles while exploring the United States during their summer vacation. They spend 27 days on the road. What is the average number of miles they travel each day?

Divide $27\overline{)4{,}374}$.

Divide the thousands. Think: $27\overline{)4}$. Not enough thousands.
Divide the hundreds. Think: $27\overline{)43}$, or $2\overline{)4}$.
Estimate 2. Multiply. Too great.
You need to correct the estimate.
Try 1. Multiply.
Subtract and compare.

$$\begin{array}{r} 1 \\ 27\overline{)4{,}374} \\ \underline{2\,7} \\ 1\,6 \end{array}$$

Divide the tens. Think: $27\overline{)167}$, or $2\overline{)16}$.
Estimate 8. Multiply. Too great.
Try 7. Multiply. Too great.
Try 6. Multiply.
Subtract and compare.

$$\begin{array}{r} 16 \\ 27\overline{)4{,}374} \\ \underline{2\,7}\downarrow \\ 1\,67 \\ \underline{1\,62} \\ 5 \end{array}$$

Divide the ones. Think: $27\overline{)54}$, or $2\overline{)5}$.
Estimate 2. Multiply.
Subtract and compare.

$$\begin{array}{r} 162 \\ 27\overline{)4{,}374} \\ \underline{2\,7} \\ 1\,67 \\ \underline{1\,62}\downarrow \\ 54 \\ \underline{54} \\ 0 \end{array}$$

The Grippo family travels an average of 162 miles each day.

Other examples:

$$52\overline{)18{,}980} \quad 365$$

$$95\overline{)26{,}272} \quad 276\ \text{R}52$$

$$74\overline{)\$264.92} \quad \$3.58$$

Divide.

1. $36\overline{)4,824}$ 2. $15\overline{)2,505}$ 3. $23\overline{)2,553}$ 4. $52\overline{)7,176}$ 5. $30\overline{)\$34.50}$

6. $57\overline{)7,048}$ 7. $28\overline{)3,679}$ 8. $34\overline{)4,877}$ 9. $40\overline{)7,318}$ 10. $21\overline{)4,908}$

11. $66\overline{)48,048}$ 12. $72\overline{)61,560}$ 13. $49\overline{)35,917}$ 14. $83\overline{)45,318}$ 15. $50\overline{)\$444.00}$

16. $22\overline{)17,281}$ 17. $60\overline{)38,864}$ 18. $45\overline{)25,375}$ 19. $78\overline{)25,196}$ 20. $91\overline{)27,138}$

21. $32\overline{)5,920}$ 22. $27\overline{)4,642}$ 23. $40\overline{)9,165}$ 24. $55\overline{)17,930}$ 25. $61\overline{)\$228.75}$

26. $6,235 \div 44$ 27. $7,849 \div 35$ 28. $32,925 \div 53$ 29. $\$140.58 \div 66$ 30. $58,660 \div 70$

31. $\frac{6,236}{39}$ 32. $\frac{\$266.40}{80}$ 33. $\frac{16,853}{19}$ 34. $\frac{8,496}{26}$ 35. $\frac{28,669}{49}$

Solve.

36. Alison Grippo buys a souvenir spoon in each of the 35 states that her family visits. She spends a total of $100.80. What is the cost of each spoon?

37. Alison and her family are among the 20,000 people who visit the Los Angeles Farmers Market each day. How many people visit in the month of July? (HINT: There are 31 days in July.)

38. Use the following information to write and solve your own word problem. The Grippos drive 2,600 miles from Los Angeles, California, to Portland, Oregon. Their car uses 1 gallon of gas for every 40 miles. The cost of gas is $1.35 per gallon.

FOCUS: MENTAL MATH

Compute mentally. Write the division examples in each row that have

1. 6 as a quotient. $60\overline{)420}$ $20\overline{)100}$ $40\overline{)240}$ $30\overline{)276}$

2. a remainder less than 4. $60\overline{)365}$ $30\overline{)180}$ $80\overline{)644}$ $40\overline{)283}$

3. 10 as the remainder. $60\overline{)490}$ $90\overline{)450}$ $20\overline{)130}$ $50\overline{)370}$

Dividing with Zero in the Quotient

Some historians believe that Marco Polo brought noodles from Asia back to Italy more than 700 years ago.

In honor of Marco Polo, Jody's hometown organizes a noodle-eating festival. Jody is in charge of setting up tables. She can seat 25 people at each table. How many tables are needed if 2,625 people are expected at the festival?

Divide $25\overline{)2,625}$.

Divide the thousands. Think: $25\overline{)2}$. Not enough thousands.

Divide the hundreds.	**Divide the tens.**	**Divide the ones.**
Think: $25\overline{)26}$.	Think: $25\overline{)12}$.	Think: $25\overline{)125}$.
Estimate 1.	Not enough tens.	Estimate 6.
Multiply.	Write 0.	Multiply. Too great.
Subtract and compare.	Multiply.	Try 5.
	Subtract and compare.	Multiply.
		Subtract and compare.

$$
\begin{array}{r}
1 \\
25\overline{)2,625} \\
25 \\
\hline
1
\end{array}
\qquad
\begin{array}{r}
10 \\
25\overline{)2,625} \\
25\downarrow \\
\hline
12 \\
0 \\
\hline
12
\end{array}
\qquad
\begin{array}{r}
105 \\
25\overline{)2,625} \\
25\downarrow \\
\hline
12 \\
0\downarrow \\
\hline
125 \\
125 \\
\hline
0
\end{array}
$$

To seat everyone, 105 tables are needed.

Other examples:

$$
\begin{array}{r}
200 \text{ R3} \\
36\overline{)7,203} \\
72 \\
\hline
0 \\
0 \\
\hline
03 \\
0 \\
\hline
3
\end{array}
\qquad
\begin{array}{r}
40 \\
14\overline{)560} \\
56 \\
\hline
0 \\
\end{array}
\qquad
\begin{array}{r}
\$0.09 \\
84\overline{)\$7.56} \\
0 \\
\hline
756 \\
756 \\
\hline
0
\end{array}
$$

Divide.

1. $25\overline{)503}$
2. $58\overline{)2,900}$
3. $22\overline{)\$15.40}$
4. $76\overline{)6,096}$
5. $13\overline{)395}$

6. $17\overline{)1,785}$
7. $23\overline{)5,520}$
8. $66\overline{)29,756}$
9. $41\overline{)16,771}$
10. $38\overline{)\$116.28}$

11. $14\overline{)8,413}$
12. $38\overline{)\$1.14}$
13. $12\overline{)\$1.44}$
14. $39\overline{)31,200}$
15. $20\overline{)6,017}$

16. $21\overline{)6,342}$
17. $80\overline{)16,876}$
18. $22\overline{)\$110.00}$
19. $93\overline{)9,395}$
20. $56\overline{)18,480}$

21. $26\overline{)\$1.82}$
22. $19\overline{)9,614}$
23. $39\overline{)31,551}$
24. $50\overline{)27,545}$
25. $44\overline{)17,632}$

26. $\$156.60 \div 54$
27. $\$80.34 \div 26$
28. $8,910 \div 33$
29. $6,899 \div 43$
30. $31,310 \div 62$

31. $7,714 \div 11$
32. $32,435 \div 36$
33. $9,021 \div 25$
34. $9,817 \div 20$
35. $33,182 \div 47$

36. $\frac{9,424}{31}$
37. $\frac{7,296}{27}$
38. $\frac{9,250}{37}$
39. $\frac{39,385}{65}$
40. $\frac{12,056}{17}$

Solve.

41. The highlight of the festival is a noodle-making contest. One expert makes 1,750 noodles in 35 seconds. How many noodles does this champion make per second?

42. Representatives from 11 different noodle companies have been invited to the festival. Each representative brings 50 boxes of noodles. How many boxes of noodles are brought to the festival?

★43. At the noodle bar, a 16-ounce box of spirogyro noodles costs $0.80. A 24-ounce box costs $1.68. Which is the better buy?

CALCULATOR

Radio station WMTH is giving away $1,000,000.00 at the rate of $50.00 every hour on the hour. Will it take more than 2 years to give the money away? more than 3 years? Use your calculator to find the answer.

More Practice, page 431

PROBLEM SOLVING
Checking That the Solution Answers the Question

Pay special attention to the question that is asked in a problem. Be sure that you answer it.

Read each problem carefully.

1. In 1973, two teenagers were the first to pedal across the United States on a bicycle built for two. Their trip covered 4,837 miles. It began on February 4 and ended on June 5. About how many months did their trip take?

2. In 1973, two teenagers were the first to pedal across the United States on a bicycle built for two. Their trip covered 4,837 miles. It began on February 4 and ended on June 5. To the nearest hundred miles, about how many miles did they travel each month?

Both of these problems give you the same information, but each problem asks a different question. The answer to each question will be different also. The first problem asks a question about time; the answer is about 4 months. The second problem asks a question about distance; the answer is about 1,200 miles.

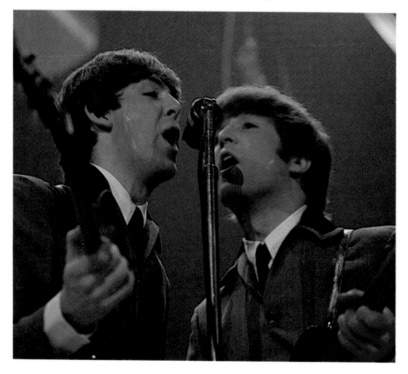

Which sentence answers the question? Write the letter of the correct answer.

1. One of the most popular songs of all time is "Yesterday" by Paul McCartney and John Lennon. During an 8-year period, 1,186 versions of the song were recorded. If the same number of versions were recorded each year, about how many versions were recorded each year?

   a. about 148 versions
   b. about 9,488 versions
   c. about 8 versions

2. Cats have an amazing homing instinct. Once, a cat left behind in California found its way back home to Oklahoma. The trip took 14 months and covered about 1,400 miles. If the cat left in September, in which month did it return to its home?

   a. 100 miles
   b. 1,200 miles
   c. November

Solve.

3. The world's largest ball of string weighs about 20,000 pounds and is almost 13 feet wide. It took the owner 28 years to gather the string. Estimate the average number of pounds of string that he collected each year.

4. Karen Stevenson won a contest by using a toothpick to eat the most baked beans in the shortest amount of time. Karen ate 2,780 beans in 30 minutes. If she could keep eating at that rate, how many beans would she eat in 1 hour?

5. The fastest-selling record in history was *John Fitzgerald Kennedy—A Memorial Album*. After it went on sale for $0.99, the record sold 4,000,000 copies in 6 days. Estimate how much money the record earned in that time.

6. The largest marching band ever assembled had 2,560 members. They marched on August 31, 1982, in a state parade in Malaysia. If there were about 30 marchers in each row, estimate how many rows there were.

7. Fifteen-year-old Patty Wilson set a world record in women's distance running in 1978. She ran 1,310 miles in 42 days. If she ran the same number of miles each day, estimate the number of miles she ran each day.

8. Otto E. Funk walked 4,165 miles from New York City to San Francisco while playing a violin. His walk began on December 14, 1928, and ended 183 days later, on June 16, 1929. For how many months did he walk?

LOGICAL REASONING

In the stamp club, 12 of the members collect U.S. stamps. 9 members collect foreign stamps. 5 of these members collect both U.S. and foreign stamps. How many members does the club have?

This Venn diagram can be used to solve the problem. There are

7 members who collect *only* U.S. stamps.

5 members who collect both U.S. and foreign stamps.

4 members who collect *only* foreign stamps.

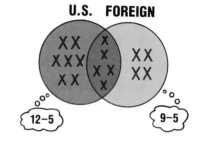

So, there are 7 + 5 + 4 = 16 members in the club.

Draw Venn diagrams and complete.

1. For the track and field team, 16 members enter track events. 11 members enter field events. 6 of these enter both track and field events.

The team has ▧ members.

2. For the coin-collecting club, 13 members collect U.S. coins. 14 members collect foreign coins. 10 of these members collect both U.S. and foreign coins.

The club has ▧ members.

3. For language classes, 16 students take Spanish. 19 students take French. 4 of these students take both Spanish and French.

▧ students take Spanish and/or French.

4. At the class picnic, 19 students played soccer. 12 students played volleyball. 11 of these students played both soccer and volleyball.

▧ students played soccer and/or volleyball.

★5. For the baseball club, 14 students collect cards from the National League. 3 students collect only National League cards. 18 students belong to the club.

▧ students collect only American League cards.

★6. At the swim meet, 16 members swam in relays. 5 members swam in relays and individual events. 25 members swam in the meet.

▧ members swam only in individual events.

GROUP PROJECT

Fabulous Facts

The problem: Fabulous facts can be found in ordinary daily events. There might be a fabulous fact to be found in your home. Think about how many hours of TV you watch each day. How much TV do your classmates watch? How many hours do you and your classmates spend watching TV in a year? Use the chart to discover a fabulous fact in your classroom.

Key Questions

- If there were 30 students in your class, and you all watched the same TV program, you'd chart this as 30 hours of TV watching for that program alone—and that's for only one week. What if you watch different programs?
- How many hours of regular programming do you watch in a week? How many hours do your classmates spend?
- How much time do you spend watching sports events and spectacles in a week? your classmates?
- How many hours of movies do you and your classmates watch in a week?

	Mon.	Tues.	Wed.	Thurs.	Fri.	Sat.	Sun.
Me							
Classmates							

Find your weekly total.

Add your classmates' totals for the week to your total. Find the total hours of TV watched for the year. Is this a fabulous fact?

CHAPTER TEST

Divide. (pages 166, 172, 174, and 180)

1. $20\overline{)80}$

2. $30\overline{)60}$

3. $\frac{92}{32}$

4. $\frac{277}{91}$

5. $86\overline{)586}$

6. $29\overline{)68}$

7. $27\overline{)1,134}$

8. $83\overline{)5,624}$

9. $54\overline{)2,259}$

10. $\frac{488}{13}$

11. $\frac{3,116}{19}$

12. $\frac{9,659}{87}$

13. $7,584 \div 75$

14. $3,791 \div 54$

15. $1,078 \div 37$

16. $36\overline{)7,596}$

17. $62\overline{)8,194}$

18. $\frac{29,782}{73}$

19. $\frac{26,952}{33}$

20. $46\overline{)44,436}$

21. $55\overline{)3,610}$

22. $62\overline{)43,555}$

23. $28\overline{)\$0.85}$

24. $\$95.00 \div 19$

25. $\$4.96 \div 84$

Estimate. Write the letter of the correct answer.
(pages 168 and 178)

26. $12\overline{)373}$ **a.** 1   **b.** 20   **c.** 30

27. $34\overline{)1,726}$ **a.** 5   **b.** 40   **c.** 50

28. $3,816 \div 42$ **a.** 80   **b.** 90   **c.** 900

29. $\frac{8,956}{73}$ **a.** 80   **b.** 100   **c.** 120

188

Solve. Use the information to answer the question.
(pages 177 and 185)

30.

The ancient Indian city of Mohenjo-Daro had giant public pools. The same pools still exist today. Scientists are interested in how fast the pools could be filled by using only the waters from the river. Which questions would help the scientists find this information? For each question, is an exact answer needed, or will an estimate be enough? Write *need not answer, need exact answer,* or *estimate.*

a. How much water will fill the pools?

b. How old are the pools?

c. How fast does the river flow?

d. Who built each pool?

31. The American Museum of Natural History in New York City is the largest museum in the world. Fran's class holds 2 bake sales to raise money for a trip to the museum. They must pay for admission, lunch, and transportation. Estimate whether they will have enough money to pay for the trip.

- There are 18 students.
- Lunch will cost $3.75 for a full meal, or $2.50 for a hamburger.
- Admission is a suggested donation of $0.75.
- Bus fare is $0.90 one way or $1.50 round trip.
- The first bake sale earned $37.45, and the second earned $78.27.

Write the letter of the correct answer. (page 185)

32. Each of the blocks used to build Stonehenge weighed about 50.2 tons. If 60 stones were used, how much did the entire monument weigh?

a. 1 block
b. 200.8 years
c. 3,012 tons

33. The blocks used to build the Cheops pyramid weighed 2.8 T each. Cheops is 480 ft 11 in. high. How much would the top 10 blocks weigh together?

a. 10 T
b. 28 T
c. 2.8 T

BONUS

1. $221\overline{)15,691}$

2. $349\overline{)28,765}$

3. $415\overline{)63,755}$

4. $621\overline{)21,221}$

5. $376\overline{)36,096}$

6. $31\overline{)32,348}$

RETEACHING

Sometimes you need to correct your estimate when you divide.

Divide $97\overline{)7{,}234}$.

Divide the thousands. Think: $97\overline{)7}$. Not enough thousands.
Divide the hundreds. Think: $97\overline{)72}$. Not enough hundreds.

Divide the tens.
  Think: $97\overline{)723}$, or $9\overline{)72}$.
  Estimate 8.

$$\begin{array}{r} 8 \\ 97\overline{)7{,}234} \\ \underline{7\,76} \end{array}$$
Multiply. Too great. You need to correct the estimate.

Try 7.

$$\begin{array}{r} 7 \\ 97\overline{)7{,}234} \\ \underline{6\,79} \\ 44 \end{array}$$
Multiply. Subtract and compare.

Divide the ones.
  Think: $97\overline{)44}$, or $9\overline{)44}$.
  Estimate 4.

$$\begin{array}{r} 74\ \text{R}56 \\ 97\overline{)7{,}234} \\ \underline{6\,79} \\ 444 \\ \underline{388} \\ 56 \end{array}$$
Multiply.
Subtract and compare.
Write the remainder.

Multiply.

1. $45\overline{)2{,}482}$ **2.** $33\overline{)2{,}636}$ **3.** $63\overline{)5{,}459}$ **4.** $54\overline{)1{,}511}$

5. $68\overline{)2{,}922}$ **6.** $45\overline{)1{,}484}$ **7.** $23\overline{)964}$ **8.** $33\overline{)493}$

9. $14{,}936 \div 37$ **10.** $4{,}493 \div 65$ **11.** $29{,}988 \div 73$ **12.** $2{,}035 \div 52$

13. $52{,}242 \div 83$ **14.** $5{,}432 \div 61$ **15.** $3{,}693 \div 44$ **16.** $15{,}836 \div 22$

17. $\$7.44 \div 12$ **18.** $\$10.88 \div 18$ **19.** $\$25.52 \div 29$ **20.** $\$22.56 \div 48$

$97\overline{)723}$ $45\overline{)2{,}482}$ $97\overline{)44}$ $97\overline{)7{,}234}$ $9\overline{)72}$

ENRICHMENT

Integers

Integers can be used to represent such things as distances above and below sea level.

A mountain is 2 km above sea level.
The height is written as $^+2$ km.
An ocean trench is 2 km below sea level.
The depth is written as $^-2$ km.
Sea level is written as 0.

Write an integer for each.

1. 4 km above sea level

2. 6° above zero

3. a 7-yard gain

4. a loss of $7.00

5. a 4-point penalty

6. a 6-foot fall

Integers can also be used to represent opposites.

EXAMPLES: 4° above zero → $^+4$ 4° below zero → $^-4$
a savings of $6.00 → $^+6$ a loss of $6.00 → $^-6$

You can show integers on a number line.

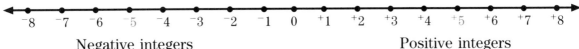

Negative integers   Positive integers

You read $^-5$ as "negative 5."   You read $^+5$ as "positive 5."
$^-5$ is 5 units to the left of 0. $^+5$ is 5 units to the right of 0.

Zero is neither a positive integer nor a negative integer.

Write an integer and its opposite for each.

7. a loss of $5.00

8. 9 feet below sea level

9. 7° above zero

10. a 5-point loss

11. 8 meters below sea level

12. a 4-foot climb

Solve.

13. A child digs a hole 5 feet deep in the sand. If 0 represents the surface, what integer would you use to represent the depth of the hole?

TECHNOLOGY

A PRINT statement is a BASIC instruction that tells the computer to type on the screen. If you are computing with the computer, you can use parentheses to tell the computer which operations to do first. The computer uses * for multiplication and / for division.

1. Write what the computer types when you type this instruction and then press RETURN or ENTER.

   PRINT 48 / (2 * 4)

2. Write what the computer types when you RUN this program.

   10   PRINT (6 + 6) / 6
   20   PRINT (2 + 2) / (2 − 1)
   30   PRINT 12 − (2 * 3)
   40   PRINT (5 − 3) * 4
   50   PRINT "WHO DO WE APPRECIATE"

A computer can also store the value of a number.

   10   LET X = 17
   20   PRINT X

When this program is RUN, the LET statement stores 17 in a **variable** called X. A variable is a location that the computer uses to store a number. If you want to see the value of a variable, you must tell the computer to print it.

3. Write what is printed when this program is RUN.

   10   LET A = 5 * 5
   20   LET B = 30 / 6
   30   PRINT B
   40   PRINT A

You have just seen two programs that use variables X, A, and B to store variables. You can make up your own variable names. Here are some rules for naming variables.

The name of a variable can be a single letter.
B   W   K   I
The name of a variable can be any two letters.
AP   RL   WT
The name of a variable can be one letter and one number.
V2   K9   T3
But the name of a variable cannot start with a number.
6F   14   8W

4. Write the variables that you can use to store numbers.

A	TR	P4	DD	HI
67	G5	14	H	MA
R2	D2	LF	W	P

5. Finish this program. It should find and print the difference between 15 and 9.

```
10  ▨
20  PRINT DF
```

6. Write a program that stores the product of 5 and 6 in a variable and prints it out.

Here is a program that prints the average of three numbers.

```
10  LET A = 15
20  LET B = 9
30  LET C = 6
40  LET AV = (A + B + C) / 3
50  PRINT AV
```

In line 40, the program first adds the values of A, B, and C to find 30. Then it divides 30 by 3 to find 10, and stores 10 in AV.

7. What is printed when this program is RUN?

8. Rewrite this program so that it prints the average of 22 and 8. Copy the whole program.

CUMULATIVE REVIEW

Write the letter of the correct answer.

1. $228.06 ÷ 7

   a. $31.48   **b.** $32.00
   c. $32.58   **d.** not given

2. 239 ÷ 7

   a. 34   **b.** 34 R1
   c. 35   **d.** not given

3. 53,687 ÷ 7

   a. 7,659 R4   **b.** 7,669 R4
   c. 7,770 R1   **d.** not given

4. 25 ÷ ■ = 1

   a. 0   **b.** 25
   c. 1   **d.** not given

5. Estimate: 18,593 ÷ 6.

   a. 1,500   **b.** 2,000
   c. 3,000   **d.** not given

6. Which number is divisible by 5?

   a. 45,306   **b.** 51,355
   c. 56,987   **d.** not given

7. 479 × 326

   a. 156,154   **b.** 158,152
   c. 166,254   **d.** not given

8. Write in standard form:
   seven and three hundred
   eighty-seven thousandths.

   a. 7.387   **b.** 70.387
   c. 7,387   **d.** not given

9. 78.42 − 63.28

   a. 14.44   **b.** 15.14
   c. 17.04   **d.** not given

10. $5,013.00 − $3,698.79

    a. $1,314.21   **b.** $1,324.51
    c. $2,111.45   **d.** not given

11. 3.7 × 0.04

    a. 1.48   **b.** 14.8
    c. 148   **d.** not given

12. 3,619 + 4,578 + 1,255 + 543

    a. 9,755   **b.** 9,995
    c. 10,215   **d.** not given

13. A convoy of trucks must carry 150
    new cars to the showroom. If each
    truck can carry 8 cars, how many
    trucks are needed?

    a. 18   **b.** 19
    c. 20   **d.** not given

14. There are 279 people in the
    woodworking classes, and 31
    people in each class. Let n = the
    number of classes. Choose the
    correct number sentence.

    a. 279 × 31 = n **b.** 279 − n = 31
    c. 279 ÷ 31 = n **d.** not given

Often when you see a streak of lightning, it is followed by thunder. Have you ever noticed that you do not hear the thunder immediately? That is because sound moves more slowly than light. How could you find the distance between you and a storm?

7 DIVIDING WITH DECIMALS
Metric Measurement

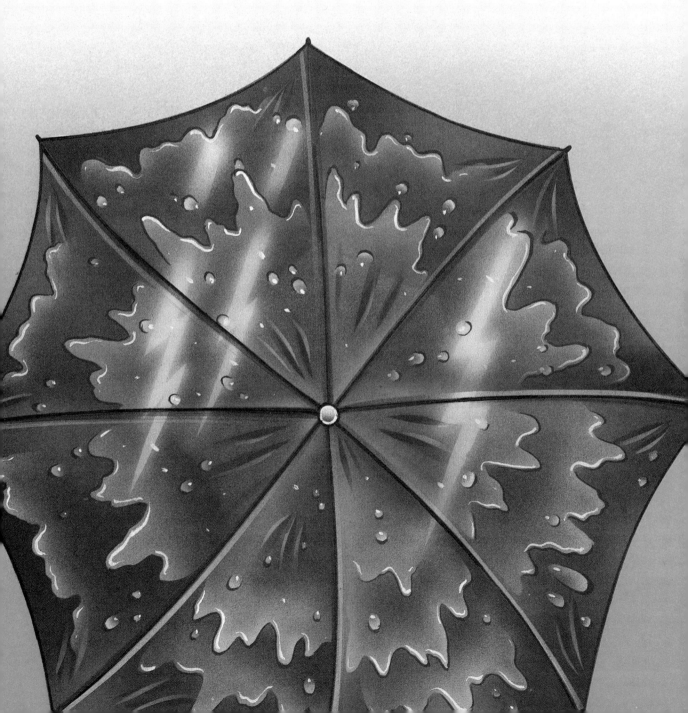

Multiplying and Dividing Decimals by 10; 100; and 1,000

A. Generally, 10 centimeters of snow contain as much moisture as 1 centimeter of rain. How much snow would there have to be to equal the moisture in 1.37 centimeters of rain?

Find 10×1.37.

$1 \times 1.37 = 1.37$	To multiply by
$10 \times 1.37 = 13.7$	10, move the decimal point one place to the right.
$100 \times 1.37 = 137.$	100, move the decimal point two places to the right.
$1,000 \times 1.37 = 1,370.$	1,000, move the decimal point three places to the right.

Write zeros in the product, as needed, in order to place the decimal point correctly.

There would have to be 13.7 centimeters of snow to equal the moisture in 1.37 centimeters of rain.

B. Find $52.6 \div 10$.

$52.6 \div 1 = 52.6$	To divide by
$52.6 \div 10 = 5.26$	10, move the decimal point one place to the left.
$52.6 \div 100 = 0.526$	100, move the decimal point two places to the left.
$52.6 \div 1,000 = 0.0526$	1,000, move the decimal point three places to the left.

Write zeros in the quotient, as needed, in order to place the decimal point correctly.

Checkpoint Write the letter of the correct answer.

Multiply or divide.

1. $269.3 \div 10$ **a.** 0.2693   **b.** 2.693   **c.** 26.93   **d.** 2,693

2. $1,000 \times 0.999$ **a.** 0.00099   **b.** 99.9   **c.** 999   **d.** 990

Multiply or divide.

1. $38 \div 10$ **3.8** **2.** $72.48 \div 100$ **3.** $251.74 \div 100$ **4.** $9,612.5 \div 1,000$

5. 10×456.2 **6.** 100×392 **7.** 100×84.87 **8.** $1,000 \times 7.169$

9. $5,348 \div 1,000$ **10.** $1,000 \times 68.412$ **★11.** $2,815.4 \div 10,000$ **★12.** $0.0564 \times 10,000$

Copy and complete each chart.

	×	10	100	1,000
13.	45.62	■	■	■
14.	1.386	■	■	■
15.	■	371.5	■	■
16.	■	■	■	9,351
17.	■	■	748.2	■

	÷	10	100	1,000
18.	348.6	■	■	■
19.	73.915	■	■	■
20.	0.4157	■	■	■
21.	■	23.48	■	■
22.	■	■	■	0.005682

Solve.

23. Scientists at the Mount Jackson Weather Station found that a total of 53.5 centimeters of snow had fallen at their station during a 10-week period. What was the average weekly snowfall?

24. Snowstorms in Howe usually leave 3.4 centimeters of snow on the ground. Recently 10 times the usual amount of snow fell on Howe. How much snow is that?

25. The average height of the waves at the beach in Palmville is 1.5 feet. After a recent hurricane, the waves were 10 times the average height. How high were the waves after the hurricane?

★26. A hurricane is 650 miles south of Miami, Florida. It is moving toward Florida at a speed of 10 miles per hour. How far will it be from Florida in 10 hours? How long will it take to reach Florida?

ANOTHER LOOK

Find the product.

1. $\begin{array}{r} 0.07 \\ \times\ 31.7 \\ \hline \end{array}$ **2.** $\begin{array}{r} 0.08 \\ \times\ 13 \\ \hline \end{array}$ **3.** $\begin{array}{r} 41.2 \\ \times\ 0.06 \\ \hline \end{array}$ **4.** $\begin{array}{r} 9.8 \\ \times\ 7.4 \\ \hline \end{array}$ **5.** $\begin{array}{r} 0.03 \\ \times\ 0.5 \\ \hline \end{array}$

6. 0.42×46.6 **7.** 0.05×3.1 **8.** 0.6×8.12 **9.** 0.04×0.3

Dividing Decimals by Whole Numbers

A. Chicago is often called the "windy city." On a particularly windy day, the wind gusted to 49.44 kilometers per hour. This figure is 3 times the average speed of the wind in Chicago. What is the average wind speed in Chicago?

Find 49.44 ÷ 3.

Divide the whole number.	Place the decimal point. Divide the tenths.	Divide the hundredths.

$$
\begin{array}{r}
1\,6 \\
3\overline{)4\,9.4\,4} \\
3\!\downarrow \\
\overline{1\,9} \\
1\,8 \\
\overline{1}
\end{array}
\qquad
\begin{array}{r}
1\,6.4 \\
3\overline{)4\,9.4\,4} \\
3 \\
\overline{1\,9} \\
1\,8 \\
\overline{1\,4} \\
1\,2 \\
\overline{2}
\end{array}
\qquad
\begin{array}{r}
1\,6.4\,8 \\
3\overline{)4\,9.4\,4} \\
3 \\
\overline{1\,9} \\
1\,8 \\
\overline{1\,4} \\
1\,2 \\
\overline{2\,4} \\
2\,4 \\
\overline{0}
\end{array}
$$

The average wind speed in Chicago is 16.48 kilometers per hour.

Another example:

$$
\begin{array}{r}
0.212 \\
18\overline{)3.816}
\end{array}
$$

> When the quotient is less than 1, write a zero in the ones place.

Checkpoint Write the letter of the correct answer.

Divide.

1. $5\overline{)3.325}$ **2.** $7\overline{)65.66}$ **3.** $8.056 \div 38$ **4.** $17\overline{)15.946}$

a. 0.557	**a.** 9.09	**a.** 0.201	**a.** 0.902
b. 0.665	**b.** 9.38	**b.** 0.21	**b.** 0.0938
c. 6.65	**c.** 93.8	**c.** 0.212	**c.** 0.938
d. 665	**d.** 938	**d.** 2.12	**d.** 938

Divide.

1. $8\overline{)5.36}$ 2. $5\overline{)9.5}$ 3. $8\overline{)6.48}$ 4. $6\overline{)9.18}$ 5. $3\overline{)4.35}$

6. $8\overline{)7.856}$ 7. $4\overline{)3.484}$ 8. $20\overline{)7.260}$ 9. $30\overline{)4.560}$ 10. $50\overline{)6.850}$

11. $13\overline{)42.12}$ 12. $71\overline{)9.94}$ 13. $24\overline{)56.64}$ 14. $39\overline{)51.48}$ 15. $98\overline{)23.52}$

16. $49\overline{)66.346}$ 17. $12\overline{)51.36}$ 18. $27\overline{)87.966}$ 19. $29\overline{)91.611}$ 20. $14\overline{)3.262}$

21. $39.33 \div 9$ 22. $2.38 \div 7$ 23. $7.28 \div 4$ 24. $21.265 \div 5$

25. $1.644 \div 3$ 26. $9.264 \div 4$ 27. $12.95 \div 37$ 28. $97.58 \div 41$

29. $91.26 \div 39$ 30. $84.422 \div 26$ 31. $91.117 \div 43$ 32. $27.768 \div 78$

Solve.

33. The forecasters at Rae's weather bureau spent $124.00 to buy 8 new thermometers. What was the price of each thermometer?

34. Rae's instruments measure a wind speed of 44.13 kilometers per hour, 3 times faster than the wind speed two hours ago. What was the wind speed two hours ago?

35. A storm must have winds of 118.5 kilometers per hour to be called a hurricane. The wind at Rae's station is blowing at a speed of 67.8 kilometers per hour. How much faster must the wind blow to reach hurricane speed?

★36. One of the heaviest rainfalls ever recorded in a 24-hour period was 186.99 centimeters. If the rainfall was constant, how much rain fell during each hour of the period? Round your answer to the nearest tenth.

CHALLENGE

Copy the chart. Match a divisor to each dividend so that all of the quotients will be the same.

Divisors	6	2	4	5	3
Dividends	1.6	3.2	4.8	2.4	4.0

Dividing by Whole Numbers—Using Zeros

A. If a tornado covers a distance of 52.16 kilometers in 2 hours, how fast does it travel per hour?

Find 52.16 ÷ 2.

Sometimes you have to write zeros in the quotient.

Divide the whole number.	Place the decimal point. Divide the tenths.	Divide the hundredths.
$\begin{array}{r} 2\,6 \\ 2\overline{)5\,2.1\,6} \\ 5\,2 \end{array}$	$\begin{array}{r} 2\,6.0 \\ 2\overline{)5\,2.1\,6} \quad \boxed{\text{Write 0.}} \\ 5\,2\downarrow \\ \hline 1 \\ 0 \\ \hline 1 \end{array}$	$\begin{array}{r} 2\,6.0\,8 \\ 2\overline{)5\,2.1\,6} \\ 5\,2\downarrow \\ \hline 1 \\ 0\downarrow \\ \hline 1\,6 \\ 1\,6 \\ \hline 0 \end{array}$

The tornado travels 26.08 kilometers per hour.

B. You may have to write zeros in the dividend.

Divide the whole number.	Place the decimal point. Divide the tenths.	Divide the hundredths.	Write a 0 in the dividend. Divide the thousandths.
$\begin{array}{r} 2 \\ 15\overline{)3\,4.8\,6} \\ 3\,0 \\ \hline 4 \end{array}$	$\begin{array}{r} 2.3 \\ 15\overline{)3\,4.8\,6} \\ 30\downarrow \\ \hline 4\,8 \\ 4\,5 \\ \hline 3 \end{array}$	$\begin{array}{r} 2.3\,2 \\ 15\overline{)3\,4.8\,6} \\ 3\,0 \\ \hline 4\,8 \\ 4\,5\downarrow \\ \hline 3\,6 \\ 3\,0 \\ \hline 6 \end{array}$	$\begin{array}{r} 2.3\,2\,4 \\ 15\overline{)3\,4.8\,6\,0} \quad \boxed{\text{Write 0.}} \\ 3\,0 \\ \hline 4\,8 \\ 4\,5 \\ \hline 3\,6 \\ 3\,0\downarrow \\ \hline 6\,0 \\ 6\,0 \\ \hline 0 \end{array}$

Checkpoint Write the letter of the correct answer.

Divide.

1. $15\overline{)47.43}$ **a.** 3.086   **b.** 3.16   **c.** 3.162   **d.** 3162

2. $0.156 \div 6$ **a.** 0.021   **b.** 0.026   **c.** 0.25   **d.** 26

200

Divide.

1. $4\overline{)1.32}$ 2. $7\overline{)5.04}$ 3. $2\overline{)1.24}$ 4. $9\overline{)35.73}$ 5. $3\overline{)12.09}$

6. $11\overline{)47.41}$ 7. $16\overline{)60.48}$ 8. $13\overline{)7.54}$ 9. $37\overline{)39.22}$ 10. $21\overline{)1.47}$

11. $4\overline{)21.8}$ 12. $5\overline{)44.02}$ 13. $6\overline{)12.63}$ 14. $5\overline{)4.88}$ 15. $8\overline{)9.16}$

16. $65\overline{)1.56}$ 17. $16\overline{)48.24}$ 18. $95\overline{)19.38}$ 19. $20\overline{)26.34}$ 20. $15\overline{)30.54}$

21. $84.62 \div 20$ 22. $3.24 \div 8$ 23. $90.51 \div 30$ 24. $5.47 \div 5$

25. $1.22 \div 4$ 26. $31.93 \div 31$ 27. $4.71 \div 6$ 28. $47.33 \div 5$

29. $25.16 \div 8$ 30. $0.19 \div 5$ 31. $1.89 \div 15$ 32. $0.789 \div 6$

Solve. For Problem 35, use the Infobank.

33. A tornado's central winds, or *updraft*, are blowing at a speed of 308.15 kilometers per hour. This speed is twice as fast as the tornado's updraft an hour ago. How fast was the tornado's updraft an hour ago?

34. The distance around a typical tornado is 2.4 kilometers. The distance around a typical hurricane is 228 times as great. What is the distance around a typical hurricane?

35. Hurricanes, like tornadoes, cause great damage. Use the information on page 418 to write and solve your own word problem.

★36. One tornado sweeps through 138.57 kilometers of Texas in 3 hours. Another tornado covers 87.58 kilometers of Oklahoma in 2 hours. Which tornado travels at a faster speed?

MIDCHAPTER REVIEW

Multiply or divide.

1. $27 \div 10$ 2. 32.328×100 3. $6.738 \div 100$ 4. $2.83 \times 1{,}000$ 5. $54.835 \div 1{,}000$

6. $3\overline{)94.62}$ 7. $2\overline{)0.462}$ 8. $6\overline{)72.12}$ 9. $8\overline{)17.24}$ 10. $5\overline{)0.41}$

11. $24\overline{)29.52}$ 12. $12\overline{)9.396}$ 13. $14\overline{)0.868}$ 14. $16\overline{)24.24}$ 15. $15\overline{)7.23}$

16. $44.36 \div 2$ 17. $21.44 \div 80$ 18. $1.53 \div 6$ 19. $3.95 \div 79$

PROBLEM SOLVING
Guessing and Checking

Sometimes the best way to solve a problem is to make a guess. Read the problem to find clues. Make a guess. Check to see whether it answers the question. If not, try again.

> In a single day in 1979, a record number of inches of rain fell on Alvin, Texas. The record was a 2-digit number. The sum of its digits is 7. It is not evenly divisible by 2 or 5, and it is less than 60. What is the number?

The first two clues in the problem tell you that the missing number has two digits whose sum is 7. List each 2-digit number whose two digits add up to 7.

16   25   34   43   52   61   70

The next clue in the problem tells you that the number is not divisible by 2 or 5. So, you can eliminate 16, 25, 34, 52, and 70 because they *are* divisible by either 2 or 5. That leaves 43 and 61.

The last clue in the problem tells you that the number is less than 60.
So, your answer is 43.

Now check your answer:
- Is it a 2-digit number? yes
- Is the sum of the digits 7? yes
- Is the number *not* evenly divisible by 2 or 5? yes
- Is the number less than 60? yes

So, in a single day, a record 43 inches of rain fell on Alvin, Texas.

Solve. If you use the guess-and-check method, show
your guesses and checks.

1. A snowstorm is considered a
blizzard when its winds reach a
certain number of miles per hour.
This is a 2-digit number. The sum of
the digits is 8. The difference
between the digits is 2. The number
is evenly divisible by 5. At how many
miles per hour is a snowstorm
considered a blizzard?

2. In 1982, a 2-day blizzard blanketed
Denver, Colorado. The number of
inches of snow that fell is a 2-digit
number that is evenly divisible by 2
and 3. The sum of its digits is 6. If
you multiply the number by 3, the
product has a 2 in the ones place.
How many inches of snow fell
during the blizzard?

3. The average low temperature in
Dublin, Ireland, is 1° lower than the
average low temperature in London,
England. The product of the two
temperatures is 72. What are the two
temperatures?

4. The average high temperature in
Hamburg, West Germany, is 1° higher
than the average high temperature in
Copenhagen, Denmark. The sum of
the two temperatures is 183. What
are the two temperatures?

5. Marcia bought a thermometer for
$1.79. She paid for it with 14 coins.
What were they?

6. Nicholas spent $4.96 for a book
about weather. He paid with 4 dollar
bills and 6 coins. What were the
coins?

7. In a single month in 1942, a record
number of inches of rain fell on Puu
Kukui, Maui, Hawaii. This is a 3-digit
number. The sum of the digits is 8.
The digit in the tens place is 0. The
number is *not* evenly divisible by 2,
4, or 5. If you multiply the number
by 3, the product is less than 1,000.
What is the record number?

★8. Santiago, Chile, averages 6 times
more rain in October than in
January. Both the number of inches
of rain that fell in October and the
number that fell in January were less
than 1. The numbers are decimals, in
tenths. The difference between them
is 0.5. Their sum is less than 0.9.
How much rain fell in October and
in January?

Centimeter and Millimeter

A. As many as 100 tiny ice crystals can cling together to form a snowflake. Some snowflakes can measure as long as 2 centimeters.

A **centimeter (cm)** is a metric unit of length used to measure small objects or distances. Find 2 cm on this metric ruler.

B. Each centimeter is divided into ten units called **millimeters (mm)**.

1 centimeter (cm) = 10 millimeters (mm)

A millimeter is used to measure the length of very small objects. The width of a dime is about a millimeter. Measure the length of this icicle to the nearest centimeter and to the nearest millimeter.

1 mm

The icicle measures 6 centimeters to the nearest centimeter.

The icicle measures 58 millimeters to the nearest millimeter.

This measurement can also be expressed as 5 cm 8 mm, or 5.8 cm.

C. You can find the distance around an object by measuring its sides and then finding the sum of the measures. Find the distance around this shape.

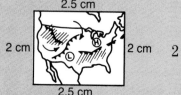

2.5 cm

2 cm

2 cm

2.5 cm

$2 + 2.5 + 2 + 2.5 = 9$

The distance around the shape is 9 cm.

Measure the length of the piece of string to the nearest

1. ▧ cm.   **2.** ▧ mm.   **3.** ▧ cm ▧ mm.

Draw a line that measures

4. 3 cm.   **5.** 10 cm.   **6.** 17 cm.   **7.** 24 mm.   **8.** 6 mm.
9. 12 mm.   **10.** 8.2 cm.   **11.** 0.7 cm.   **12.** 31.2 cm.   **13.** 14.8 cm.

Measure to find the distance around each shape.

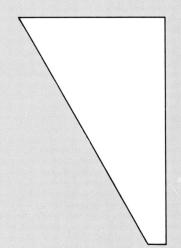

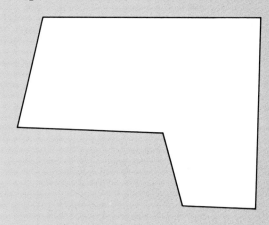

FOCUS: ESTIMATION

How long is each item? Copy and complete the table. First estimate each length to the nearest centimeter. Then measure.

Object	Estimate	Measure
A book	▧	▧
An eraser	▧	▧
Your index finger	▧	▧
A piece of chalk	▧	▧

More Practice, page 432

Meters and Kilometers

A. **Meters (m)** and **kilometers (km)** are metric units of length.

> 1 centimeter (cm) = 10 millimeters (mm)
> 1 meter (m) = 100 centimeters (cm)
> 1 kilometer (km) = 1,000 meters (m)

The length of a baseball bat is about 1 meter.

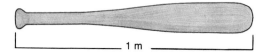

The distance a person can walk in 15 minutes is about 1 kilometer.

B. To rename larger units with smaller units, you can multiply.

2 cm = ▨ mm	5 m = ▨ cm	9 km = ▨ m
2 × 10 = 20	5 × 100 = 500	9 × 1,000 = 9,000
2 cm = 20 mm	5 m = 500 cm	9 km = 9,000 m

C. To rename smaller units with larger units, you can divide.

7 mm = ▨ cm	4 cm = ▨ m	8 m = ▨ km
7 ÷ 10 = 0.7	4 ÷ 100 = 0.04	8 ÷ 1,000 = 0.008
7 mm = 0.7 cm	4 cm = 0.04 m	8 m = 0.008 km

Checkpoint Write the letter of the correct answer.

Complete.

1. 16 cm = ▨ mm   **a.** 0.16   **b.** 1.6   **c.** 160   **d.** 1,600

2. 756 mm = ▨ m   **a.** 0.0756   **b.** 0.756   **c.** 75.6   **d.** 756,000

3. 53 km = ▨ m   **a.** 0.053   **b.** 0.53   **c.** 530   **d.** 53,000

Which unit would you use to measure?
Write *millimeter, centimeter, meter,* or *kilometer.*

1. the length of a soccer field

2. the distance to the sun

3. the length of a pen

4. the length of a thumbnail

5. the height of a tree

6. a day's car ride

7. the length of a straw

8. the length of an ant

Choose the appropriate unit. Write *millimeter, centimeter, meter,* or *kilometer.*

9. A car travels at 55 ▨ an hour.

10. A man is 2 ▨ tall.

11. A house is 10 ▨ high.

12. A plane flies 500 ▨ in a day.

13. A paper clip is 4 ▨ long.

14. A pair of scissors is 5 ▨ long.

Complete.

15. 48 km = ▨ m

16. 0.01 m = ▨ cm

17. 27 km = ▨ m

18. 11 m = ▨ km

19. 93 cm = ▨ m

20. 38.2 m = ▨ km

21. 3 km = ▨ m

22. 47.2 m = ▨ cm

23. 209.5 km = ▨ m

24. 4,967 m = ▨ km

25. 6,035 mm = ▨ cm

26. 0.386 cm = ▨ mm

27. 0.32 m = ▨ cm

28. 7,004 m = ▨ km

29. 835.7 cm = ▨ mm

Solve. For Problem 31, use the Infobank.

30. Weather satellites orbit Earth at an altitude of 1,400,000 meters. What is this altitude in kilometers?

31. Use the information on page 418 to solve. Calculate the widths, in kilometers, of the three lowest bands of Earth's atmosphere.

ANOTHER LOOK

Add.

1.	**2.**	**3.**	**4.**	**5.**
35,984	52,897	10,459	23,845	97,002
21,065	1,250	5,682	1,329	21,570
+ 10,120	+ 35,381	+ 374	+ 47	+ 53

More Practice, page 433

PROBLEM SOLVING
Identifying Extra/Needed Information

One year, 108.3 cm of rain fell in Rio de Janeiro, Brazil. That was 28.9 centimeters less rain than fell in Brasília, Brazil's capital city. Rio de Janeiro's rainfall was 3 times as great as that of Santiago, Chile. What was the rainfall in that year in Santiago, Chile.

A problem may contain more information than you need to answer the question that is asked. If the problem you are reading seems to contain extra information, follow these steps.

— **1.** Study the question.
What was the rainfall in Santiago?

— **2.** List the information in the problem.
    a. Rio de Janeiro's rainfall was 108.3 cm.
    b. Rio de Janeiro's rainfall was 28.9 cm less than that of Brasília.
    c. Rio de Janeiro's rainfall was 3 times as great as that of Santiago, Chile.

— **3.** Cross out the information that will not help you answer the question. (Cross out *b.*)

— **4.** Use the information that is left to solve the problem.

Solve: $108.3 \div 3 = 36.1$.

There were 36.1 centimeters of rainfall that year in Santiago, Chile.

Write the letter of the information that is not needed.

1. Bridgeport, Connecticut, lies 87 km northeast of New York City. In 1976, Hurricane Belle traveled from New York to Bridgeport in about 3 hours. In Bridgeport, its winds were recorded at 124 km per hour. At what speed did Hurricane Belle travel from New York to Bridgeport?

   a. Bridgeport is 87 km from New York City.
   b. Belle's winds were recorded at 124 km per hour.
   c. Belle took about 3 hours to reach Bridgeport from New York.

2. At 1:00 P.M., a typhoon was reported heading toward Manila at 23.3 km per hour. It hit Manila 4 hours later. At 6:00 P.M., it struck Cavita, 16 km away. How far from Manila was the typhoon at the time of the first report?

   a. The typhoon was first reported at 1:00 P.M.
   b. It was moving at 23.3 km per hour.
   c. Cavita, 16 km from Manila, was hit at 6:00 P.M.

Solve.

3. Canton, China, had 161.5 cm of rain in one year. Shanghai had 53 cm of rain in the same year. Tientsin had only 29.2 cm of rain. That year, about how many times as great as Shanghai's rainfall was Canton's?

4. The greatest amount of rain in a 30-day period fell 125 years ago in a town in India. The town was drenched with 929.99 cm of rain. To the nearest tenth of a centimeter, what was the average daily rainfall?

5. In 1982, 54.6 cm of snow fell on Springfield, Missouri. Almost 9.4 cm more than that fell on Kansas City. Marquette, Michigan, had 8 times as much snow as Springfield. How much snow fell on Marquette?

6. The greatest snowfall in one year measured 31.1 meters. It fell on Mount Rainier, Washington. Mount Rainier is 4,393 meters high. To the nearest tenth of a meter, what was the average monthly snowfall?

7. In Honolulu, Hawaii, the average wind speed is 18.8 km per hour. That is almost twice the average for Chattanooga, Tennessee. On Mount Washington, in New Hampshire, the average wind speed is 3 times as great as that in Honolulu. What is the wind speed on Mount Washington?

8. Probably the heaviest recorded rainfall fell on the island of Guadeloupe, in 1970. In one minute, 3.84 cm of rain fell. At that rate, 230.9 cm of rain would fall in one hour. How many centimeters of rain fell per second?

Liters and Milliliters

A. Maria collects rainwater in a 1-liter jug for her school science project. One rainy day, she collects 364 milliliters of water.

Milliliters (mL) and **liters (L)** are metric units of capacity.

> 1 liter (L) = 1,000 milliliters (mL)

A carton of milk holds about 1 liter.

An eyedropper holds about 1 milliliter.

B. To rename larger units with smaller units, you can multiply.

$4\text{ L} = \ \text{mL}$
$4 \times 1{,}000 = 4{,}000$
$4\text{ L} = 4{,}000\text{ mL}$

C. To rename smaller units with larger units, you can divide.

$8\text{ mL} = \ \text{L}$
$8 \div 1{,}000 = 0.008$
$8\text{ mL} = 0.008\text{ L}$

Checkpoint Write the letter of the correct answer.

Complete.

1. 78 L = ▨ mL

a. 0.078
b. 780
c. 7,800
d. 78,000

2. 863 mL = ▨ L

a. 0.0863
b. 0.863
c. 8.63
d. 863,000

3. 119 L = ▨ mL

a. 0.119
b. 1.19
c. 11,900
d. 119,000

4. 47 mL = ▨ L

a. 0.0047
b. 0.047
c. 0.47
d. 47,000

Which unit would you use to measure the capacity?
Write *milliliter* or *liter*.

1. a thermos

2. a teaspoon

3. a canteen

4. a bucket

5. a thimble

6. a carton of juice

7. an aquarium

8. a test tube

9. a raindrop

10. a water-storage tank

11. a gas tank in a car

12. an oil drum

13. a water glass

14. a water trough

15. an ink bottle

Complete.

16. 6 L =   mL

17. 12 L =   mL

18. 0.91 L =   mL

19. 3.3 L =   mL

20. 27 mL =   L

21. 789 mL =   L

22. 15 mL =   L

23. 357 mL =   L

24. 0.4 L =   mL

25. 269 L =   mL

26. 128 mL =   L

27. 4.087 L =   mL

28. 1 mL =   L

29. 10 L =   mL

30. 3,974 mL =   L

31. 23.087 L =   mL

32. 578 mL =   L

33. 0.56 L =   mL

34. 141 L =   mL

35. 0.099 L =   mL

36. 2 mL =   L

37. 79.002 mL =   L

38. 8.091 L =   mL

39. 2,965 mL =   L

Solve.

40. Scientists find 6 milliliters of pollutants in a 30-liter
tank of rainwater. How many milliliters of
pollutants are there in each liter of rainwater.

41. After a six-month cleanup, scientists found only 2
milliliters of pollutants in a 30-liter tank of
rainwater. How many milliliters of pollutants were
there now in each liter? How many fewer milliliters
is this than the earlier measurement?

CHALLENGE

Jim has a large vat full of water. He wants to measure
out 1 liter of the water, but he does not have a 1-liter
container. He does have a 4-liter container and a 7-liter
container. How can he measure exactly 1 liter of the
water?

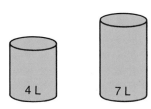

More Practice, page 233

Kilograms, Grams, and Milligrams

A. Although hailstones can be as large as softballs, most of them are the size of peas. The largest hailstone ever found fell on Coffeyville, Kansas, in 1970. The hailstone had a mass of 800 grams.

Kilograms (kg), grams (g), and **milligrams (mg)** are metric units of mass.

> 1 gram (g) = 1,000 milligram (mg)
> 1 kilogram (kg) = 1,000 gram (g)

A feather has a mass of about 1 gram.

An iron has a mass of about 1 kilogram.

A grain of sand has a mass of about 1 milligram.

B. To rename larger units with smaller units, you can multiply.

4 g = ▨ mg   6 kg = ▨ g
4 × 1,000 = 4,000   6 × 1,000 = 6,000
4 g = 4,000 mg   6 kg = 6,000 g

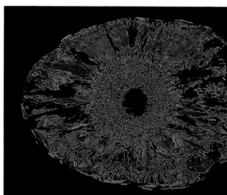

C. To rename smaller units with larger units, you can divide.

7 mg = ▨ g   8 g = ▨ kg
7 ÷ 1,000 = 0.007   8 ÷ 1,000 = 0.008
7 mg = 0.007 g   8 g = 0.008 kg

Checkpoint Write the letter of the correct answer.

Complete.

1. 38 g = ▨ kg
   a. 0.0038
   b. 0.038
   c. 3.8
   d. 3,800

2. 938 mg = ▨ g
   a. 0.938
   b. 9.38
   c. 93,800
   d. 938,000

3. 78 g = ▨ mg
   a. 0.078
   b. 7.8
   c. 7,800
   d. 78,000

4. 49 kg = ▨ g
   a. 0.049
   b. 490
   c. 4,900
   d. 49,000

Which unit would you use to measure the mass? Write
milligram, gram, or *kilogram.*

1. A car **2.** A snowflake **3.** An adult

4. A pencil **5.** A flash cube **6.** A book

7. An airplane **8.** A bicycle **9.** An apple

10. A needle **11.** A chair **12.** A house

13. A bee **14.** A desk **15.** An elephant

16. A telephone **17.** An ice cube **18.** A tractor

Complete.

19. 19 kg = ▉ g   **20.** 1.2 kg = ▉ g   **21.** 47 g = ▉ mg   **22.** 25 kg = ▉ g

23. 64 g = ▉ kg   **24.** 35 mg = ▉ g   **25.** 345 mg = ▉ kg   **26.** 29.4 g = ▉ kg

27. 0.01 g = ▉ mg   **28.** 498 g = ▉ kg   **29.** 3,732 mg = ▉ kg   **30.** 34.3 g = ▉ mg

31. 0.045 kg = ▉ g   **32.** 1,095 mg = ▉ g   **33.** 278 kg = ▉ mg   **34.** 56.4 g = ▉ kg

35. 89.9 kg = ▉ g   **36.** 576 g = ▉ mg   **37.** 780 mg = ▉ kg   **38.** 52.9 kg = ▉ g

39. 7,897 mg = ▉ g   **40.** 8.9 kg = ▉ mg   **41.** 0.11 g = ▉ mg   **42.** 9 kg = ▉ mg

Solve.

43. The mass of a hailstone is about 1 gram. After one hailstorm, 9.8 kg of hailstones were scraped off the roof of a building. About how many hailstones fell on the roof?

44. A scientist is flying in an airplane that, including its cargo, has a total mass of 3,500 kg. The plane releases 1,200 kg of rain-making chemicals into the clouds. How much does the plane now weigh?

CHALLENGE

Lou uses a balance scale to find the mass of a sea gull that has been injured. He balances the scale by placing the bird in one tray, and placing known masses in the other tray. The bird's mass is 1.5 kg. Look at the table. Find at least four combinations of masses that Lou can use to find 1.5 kg.

KNOWN MASSES

Mass	Number available
1 kg	1
0.75 kg	1
0.5 kg	2
0.25 kg	4
0.01 kg or 10 g	10

PROBLEM SOLVING
Solving Multistep Problems/Making a Plan

Making a plan can help you solve complicated problems. Write down the steps you need to take to solve the problem.

The year after the drought, Bayville had a total of 37.58 cm of rain. That is 2 times the amount of rain that fell during the drought year of 1983. In 1982, Bayville had 1.5 times the amount of rain that fell in 1983. What is the total amount of rainfall for the years 1982, 1983, and 1984?

Needed data: How much rain fell in 1983?
                 How much rain fell in 1982?

Plan
Step 1: Find the amount of rainfall in 1983.
Step 2: Find the amount of rainfall in 1982.
Step 3: Find the total amount of rainfall for the years 1982, 1983, and 1984.

Step 1:

amount of rainfall in 1984	÷	2	=	amount of rainfall in 1983
37.58 cm	÷	2	=	18.79 cm

Step 2:

amount of rainfall in 1983	×	1.5	=	amount of rainfall in 1982
18.79 cm	×	1.5	=	28.185 cm

Step 3:

amount of rainfall in 1982	+	amount of rainfall in 1983	+	amount of rainfall in 1984	=	total amount of rainfall
28.185 cm	+	18.79 cm	+	37.58 cm	=	84.555 cm

The total amount of rainfall was 84.555 cm.

Write the missing step or steps in the plan.

1. Bayville has a seawall 11.7 m high. Water at the seawall is usually 8.2 m deep. Before a storm, a stack of sandbags 1 m high is placed on the seawall. The water rises 4 m. Will the water go over the sandbags on the seawall?

Step 1: Find the height of the seawall and the sandbags.
Step 2: Find the height of the water during the storm.
Step 3:

2. Ed and Jo leave Crewe on Friday night. There are 28 cm of snow on the ground. The average daily snowfall is 3 cm. The day Ed and Jo arrived, there were 13 cm of snow on the ground. On which day of the week did Ed and Jo arrive at Crewe?

Step 1: Find out how much snow fell while they were there.
Step 2:
Step 3:

Make a plan for each problem. Solve. You may be able to make more than one plan to solve some problems.

3. Taking a shower instead of a bath saves 40 liters of water. Using the dishwasher instead of washing dishes by hand saves 36 liters. Each of the 5 people in a family takes 1 shower each day, and the dishwasher is run 2 times daily. How many liters of water are saved each day?

4. A weather balloon measures 5 m wide. As it rises, it expands. At a height of 27,000 m, it bursts. The balloon's width expands 0.5 m for every 1,000 m it rises. What was the width of the weather balloon just before it burst at 27,000 m?

5. A blizzard dumped 16.8 cm of snow on Bayville in 7 days. The snow fell at the same rate during the 7 days. How many centimeters of snow fell in the first 5 hours of the storm?

6. When the ice freezes to a thickness of 10 cm, officials let 50 people skate on Bayview Pond at one time. How many cm thick would the ice have to be if only 5 people wanted to skate?

★7. The water at the Bayville seawall is 8.3 m deep before a storm. The water level rises 1.4 m every hour. People add 0.8 m of sandbags to the 11.7 m wall each hour. The storm lasts 4 hours. What is the difference in height between the water level and the seawall after 4 hours?

★8. John measures how much rain has fallen during each hour of a rainstorm. In the first hour, 1.5 cm fell. In the second hour, 1.25 cm fell, and 1.3 cm fell in the third hour. If the rain continues to fall at the same rate, how much rain will have fallen after 5 hours of the rainstorm?

READING MATH

Read this paragraph.

name of a person jerry went to the librarian and asked beginning of exact words where can i find the card catalog pause please question end of exact words

A few symbols would make these sentences easier to read and understand.

Jerry went to the librarian and asked, "Where can I find the card catalog, please?"

Symbols such as commas, capital letters, and question marks make sentences easier to read. They also save space. Math has symbols, too. You already know the symbols $>$, $+$, $\neq$, $<$, $\times$, $\div$.

1. Tell what they mean.

2. Write three other symbols, and tell what they mean.

Use these symbols to rewrite each of the following problems: $>$, $<$, $=$, $^\circ$, $+$, $-$, $\times$, $\div$, \$, $\neq$.

3. $0.07 \ \square \ 70$

4. $32\square$ F = freezing

5. $637 \ \square \ 729$

6. $140 \ \square \ 35 = 4$

7.
$$\begin{array}{r} 372 \\ \square \quad 6 \\ \hline 2{,}232 \end{array}$$

8.
$$\begin{array}{r} 49 \\ \square \quad 4 \\ \hline 196 \end{array}$$

9.
$$\begin{array}{r} \$7.12 \\ +\ 3.81 \\ \hline \square 10.93 \end{array}$$

10.
$$\begin{array}{r} 792 \\ \square 312 \\ \hline 480 \end{array}$$

Use symbols to write a number sentence that will solve each word problem. Then solve each problem.

11. The earliest English colony in North America, Roanoke, was started 22 years before the founding of Jamestown. Jamestown was founded in 1607. When was Roanoke founded?

12. John Adams became our second President in 1797. His son John Quincy Adams became President in 1825. How many years after his father became the President did John Quincy Adams begin his presidency?

GROUP PROJECT

Planning for the Summer

The problem: This summer you want to earn enough money to buy and to do certain things. You have plenty of time, so start now! To make the most of your vacation, plan a budget for the summer. Make a chart like the one below to help you.

Key Questions

- How do you plan to earn money?
- How will you budget your money?
- What expenses will you have?
- How much money do you plan to spend on clothes?
- What kind of activities are you planning?
- How much will they cost?

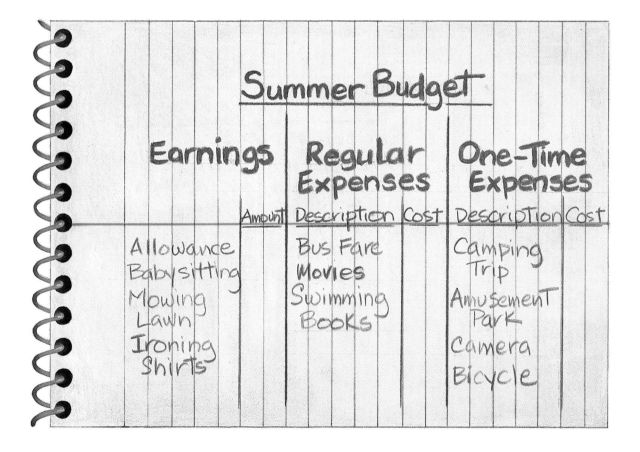

CHAPTER TEST

Multiply or divide. (page 196)

1. $77.28 \div 10$　　**2.** $912.04 \div 100$　　**3.** $0.105 \div 100$　　**4.** $36.945 \div 1{,}000$

5. $602.046 \times 1{,}000$　　**6.** $8.47 \times 1{,}000$　　**7.** 35.04×100　　**8.** 7.045×10

Divide. (pages 196, 198, and 200)

9. $8\overline{)21.12}$　　**10.** $17\overline{)162.52}$　　**11.** $61\overline{)6.344}$　　**12.** $19\overline{)14.44}$

13. $4\overline{)1.232}$　　**14.** $6\overline{)0.018}$　　**15.** $24\overline{)1.992}$　　**16.** $67\overline{)404.01}$

17. $6\overline{)0.357}$　　**18.** $95\overline{)3.8}$　　**19.** $75\overline{)4.95}$　　**20.** $16\overline{)24.8}$

21. $104.58 \div 45$　　　　**22.** $404.01 \div 67$　　　　**23.** $184.9 \div 43$

Draw a line for each measurement. (page 204)

24. 16 mm　　　　**25.** 13 cm　　　　**26.** 7.6 cm

Which unit would you use to measure? (pages 210 and 212)
Write *millimeter, centimeter, meter,* or *kilometer.*

27. the length of a chalkboard　　　　**28.** the height of a blade of grass

Write *milliliter* or *liter.*

29. the capacity of a barrel　　　　**30.** the capacity of a thimble

Write *milligram, gram,* or *kilogram.*

31. the mass of a button　　　　**32.** the mass of a tractor

Complete. (pages 210 and 212)

33. $0.62 \text{ kg} = \blacksquare \text{ mg}$　　　　**34.** $74 \text{ mL} = \blacksquare \text{ L}$

35. $6.2 \text{ m} = \blacksquare \text{ cm}$　　　　**36.** $2{,}006 \text{ mL} = \blacksquare \text{ L}$

Solve. (pages 200–201, 208–209, and 214–215)

37. The forest service measured 136.75 cm of snow after a 5-d snowstorm on Razorback Mountain. During that storm, the average daily snowfall was 2.63 cm more than the previous storm's daily average. What was the average daily snowfall for the previous storm?

38. The water level at the Razorback Mountain reservoir increased by 4.76 cm after 7 d of warm temperatures and melting snow. Of that increase, 0.19 cm was due to a rainstorm. What was the average daily rise in the water level from snow melt alone?

39. The heaviest snowstorm of the year on Razorback Mountain dropped 157.62 cm of snow. That is 6 times the amount of the first snowfall of the year. How much snow fell in 1 h during the first snowstorm of the year?

40. After a long drought, Razorback finally received fresh snow. The number of centimeters of snow that fell is a 2-digit number. Both digits are prime numbers. Their sum is 6. If you multiply the second digit by 4, the product has 0 in the ones place. How many centimeters of snow did Razorback Mountain receive?

BONUS

Choose the better price.

1. 5 onions for $1.95 or 3 for $1.47

2. 6 kiwi fruits for $2.10 or 7 for $2.24

3. 4 tires for $210.40 or 5 for $265.50

RETEACHING

A. Sometimes when you divide decimals, you have to write zeros in the quotient or in the dividend.

Find 65.35 ÷ 5.

Divide the whole number.

$$
\begin{array}{r}
13 \\
5\overline{)65.35} \\
65
\end{array}
$$

Place the decimal point. Divide the tenths.

$$
\begin{array}{r}
13.0 \\
5\overline{)65.35} \\
65 \\
3 \\
0 \\
\overline{3}
\end{array}
$$

Write 0.

Write 0.

Divide the hundredths.

$$
\begin{array}{r}
13.07 \\
5\overline{)65.35} \\
65 \\
3 \\
0 \\
\overline{35} \\
35
\end{array}
$$

B. Find the quotient of 48.24 ÷ 5

Divide the whole number.

$$
\begin{array}{r}
9 \\
5\overline{)48.24} \\
45 \\
\overline{3}
\end{array}
$$

Place the decimal point. Divide the tenths.

$$
\begin{array}{r}
9.6 \\
5\overline{)48.24} \\
45 \\
3\,2 \\
3\,0 \\
\overline{2}
\end{array}
$$

Divide the hundredths.

$$
\begin{array}{r}
9.64 \\
5\overline{)48.24} \\
45 \\
3\,2 \\
3\,0 \\
24 \\
20 \\
\overline{4}
\end{array}
$$

Write a 0 in the dividend. Divide the thousandths.

$$
\begin{array}{r}
9.648 \\
5\overline{)48.240} \\
45 \\
3\,2 \\
3\,0 \\
24 \\
20 \\
40 \\
40
\end{array}
$$

Write 0.

Divide.

1. $8\overline{)0.74}$ 2. $8\overline{)3.24}$ 3. $7\overline{)42.14}$ 4. $8\overline{)32.16}$ 5. $4\overline{)44.36}$

6. $55\overline{)4.62}$ 7. $40\overline{)283.6}$ 8. $45\overline{)9.18}$ 9. $42\overline{)3.15}$ 10. $19\overline{)77.14}$

11. $5\overline{)47.33}$ 12. $8\overline{)25.16}$ 13. $4\overline{)83.14}$ 14. $2\overline{)14.33}$ 15. $5\overline{)17.2}$

16. 4.62 ÷ 55 17. 283.6 ÷ 40 18. 1.8 ÷ 12 19. 34.2 ÷ 90

20. 15.15 ÷ 5 21. 6.49 ÷ 2 22. 116.5 ÷ 25 23. 83.42 ÷ 4

ENRICHMENT

Order of Operations

When you solve a number sentence, you need to perform the operations in a certain order to get the correct answer.

From left to right, you multiply and divide, then add and subtract.

Find $10 - 2 \times 3$.
Here are two solutions.

Which is correct?

$$\underbrace{10 - 2}_{8} \times 3$$
$$8 \times 3$$
$$24$$

$$10 - \underbrace{2 \times 3}_{6}$$
$$10 - 6$$
$$4$$

Solved by subtracting and then multiplying

Solved by multiplying and then subtracted

The second solution is correct because it follows the rule: First multiply and divide; then add and subtract.

When parentheses are used in a number sentence, do the operations inside the parentheses first. Then do the operations outside, from left to right.

$$
\begin{array}{ccc}
(5 + 7) & \times & 3 \\
\downarrow & & \downarrow \\
12 & \times & 3 \\
& \downarrow & \\
& 36 &
\end{array}
\qquad
\begin{array}{ccc}
8 & \div & (2 + 2) \\
\downarrow & & \downarrow \\
8 & \div & 4 \\
& \downarrow & \\
& 2 &
\end{array}
$$

Complete. Use the order of operations.

1. $7 \times 4 + 8$

2. $56 - 5 \times 8$

3. $14 + 8 \div 4$

4. $36 \div 6 - 4$

5. $22 + (8 \times 5)$

6. $8 - (4 + 2)$

7. $(46 - 22) \times 33$

8. $(45 + 9) \div 3$

9. $(32 + 48) \times 18$

10. $7 \times 8 + (3 + 2)$

11. $84 + (360 \div 15)$

12. $(6 + 3) \times (4 + 4)$

Work backwards from the answer and draw parentheses to show which operation was done first.

EXAMPLE: $4 \times 3 + 2 = 20 \longrightarrow 4 \times (3 + 2) = 20$, but $(4 \times 3) + 2 \neq 20$

13. $8 \times 6 - 4 = 16$

14. $36 + 9 \div 5 = 9$

15. $12 \div 2 \times 9 = 54$

16. $6 + 3 \times 7 = 63$

17. $6 + 3 \times 7 = 27$

18. $81 \div 9 + 9 = 18$

TECHNOLOGY

Here is a BASIC program that renames numbers of days as weeks.

```
10   LET D = 21
20   LET W = D / 7
30   PRINT D "DAYS EQUAL"
40   PRINT W "WEEKS"
```

1. What is printed when this program is RUN?

2. Write the line you would change to make the computer print this.

35 DAYS EQUAL
5 WEEKS

3. Write a program that stores 72 in a variable for hours. Have the computer compute the number of days and print this.

72 HOURS EQUAL
3 DAYS

Here is a short program that prints the sum of 7 and 9.

```
10   LET N1 = 7
20   LET N2 = 9
30   LET S = N1 + N2
40   PRINT S
```

When you RUN this program, it will print the number 16 on the screen. You might want to have the computer tell you which computation it is performing. To do this, add these extra lines to your program.

Instruction	The computer prints this.
40 PRINT "THE SUM OF"	THE SUM OF
50 PRINT N1	7
60 PRINT "AND"	AND
70 PRINT N2	9
80 PRINT "IS"	IS
90 PRINT S	16

Look closely at line 40. There are three pairs of quotation marks. The computer will print the letters between each set of quotation marks. Anything else must be a variable, so the computer prints the value of that variable. Here's the whole program.

```
10  LET N1 = 7
20  LET N2 = 9
30  LET S = N1 + N2
40  PRINT "THE SUM OF" N1 "AND" N2 "IS" S
```

4. Write the line you would change to make the computer print this.
THE SUM OF 18 AND 9 IS 27

5. Write the line you would change to make the computer print this.
18 PLUS 9 EQUALS 27

6. Write a program that computes the quotient of 72 divided by 9 and that prints this.
72 DIVIDED BY 9 IS 8

You can use parentheses in a LET statement the same way you did in the PRINT statement. The computer will compute inside the parentheses first and then perform the rest of the calculations.

7. What is printed when this program is RUN?

```
10  LET N = 3 * (7 − 3)
20  PRINT "THE NUMBER IS" N
```

You can have more than one set of parentheses in an expression.

```
10  LET N = (4 − 2) * (8 + 2)
20  PRINT "THE NUMBER IS" N
```

The computer computes in the first set of parentheses and then in the second set of parentheses. At the very end, it multiplies 2 * 10, and stores the answer, 20, in N.

8. What is printed when you RUN this program?

```
10  LET F = (7 − 1) − (5 − 2)
20  PRINT "MY DOG HAS" F "FLEAS"
```

CUMULATIVE REVIEW

Write the letter of the correct answer.

1. $47\overline{)497}$

   a. 10
   b. 10 R15
   c. 10 R27
   d. not given

2. $6{,}744 \div 72$

   a. 94 R6
   b. 98 R5
   c. 101 R1
   d. not given

3. $9{,}943 \div 33$

   a. 298 R8
   b. 301 R10
   c. 310
   d. not given

4. $20\overline{)200}$

   a. 10
   b. 20
   c. 100
   d. not given

5. $\$76.80 \div 32$

   a. $2.40
   b. $2.50
   c. $2.75
   d. not given

6. $54\overline{)6{,}898}$

   a. 124 R2
   b. 127 R40
   c. 127 R60
   d. not given

7. 0.5×6.85

   a. 3.425
   b. 3.445
   c. 4.755
   d. not given

8. Write in standard form:
   348 billion, 956 million, 729 thousand, 889.

   a. 34,895,672,989
   b. 348,956,729,889,000
   c. 348,956,729,889
   d. not given

9. Compare. Write $>$, $<$, or $=$ for ●.
   0.26 ● 0.260

   a. $>$
   b. $<$
   c. $=$
   d. not given

10. 358×792

    a. 63,445
    b. 238,436
    c. 283,426
    d. not given

11. Michael's teacher is 3 times as old as Michael. Michael is 12 years old. Let n = the age of Michael's teacher. Choose the correct number sentence.

    a. $3 \times n = 12$
    b. $3 + n = 12$
    c. $12 \times 3 = n$
    d. not given

12. Dana's grandfather is 78 years old. He is 6 times as old as Dana. How old is Dana?

    a. 12
    b. 13
    c. 14
    d. not given

Could you describe a watermelon to someone who has never seen one? Try.

8 NUMBER THEORY, FRACTIONS

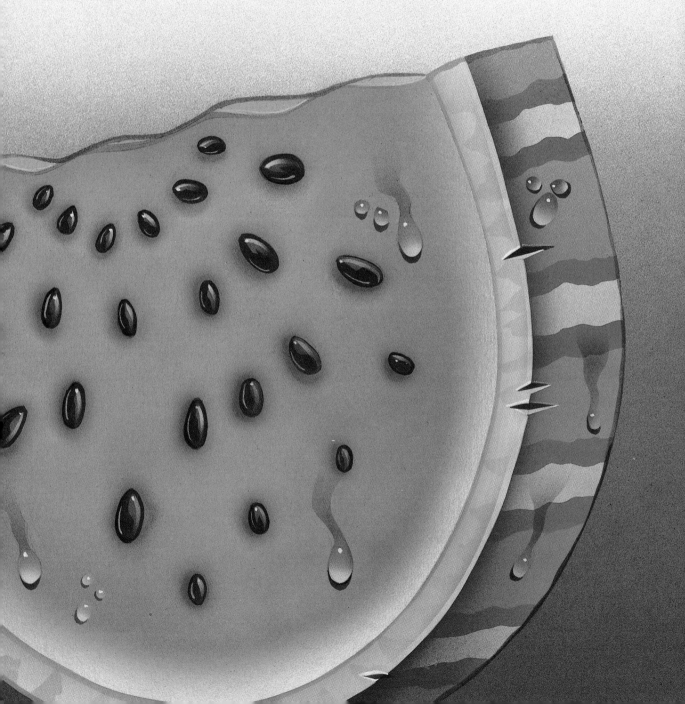

Least Common Multiples

A. The product of any number and 3 is a **multiple** of 3.

0, 3, 6, 9, 12, 15, and 18 are some multiples of 3. You can find the multiples of any number by multiplying the number by any whole number, such as 0, 1, 2, 3, and so on.

$0 \times 5 = 0$
$1 \times 5 = 5$
$2 \times 5 = 10$
$3 \times 5 = 15$ } some multiples of 5

$3 \times 16 = 48$
$5 \times 16 = 80$
$9 \times 16 = 144$
$20 \times 16 = 320$ } some multiples of 16

B. A number is a **common multiple** of two numbers if it is a multiple of both numbers.

0, 3, 6, 9, 12, and 15 are multiples of 3.
0, 2, 4, 6, 8, 10, and 12 are multiples of 2.
So, 0, 6, and 12 are some common multiples of 2 and 3.

C. The **least common multiple** of two or more numbers is the smallest number, other than 0, that is a common multiple.

Find the least common multiple of 2, 4, and 6.
List several multiples of each number.

2 ⟶ 0, 2, 4, 6, 8, 10, 12 . . .
4 ⟶ 0, 4, 8, 12, 16, 20, 24 . . .
6 ⟶ 0, 6, 12, 18, 24, 30 . . .

List the common multiples. 0, 12

Find the least common multiple. 12

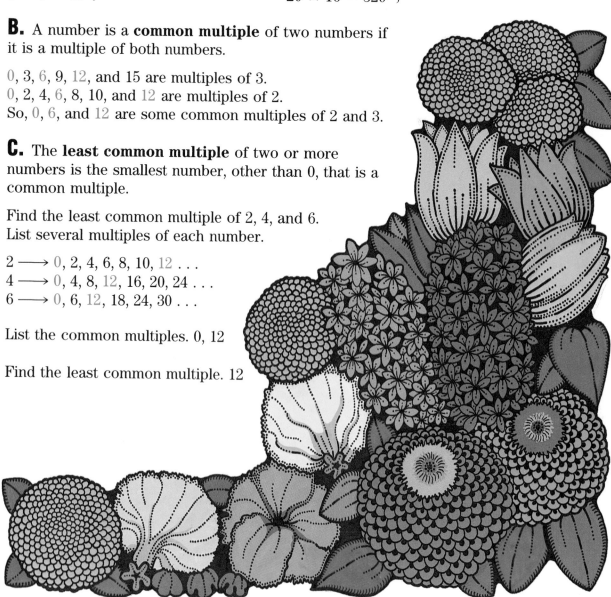

Find the first six multiples of each number.

1. 2   **2.** 4   **3.** 6   **4.** 5   **5.** 9   **6.** 8

7. 10   **8.** 12   **9.** 15   **10.** 20   **11.** 22   **12.** 25

Find the first three common multiples.

13. 2, 5   **14.** 3, 4   **15.** 4, 6   **16.** 2, 7

17. 6, 9   **18.** 8, 10   **19.** 6, 15   **20.** 12, 18

Find the least common multiple.

21. 2, 4   **22.** 4, 5   **23.** 4, 10   **24.** 5, 6

25. 6, 8   **26.** 3, 9   **27.** 8, 20   **28.** 12, 30

29. 7, 3, 9   **30.** 8, 4, 6   **31.** 9, 6, 4   **32.** 5, 3, 4

Solve.

33. Marge wants to buy the same number of tulip bulbs as she does hyacinth bulbs. Tulip bulbs are sold 6 per pack. Hyacinth bulbs are sold 8 per pack. What is the fewest number of packs of each that Marge can buy? How many of each kind of bulb will she have?

34. Marge wants to buy an equal number of marigolds, zinnias, and petunias at the sale. Use the chart to find the least number of each she could buy, and how much money she would spend.

Flower	Sale price
Marigolds	3 plants for $4.00
Zinnias	4 plants for $5.00
Petunias	8 plants for $7.50

ANOTHER LOOK

Divide.

1. $13\overline{)1,625}$ **2.** $18\overline{)3,726}$

3. $21\overline{)5,418}$ **4.** $14\overline{)3,185}$

5. $3,648 \div 12$ **6.** $2,987 \div 25$

7. $3,531 \div 11$ **8.** $2,235 \div 15$

More Practice, page 434

Greatest Common Factor

A. The **factors** of 6 are all the numbers that can be multiplied to form a product of 6.

Find the factors of 6.

$6 \times 1 = 6$ $\qquad$ $3 \times 2 = 6$ $\qquad$ $2 \times 3 = 6$

1, 2, 3, and 6 are the factors of 6.

B. What are the **common factors** of 9 and 12?

List the factors of 9.
$1 \times 9 = 9$ $\quad$ $3 \times 3 = 3$
1, 3, and 9 are factors.

List the factors of 12.
$1 \times 12 = 12$ $\quad$ $3 \times 4 = 12$ $\quad$ $2 \times 6 = 12$
1, 2, 3, 4, 6, and 12 are factors.

1 and 3 appear on both lists.
1 and 3 are common factors of 9 and 12.

C. The **greatest common factor** of two or more numbers is the greatest number that is a factor of each number.

Find the greatest common factor of 24, 32 and 36.

List the factors of 24. $\quad$ List the factors of 32.
1, 2, 3, 4, 6, 8, 12, 24. $\quad$ 1, 2, 4, 8, 16, 32.

List the factors of 36.
1, 2, 3, 4, 6, 9, 12, 18, 36.

List the common factors: 1, 2, 4.
The greatest common factor is 4.

List the factors of each number.

1. 10   **2.** 8   **3.** 18   **4.** 25   **5.** 28

6. 30   **7.** 36   **8.** 48   **9.** 56   **10.** 60

11. 64   **12.** 72   **13.** 81   **14.** 90   **15.** 100

List the common factors.

16. 2, 4   **17.** 6, 10   **18.** 8, 12   **19.** 10, 15   **20.** 18, 24

21. 12, 30   **22.** 20, 50   **23.** 14, 28   **24.** 32, 60   **25.** 35, 70

Copy and complete the table.

Numbers	Factors	Greatest common factor
26. 8 12	1, 2, 4, 8 1, 2, 3, 4, 6, 12	■
27. 24 32	■, ■, ■, ■, ■, ■, ■, ■ ■, ■, ■, ■, ■, ■	■
28. 36 40	■, ■, ■, ■, ■, ■, ■, ■, ■ ■, ■, ■, ■, ■, ■, ■, ■	■
29. 24 60	■, ■, ■, ■, ■, ■, ■, ■ ■, ■, ■, ■, ■, ■, ■, ■, ■, ■, ■, ■	■
★30. ■ ■	1, 2, 3, 4, 6, ■ 1, 2, 3, 6, 9, ■	■
★31. 36 24 42	■, ■, ■, ■, ■, ■, ■, ■, ■ ■, ■, ■, ■, ■, ■, ■, ■ ■, ■, ■, ■, ■, ■, ■, ■	■

FOCUS: MENTAL MATH

Compute mentally.

1. $(3 \times 4) \div 4$ **2.** $(4 \times 5) \div 5$ **3.** $(6 \times 7) \div 6$

4. $(8 \times 4) \div 8$ **5.** $(2 \times 8) \div 4$ **6.** $(9 \times 4) \div 6$

7. $(2 \times 6) \div 4$ **8.** $(3 \times 4) \div 6$ **9.** $(12 \times 2) \div 6$

10. $(18 \times 2) \div 9$ **11.** $(16 \times 2) \div 8$ **12.** $(15 \times 3) \div 9$

Prime and Composite Numbers

A. A **prime number** has exactly two factors: itself and 1.

List all the factors of 19. $1 \times 19 = 19$

The factors of 19 are 1 and 19.
19 is a prime number.

Some other prime numbers are 2, 3, 5, 29, and 61.

B. A **composite number** has more than two factors.

List all the factors of 20. 1×20; 2×10; 4×5

The factors of 20 are 1, 2, 4, 5, 10, and 20.

20 is a composite number.

> Since it only has one factor, 1 is neither composite nor prime.

C. A composite number can be shown as the product of prime factors. This is called the **prime factorization** of the number. You can use a factor tree to help you write the prime factorization of 36.

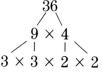

The prime factorization of 36 is $2 \times 2 \times 3 \times 3$.

Here are two more factor trees for 36.

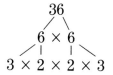

Notice that the prime factors of 36 are always the same. The order of the factors is not important.

List all the factors of each number.

1. 8 **2.** 9 **3.** 14 **4.** 28 **5.** 32 **6.** 27

Write *prime* or *composite* to describe each number.

7. 5 **8.** 23 **9.** 21 **10.** 15 **11.** 81 **12.** 49

13. 93 **14.** 29 **15.** 53 **16.** 42 **17.** 97 **18.** 103

Copy and complete each factor tree.

19.
$$20$$
$$2 \times 10$$
$$2 \times \blacksquare \times \blacksquare$$

20.
$$24$$
$$6 \times 4$$
$$2 \times \blacksquare \times \blacksquare \times \blacksquare$$

21.
$$18$$
$$2 \times \blacksquare$$
$$\blacksquare \times \blacksquare \times \blacksquare$$

Draw a factor tree to write the prime factorization of each number.

22. 27 **23.** 32 **24.** 48 **25.** 45 **26.** 53

27. 81 **28.** 125 **29.** 132 **30.** 115 **31.** 113

CHALLENGE

Here is a way to find all the prime numbers between 1 and 50.

Copy the chart of the numbers 1 through 50.

Step 1: Cross out 1.
Step 2: Circle all the numbers that are multiples of

2 except 2.   3 except 3.   5 except 5.   7 except 7.

1. Are the circled numbers prime or composite?

2. Are the uncircled numbers prime or composite?

3. Write all the prime numbers between 1 and 50.

X̶	2	3	4	5	6	7	8	9	10
11	12	13	14	15	16	17	18	19	20
21	22	23	24	25	26	27	28	29	30
31	32	33	34	35	36	37	38	39	40
41	42	43	44	45	46	47	48	49	50

Fractions

A. Grain is stored in the 10 compartments of a cargo ship. If 4 compartments are filled with wheat, what fraction of the ship's compartments are filled?

You can write a fraction to show the part of the ship that has been filled.

numerator ⟶ $\underline{4}$ ⟵ **number of parts filled**
denominator ⟶ 10 ⟵ **total number of parts**

Read: four tenths.   Write: $\frac{4}{10}$, or 0.4.

Of the cargo compartments, $\frac{4}{10}$ are filled.

B. You can use a fraction to describe parts of a set. What fraction of the set is circles?

○ ○ ○ □ □ □ △ △ △

$\underline{3}$ ⟵ number of circles
9 ⟵ total number of shapes

Read: three ninths   **Write:** $\frac{3}{9}$.

C. You can use a number line to help you estimate the value of a fraction.

This number line shows $\frac{1}{8}$, $\frac{5}{8}$, and $\frac{7}{8}$.

Each fraction is between 0 and 1 on the number line.

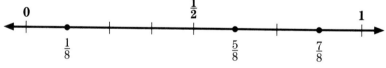

$\frac{1}{8}$ is close to 0. $\frac{5}{8}$ is close to $\frac{1}{2}$. $\frac{7}{8}$ is close to 1.

Checkpoint   Write the letter of the correct answer.

What fraction of each drawing is shaded?

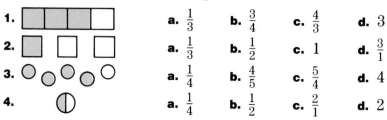

1.   **a.** $\frac{1}{3}$ **b.** $\frac{3}{4}$ **c.** $\frac{4}{3}$ **d.** 3

2.   **a.** $\frac{1}{3}$ **b.** $\frac{1}{2}$ **c.** 1   **d.** $\frac{3}{1}$

3.   **a.** $\frac{1}{4}$ **b.** $\frac{4}{5}$ **c.** $\frac{5}{4}$ **d.** 4

4.   **a.** $\frac{1}{4}$ **b.** $\frac{1}{2}$ **c.** $\frac{2}{1}$ **d.** 2

Write the fraction for the part that is shaded.

1.

2.

3.

4. ⬤ ◯ ◯
   ⬤ ◯ ◯
   ◯ ◯ ◯

5. ⬤ ◯ ◯ ◯ ⬤
   ⬤ ◯ ⬤ ◯ ◯

Write the fraction.

6. one fourth

7. three fifths

8. seven twelfths

9. six sevenths

10. five ninths

11. two eighths

Write whether the fraction is close to *0*, close to *1*, or close to $\frac{1}{2}$.

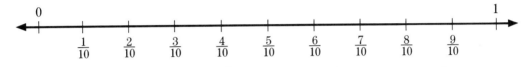

12. $\frac{1}{10}$

13. $\frac{9}{10}$

14. $\frac{6}{10}$

15. $\frac{4}{10}$

16. $\frac{2}{10}$

17. $\frac{8}{10}$

Write the decimal.

18. $\frac{3}{10}$

19. $\frac{7}{10}$

20. $\frac{1}{10}$

21. $\frac{5}{10}$

22. $\frac{4}{10}$

23. $\frac{2}{10}$

24. $\frac{6}{10}$

25. $\frac{5}{10}$

26. $\frac{8}{10}$

27. $\frac{9}{10}$

★28. $\frac{10}{10}$

Solve.

29. Many years ago, farmers did not have machines to help them harvest their crops. They could only harvest 2 acres per day. If Mr. Pitts had a 12-acre field, what fraction of it could be harvested in one day?

★30. Jason had to plant two 40-acre fields. After one day, Jason had planted 11 acres of corn and 5 acres of squash. What fraction of his farm had Jason planted with corn and squash?

FOCUS: REASONING

Alan, Brad, and Clark are brothers. Alan and the oldest brother share a set of bunk beds. Brad is younger than Alan. Who shares the bunk beds with Alan?

PROBLEM SOLVING
Choosing the Operation

Remember that the wording of a problem and the question it asks can give you hints about how to solve the problem. You can use these hints to help you choose the best operation to use.

Hints:

If you know	and you want to find	you can
• how many there are in two or more groups	how many in all	add.
• how many there are in one group • how many join it	the total number	add.
• how many there are in one group • how many more there are in a second group	how many there are in a second group	add.
• how many there are in one group • the number taken away	how many are left	subtract.
• how many there are in two groups	how much larger one group is than the other	subtract to compare.
• how many there are in one group • how many there are in part of the group	how many there are in the remaining part of the group	subtract.
• the number in each group is the same • how many there are in each group • how many groups there are	how many in all	multiply.
• the number in each group is the same • how many in all • how many in each group	how many groups	divide.
• the number in each group is the same • how many there are in all • how many groups there are	how many in each group	divide.

Choose the best operation to solve each problem. Write the letter of the correct answer.

1. The Dysons had 99 cattle in 1979. In 1984, they had 186 cattle. How many cattle did the Dysons acquire between 1979 and 1984?

   a. addition   **b.** subtraction
   c. multiplication   **d.** division

2. The Dysons' cattle eat 33,480 pounds of feed in 30 days. How much feed do they eat in 1 day?

   a. addition   **b.** subtraction
   c. multiplication   **d.** division

Solve.

3. The Dysons farm 475 acres of cornfields. If each acre yields 155.5 bushels of corn, how many bushels of corn do the Dysons harvest?

4. In 1977, Tom Dyson drove his tractor 7,181 miles. In 1978, he drove it 8,473 miles. He drove the tractor 7,811 miles in 1979, and in 1980, he drove 8,374 miles. Compare and order the mileages from the least to the greatest.

5. The Dyson's trailer carries 85 bales of hay. Each bale weighs 60.7 pounds. What is the total weight of the bales in the trailer?

6. The tractor cost the Dysons $45,000. The harvester cost $72,550. Estimate how much more was spent on the harvester than on the tractor.

7. When they first began to farm, the Dysons harvested 80 bushels of corn per day by hand. Using machines, they can harvest 3,500 bushels per day. By how much has their harvest production increased?

8. Last month, Tom drove the tractor for 27 days. After that, he repaired an axle, and noted that he had driven 3,294 miles that month. On the average, how many miles did Tom drive the tractor each day last month?

★9. In October, the Dysons sold 75 bushels of corn for $221.25. In November, they sold 75 bushels for $237.75. For how much less did the Dysons sell each bushel of corn in October?

★10. The Dysons have just sold some of their corn to a food company for $3.20 per bushel. If the price of corn falls to $2.80 per bushel next month, how much less will the Dysons receive for 10,000 bushels?

Equivalent Fractions

A. Ed divided his garden into 6 equal parts. He planted $\frac{1}{3}$ of the garden with string beans. How many sixths of the garden have string beans? You can draw a picture to find the answer.

$\frac{1}{6}$	$\frac{1}{6}$	$\frac{1}{6}$	$\frac{1}{6}$	$\frac{1}{6}$	$\frac{1}{6}$

$$\underbrace{}_{\frac{1}{3}} \quad \underbrace{}_{\frac{1}{3}} \quad \underbrace{}_{\frac{1}{3}}$$

Of the garden, $\frac{2}{6}$ is string beans.

$\frac{1}{3}$ and $\frac{2}{6}$ are **equivalent fractions.** They name the same part.

B. You can find equivalent fractions by multiplying the numerator and the denominator of a fraction by any number but zero.

$$\frac{1}{3} = \frac{1 \times 2}{3 \times 2} = \frac{2}{6} \qquad \frac{1 \times 3}{3 \times 3} = \frac{3}{9} \qquad \frac{1 \times 4}{3 \times 4} = \frac{4}{12}$$

So, $\frac{1}{3} = \frac{2}{6} = \frac{3}{9} = \frac{4}{12}$.

C. Complete.

$$\frac{2}{4} = \frac{\blacksquare}{8}$$

Think: $4 \times 2 = 8.$ So, $\frac{2 \times 2}{4 \times 2} = \frac{4}{8}.$

Checkpoint Write the letter of the correct answer.

Complete.

1. $\frac{3}{4} = \frac{\blacksquare}{8}$

a. 2
b. 3
c. 6
d. 7

2. $\frac{2}{3} = \frac{\blacksquare}{12}$

a. 2
b. 4
c. 6
d. 8

3. $\frac{3}{5} = \frac{6}{\blacksquare}$

a. 2
b. 5
c. 8
d. 10

Complete.

1. $\frac{1}{5} = \frac{1 \times 2}{5 \times 2} = \frac{\blacksquare}{\blacksquare}$

2. $\frac{1}{3} = \frac{1 \times 3}{3 \times 3} = \frac{\blacksquare}{\blacksquare}$

3. $\frac{1}{6} = \frac{1 \times 2}{6 \times 2} = \frac{\blacksquare}{\blacksquare}$

Write the next three equivalent fractions.

4. $\frac{3}{4}, \frac{6}{8}, \frac{9}{12}, \frac{\blacksquare}{\blacksquare}, \frac{\blacksquare}{\blacksquare}, \frac{\blacksquare}{\blacksquare}$

5. $\frac{1}{3}, \frac{2}{6}, \frac{3}{9}, \frac{\blacksquare}{\blacksquare}, \frac{\blacksquare}{\blacksquare}, \frac{\blacksquare}{\blacksquare}$

6. $\frac{1}{4}, \frac{2}{8}, \frac{3}{12}, \frac{\blacksquare}{\blacksquare}, \frac{\blacksquare}{\blacksquare}, \frac{\blacksquare}{\blacksquare}$

7. $\frac{2}{5}, \frac{4}{10}, \frac{6}{15}, \frac{\blacksquare}{\blacksquare}, \frac{\blacksquare}{\blacksquare}, \frac{\blacksquare}{\blacksquare}$

Complete.

8. $\frac{1}{10} = \frac{\blacksquare}{20}$

9. $\frac{1}{10} = \frac{\blacksquare}{30}$

10. $\frac{2}{4} = \frac{\blacksquare}{8}$

11. $\frac{1}{4} = \frac{\blacksquare}{12}$

12. $\frac{4}{8} = \frac{\blacksquare}{16}$

13. $\frac{4}{5} = \frac{\blacksquare}{10}$

14. $\frac{3}{6} = \frac{\blacksquare}{12}$

15. $\frac{3}{6} = \frac{\blacksquare}{18}$

16. $\frac{1}{5} = \frac{\blacksquare}{10}$

17. $\frac{2}{5} = \frac{\blacksquare}{10}$

18. $\frac{1}{5} = \frac{\blacksquare}{15}$

19. $\frac{5}{9} = \frac{10}{\blacksquare}$

20. $\frac{3}{4} = \frac{9}{\blacksquare}$

21. $\frac{6}{10} = \frac{12}{\blacksquare}$

★22. $\frac{2}{8} = \frac{1}{\blacksquare}$

Solve.

23. Pat spent the weekend planting corn. She planted $\frac{1}{7}$ of it Saturday and $\frac{4}{28}$ of it Sunday. Did she plant the same amount of corn each day?

★24. Sally spent $\frac{1}{4}$ of a day planting tomatoes. Melanie spent 8 hours tending her garden. A day is 24 hours long. Who spent more time working in the garden?

CALCULATOR

You can use a calculator to find equivalent fractions.

Complete.

$\frac{3}{25} = \frac{\blacksquare}{125}$

Divide the denominators:

Multiply the quotient by the numerator:

$\boxed{5} \ \boxed{x} \ \boxed{3} \ \boxed{=} \ \boxed{15}$

So, $\frac{3}{25} = \frac{15}{125}$.

Use your calculator to find equivalent fractions.

1. $\frac{5}{9} = \frac{\blacksquare}{162}$

2. $\frac{4}{7} = \frac{\blacksquare}{105}$

3. $\frac{3}{18} = \frac{\blacksquare}{54}$

4. $\frac{5}{12} = \frac{\blacksquare}{124}$

Simplifying Fractions

A. The Brennan family has a fruit farm. There are 12 orchards on their farm. Of these, 6 are apple orchards. This means that $\frac{6}{12}$ of the Brennan orchards are apple orchards.

Write $\frac{6}{12}$ in simplest form.

A fraction is in **simplest form** if the denominator and the numerator have no common factors. You can find equivalent fractions by dividing the numerator and the denominator of a fraction by a common factor.

$$\frac{6}{12} = \frac{6 \div 2}{12 \div 2} = \frac{3}{6} \qquad \boxed{\frac{3}{6} \text{ is not in simplest form.}}$$

Continue dividing to find another fraction equivalent to $\frac{3}{6}$.

$$\frac{3}{6} = \frac{3 \div 3}{6 \div 3} = \frac{1}{2}$$

1 and 2 have no common factor greater than 1. So, $\frac{6}{12}$ in simplest form is $\frac{1}{2}$.

B. Another way to simplify a fraction is by dividing the numerator and the denominator by their greatest common factor.

Write $\frac{6}{12}$ in simplest form.

The greatest common factor of 6 and 12 is 6. Divide both the numerator and the denominator by 6.

$$\frac{6}{12} \div \frac{6}{6} = \frac{1}{2} \qquad \frac{6}{12} \text{ in simplest form is } \frac{1}{2}.$$

Checkpoint Write the letter of the correct answer.

Find the equivalent fraction in simplest form.

1. $\frac{4}{12}$

a. $\frac{1}{4}$
b. $\frac{1}{3}$
c. $\frac{2}{6}$
d. $\frac{8}{24}$

2. $\frac{27}{30}$

a. $\frac{1}{3}$
b. $\frac{9}{10}$
c. $\frac{3}{3}$
d. 3

3. $\frac{75}{100}$

a. $\frac{1}{25}$
b. $\frac{15}{25}$
c. $\frac{7}{10}$
d. $\frac{3}{4}$

4. $\frac{12}{72}$

a. $\frac{1}{12}$
b. $\frac{3}{36}$
c. $\frac{1}{6}$
d. $\frac{2}{12}$

Divide to write the fraction in simplest form.

1. $\frac{4 \div 4}{8 \div 4} = \blacksquare$

2. $\frac{8 \div 4}{12 \div 4} = \blacksquare$

3. $\frac{8 \div 2}{10 \div 2} = \blacksquare$

4. $\frac{9 \div 9}{27 \div 9} = \blacksquare$

5. $\frac{10 \div 10}{30 \div 10} = \blacksquare$

6. $\frac{15 \div 5}{35 \div 5} = \blacksquare$

7. $\frac{45 \div 9}{54 \div 9} = \blacksquare$

8. $\frac{99 \div 11}{121 \div 11} = \blacksquare$

Complete.

9. $\frac{4}{8} = \frac{\blacksquare}{2}$

10. $\frac{2}{10} = \frac{\blacksquare}{5}$

11. $\frac{6}{27} = \frac{\blacksquare}{9}$

12. $\frac{8}{40} = \frac{\blacksquare}{5}$

13. $\frac{7}{56} = \frac{\blacksquare}{8}$

14. $\frac{6}{16} = \frac{\blacksquare}{8}$

15. $\frac{9}{24} = \frac{\blacksquare}{8}$

16. $\frac{13}{52} = \frac{\blacksquare}{4}$

17. $\frac{12}{62} = \frac{\blacksquare}{31}$

18. $\frac{63}{162} = \frac{\blacksquare}{18}$

Write the fraction in simplest form.

19. $\frac{4}{16}$

20. $\frac{6}{8}$

21. $\frac{8}{18}$

22. $\frac{10}{12}$

23. $\frac{12}{15}$

24. $\frac{18}{20}$

25. $\frac{60}{100}$

26. $\frac{7}{21}$

27. $\frac{15}{30}$

28. $\frac{14}{28}$

29. $\frac{20}{25}$

30. $\frac{4}{18}$

Solve. For Problem 32, use the Infobank.

31. Last year, the Makey's wheat crop was 2.7 metric tons. Their silo is capable of storing 5 metric tons of wheat. How many more metric tons could it hold after last year's crop was stored?

32. Use the information on page 418 to solve. Write as a fraction in simplest form the part of the total United States wheat crop produced in Kansas.

CHALLENGE

Copy the chart and the riddle. Simplify each fraction in the box. Then solve the riddle by writing the letter that matches each fraction in the blank above that fraction. One letter is done for you.

$\frac{3}{27}$	$\frac{6}{27}$	$\frac{8}{24}$	$\frac{4}{20}$	$\frac{12}{20}$	$\frac{8}{20}$	$\frac{9}{21}$	$\frac{12}{32}$	$\frac{20}{24}$	$\frac{6}{24}$	$\frac{18}{24}$	$\frac{8}{28}$
$\blacksquare$	$\blacksquare$	$\blacksquare$	$\blacksquare$	$\blacksquare$	$\blacksquare$	$\blacksquare$	$\blacksquare$	$\frac{1}{4}$	$\blacksquare$	$\blacksquare$	
A	E	I	O	U	Y	G	K	N	R	S	T

Riddle: What kind of keys can you eat?

Answer: $\underset{\frac{2}{7}}{\rule{1.5em}{0.4pt}}$ $\underset{\frac{3}{5}}{\rule{1.5em}{0.4pt}}$ $\underset{\frac{1}{4}}{\overset{R}{\rule{1.5em}{0.4pt}}}$ $\underset{\frac{3}{8}}{\rule{1.5em}{0.4pt}}$ $\underset{\frac{2}{9}}{\rule{1.5em}{0.4pt}}$ $\underset{\frac{2}{5}}{\rule{1.5em}{0.4pt}}$ $\underset{\frac{3}{4}}{\rule{1.5em}{0.4pt}}$!

Mixed Numbers

A. Look at the picture. The circles are divided into fourths. Count the fourths. How many fourths are shaded?

Five fourths are shaded. Write $\frac{5}{4}$.

How many circles are shaded?
The shaded area is one whole circle
and one fourth of the other circle. Write $1\frac{1}{4}$. $\frac{5}{4} = 1\frac{1}{4}$

B. A **mixed number** has a whole-number part and a fraction part. You can rename some fractions as mixed numbers or as whole numbers.

$$\frac{7}{3} \longrightarrow 3\overline{)7} \quad \begin{array}{l} \text{Divide the numerator by} \\ \text{the denominator. Write the} \\ \text{remainder as a fraction.} \end{array} \qquad \frac{8}{4} \longrightarrow 4\overline{)8}$$

with $\frac{2}{3\overline{)7}}$, $\underline{6}$, 1 and $\frac{2}{4\overline{)8}}$, $\underline{8}$, 0

So, $\frac{7}{3} = 2\frac{1}{3}$.

So, $\frac{8}{4} = 2$.

C. You can also rename a mixed number or a whole number as a fraction.

Multiply the whole number by the denominator.	Add the product and the numerator.	Write the sum over the denominator.
$5\frac{1}{3} \longrightarrow 3 \times 5 = 15$	$15 + 1 = 16$	$\frac{16}{3}$

Another example: $6 = \frac{6}{1}, \frac{12}{2}, \frac{18}{3}$

CHECKPOINT Write the letter of the correct answer.

Rename as a mixed number.

1. $\frac{5}{2}$ **a.** $\frac{10}{4}$ **b.** $2\frac{1}{2}$ **c.** $\frac{10}{2}$ **d.** 7

Rename as a fraction.

2. $2\frac{1}{3}$ **a.** $\frac{4}{3}$ **b.** $\frac{5}{3}$ **c.** $\frac{6}{3}$ **d.** $\frac{7}{3}$

Write a fraction and a mixed number for the part that is shaded.

1.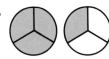

2.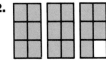

3.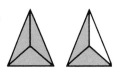

4.

5.

6.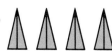

Write as a whole number or as a mixed number.

7. $\frac{8}{3}$ **8.** $\frac{14}{3}$ **9.** $\frac{11}{4}$ **10.** $\frac{16}{5}$ **11.** $\frac{15}{4}$ **12.** $\frac{9}{2}$

13. $\frac{8}{4}$ **14.** $\frac{7}{3}$ **15.** $\frac{15}{5}$ **16.** $\frac{13}{5}$ **17.** $\frac{19}{3}$ **18.** $\frac{29}{4}$

19. $\frac{7}{2}$ **20.** $\frac{16}{8}$ **21.** $\frac{18}{2}$ **22.** $\frac{15}{2}$ **23.** $\frac{21}{5}$ **24.** $\frac{36}{9}$

25. $\frac{5}{3}$ **26.** $\frac{8}{7}$ **27.** $\frac{16}{4}$ **28.** $\frac{10}{3}$ **29.** $\frac{6}{5}$ **30.** $\frac{9}{3}$

Write as a fraction.

31. $3\frac{1}{2}$ **32.** $3\frac{3}{4}$ **33.** $2\frac{2}{3}$ **34.** $6\frac{7}{8}$ **35.** 8 **36.** $10\frac{1}{4}$

37. $3\frac{2}{3}$ **38.** 9 **39.** $7\frac{3}{8}$ **40.** $4\frac{1}{2}$ **41.** $5\frac{5}{8}$ **42.** $4\frac{3}{10}$

43. $3\frac{2}{5}$ **44.** $1\frac{4}{7}$ **45.** $4\frac{3}{4}$ **46.** $11\frac{9}{10}$ **47.** $5\frac{1}{6}$ **48.** $10\frac{2}{3}$

49. $4\frac{3}{8}$ **50.** 7 **51.** $8\frac{1}{3}$ **52.** $22\frac{5}{8}$ **53.** $4\frac{6}{7}$ **54.** $11\frac{3}{4}$

CHALLENGE

Marty has $5\frac{1}{3}$ yards of ribbon. He cuts the ribbon into equal pieces. Each piece is $\frac{1}{3}$ yard long. How many pieces does Marty have?

Compare and Order Fractions and Mixed Numbers

A. Compare $\frac{1}{3}$ and $\frac{2}{3}$.

Check to see that the denominators are the same.

$\frac{1}{3}$ $\frac{2}{3}$

Compare the numerators.

$1 < 2$

So, $\frac{1}{3} < \frac{2}{3}$.

B. Compare $\frac{1}{4}$ and $\frac{1}{3}$. The denominators are not the same.

Write the equivalent fractions for $\frac{1}{4}$ and $\frac{1}{3}$ that have a common denominator.

The least common multiple of 4 and 3 is 12.

$\frac{1}{3} = \frac{4}{12}$ $\frac{1}{4} = \frac{3}{12}$

Compare the numerators.

$\frac{4}{12} > \frac{3}{12}$

So, $\frac{1}{3} > \frac{1}{4}$.

C. Compare $3\frac{1}{3}$ and $3\frac{4}{9}$.

Compare the whole numbers.

$3 = 3$

So, $3\frac{1}{3} < 3\frac{4}{9}$.

Write an equivalent fraction with a common denominator.

$\frac{1}{3} = \frac{3}{9}$

Compare the fractions.

$\frac{3}{9} < \frac{4}{9}$

D. You can order $\frac{3}{4}$, $\frac{5}{8}$, and $\frac{2}{3}$ by comparing these fractions.

Write equivalent fractions that have a common denominator.

$\frac{3}{4} = \frac{18}{24}$

$\frac{5}{8} = \frac{15}{24}$

$\frac{2}{3} = \frac{16}{24}$

Compare to find the greatest fractions.

$\frac{18}{24}, \frac{15}{24}, \frac{16}{24}$

$\frac{18}{24} > \frac{16}{24}$

$\frac{18}{24} > \frac{15}{24}$

Continue comparing.

$\frac{16}{24} > \frac{15}{24}$

Once you have compared all of the fractions, you can write them in either order

from the least to the greatest: $\frac{5}{8}, \frac{2}{3}, \frac{3}{4}$; or,

from the greatest to the least: $\frac{3}{4}, \frac{2}{3}, \frac{5}{8}$.

Compare. Write >, <, or = for ●.

1. $\frac{1}{3}$ ● $\frac{2}{3}$

2. $\frac{3}{5}$ ● $\frac{4}{5}$

3. $\frac{5}{9}$ ● $\frac{2}{9}$

4. $\frac{2}{10}$ ● $\frac{7}{10}$

5. $\frac{3}{4}$ ● $\frac{1}{4}$

6. $\frac{4}{7}$ ● $\frac{3}{4}$

7. $\frac{6}{8}$ ● $\frac{3}{4}$

8. $\frac{1}{3}$ ● $\frac{2}{6}$

9. $\frac{1}{2}$ ● $\frac{5}{7}$

10. $\frac{2}{9}$ ● $\frac{1}{8}$

11. $3\frac{3}{9}$ ● $2\frac{1}{9}$

12. $5\frac{3}{8}$ ● $2\frac{7}{8}$

13. $4\frac{6}{8}$ ● $5\frac{4}{8}$

14. $7\frac{1}{7}$ ● $1\frac{6}{7}$

15. $2\frac{2}{3}$ ● $2\frac{3}{4}$

16. $\frac{2}{5}$ ● $\frac{4}{10}$

17. $\frac{5}{8}$ ● $\frac{3}{8}$

18. $2\frac{1}{2}$ ● $2\frac{3}{8}$

19. $\frac{5}{6}$ ● $\frac{4}{5}$

20. $1\frac{1}{3}$ ● $1\frac{1}{4}$

Write in order from the least to the greatest.

21. $\frac{3}{5}, \frac{2}{8}, \frac{1}{2}$

22. $\frac{1}{3}, \frac{2}{3}, \frac{1}{7}$

23. $\frac{2}{8}, \frac{2}{4}, \frac{2}{6}$

24. $\frac{4}{5}, \frac{2}{3}, \frac{3}{4}$

Write in order from the greatest to the least.

25. $\frac{3}{7}, \frac{1}{2}, \frac{2}{3}$

26. $\frac{3}{4}, \frac{1}{7}, \frac{2}{9}$

27. $\frac{1}{2}, \frac{4}{6}, \frac{2}{6}$

28. $\frac{3}{5}, \frac{3}{4}, \frac{3}{8}$

Solve.

29. During the first week of harvesting, one group of workers picked $\frac{2}{5}$ of the apple crop. A second group picked $\frac{3}{10}$ of the pear crop. Which group picked more of its crop?

30. Jesse makes his special juice. The recipe calls for $\frac{1}{2}$ gallon apple juice, $\frac{1}{6}$ gallon grape juice, and $\frac{1}{3}$ gallon pear juice. Order the ingredients from the least to the greatest.

MIDCHAPTER REVIEW

Find the least common multiple.

1. 2, 17

2. 5, 6

3. 9, 2

4. 7, 3

Find the greatest common factor.

5. 12, 18

6. 16, 24

7. 25, 24, 32, 48

8. 27, 45, 36

Complete.

9. $\frac{2}{3} = \frac{\blacksquare}{6}$

10. $\frac{3}{10} = \frac{\blacksquare}{20}$

11. $\frac{5}{9} = \frac{\blacksquare}{45}$

12. $\frac{2}{7} = \frac{\blacksquare}{21}$

Write the fraction in simplest form.

13. $\frac{8}{10}$

14. $\frac{9}{27}$

15. $\frac{17}{68}$

16. $\frac{15}{255}$

Write as a whole number or a mixed number.

17. $\frac{12}{4}$

18. $\frac{10}{9}$

19. $\frac{35}{6}$

20. $\frac{98}{7}$

PROBLEM SOLVING
Practice

Write the letter of the solution.

1. One farm contains 430 acres of land. Its owner plans to sell some of it at $4,000 per acre. How much money will the owner receive if he sells half his land?

  a. $860,000
  b. 215 acres
  c. Yes, he does.

2. The Lees and the Sands each owned 250-acre farms. They each bought another 75 acres. Today, the Lees have 354 acres. Who owns more land?

  a. 29 acres more
  b. the Sands
  c. the Lees

Write a number sentence, and solve.

3. In 1982, there were 34 fewer farms in the United States than in 1981. If there were 2,400 farms in 1982, how many farms were there in 1981?

4. The Parker farm has 3 times as many cattle as the Steinway farm. The Parkers have 90 cattle. How many cattle do the Steinways have?

Write the letter of the best estimate.

5. An Alaskan chicken farmer sells 2,416 eggs in one month. The average hen on the farm lays about 16 eggs per month. How many hens does the farmer have?

  a. 15   **b.** 150
  c. 1,500   **d.** 15,000

6. In 1983, a bushel of corn cost $3.30. A bushel of wheat cost $3.54. If a farmer sold 7,500 bushels of corn and 2 times as much wheat, about how much money would the farmer have received?

  a. $28,000   **b.** $53,000
  c. $75,000   **d.** $773,000

List the information in the problem. Cross out the information that won't help you answer the question.

7. Jason Parker is loading bales of hay onto a mechanical arm. The arm moves the 50-pound bales up to the hayloft. Jason loads 1 bale every 5 seconds. How many bales will he load in 60 seconds?

8. Between 1970 and 1980, world production of eggs increased by 5,568,000 metric tons. In 1980, production was 26,700,000 metric tons. A metric ton equals 1,000 kilograms. How many metric tons of eggs were produced in 1970?

Solve.

9. One farmer earned $12,600 from crop sales in 1984. Each year, she earns 0.04 times more than she did the previous year. How much will she earn in 1987?

10. Dan Parker wants to take 7 horses to Toni Steinway's farm. His trailer can hold only 2 horses. How many trips will Dan Parker have to make?

11. Between 1980 and 1983, the number of farms in Iowa decreased from 119,000 to 115,000. If the decrease in farms continues at the same rate, how many farms will there be in 1989?

12. One acre of Toni Steinway's farmland is worth $2,000. When she bought it, she paid only $100 per acre. If Toni sold 80 acres of her land, how much profit would she make?

13. Last year, the Parkers earned $11,885 after expenses. The Steinways earned $2,100 more. The Jacksons earned as much from their large farm as the Parkers and the Steinways together. How much did the Jacksons earn?

14. The United States is the world's largest corn exporter. In one year, it exported 54,856,000 metric tons of corn. That same year, Argentina, exporting 9,112,000 metric tons, was second. Estimate how much larger the amount of corn the United States exported was than the amount exported by Argentina.

15. Nuts are an important agricultural product. In 1983, the United States produced 141.8 million pounds of pecans. In 1982, pecan production totaled 107.6 million pounds. How many pounds of pecans did the United States produce from 1982 through 1983?

16. In 1980, Canada produced 17,637,000 bushels of rye. The United States, producing 16,259,000 bushels, was close behind. In the United States, Nebraska produced 749,000 bushels, and Oklahoma produced 744,000 bushels. How many more bushels did Nebraska produce than Oklahoma?

17. A healthy apple tree can produce about 30 bushels of fruit. Recently, a fast-working farmer picked a record 270 bushels in 9 hours. If he completely picked one tree per hour, how many trees did he work on?

18. By 1624, English settlers in Jamestown had 20,000 cattle. That number was about 0.003 of the number of cattle there were in the United States by 1900. About how many cattle were there in the United States by 1900?

Adding Like Fractions

A. Jean is raising a dairy cow named Bon Bon for her 4-H project. Jean makes cheese from the cow's milk. On Monday, Jean gives $\frac{3}{8}$ wheel of cheese to friends. On Thursday, she gives $\frac{4}{8}$ wheel to her 4-H club leader. How much cheese does Jean give away? You can show this in a picture.

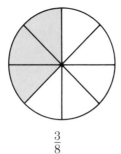
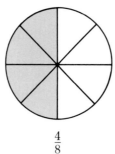

$$\frac{3}{8} \qquad\qquad \frac{4}{8}$$

Count the shaded parts. There are $\frac{7}{8}$.

Add $\frac{3}{8} + \frac{4}{8}$.

To add fractions with like denominators, add the numerators. Write the sum over the denominator.

$$\frac{3}{8} + \frac{4}{8} = \frac{3+4}{8} = \frac{7}{8}$$

Jean gives away $\frac{7}{8}$ wheel of cheese.

B. Sometimes you need to rename the sum as a mixed number in simplest form.

$$\frac{3}{4} + \frac{3}{4} = \frac{3+3}{4} = \frac{6}{4} = 1\frac{2}{4}, \text{ or } 1\frac{1}{2}$$

Checkpoint   Write the letter of the correct answer.

Add. The answer must be in simplest form.

1. $\frac{2}{4} + \frac{1}{4}$ **2.** $\frac{3}{5} + \frac{4}{5}$ **3.** $\frac{5}{9} + \frac{7}{9}$ **4.** $\frac{9}{10} + \frac{2}{10}$

a. $\frac{3}{16}$ **a.** $\frac{7}{25}$ **a.** $\frac{12}{81}$ **a.** $\frac{1}{10}$

b. $\frac{3}{8}$ **b.** $\frac{7}{10}$ **b.** $1\frac{3}{9}$ **b.** $\frac{7}{10}$

c. $\frac{3}{4}$ **c.** $\frac{7}{5}$ **c.** $\frac{4}{3}$ **c.** $\frac{11}{10}$

d. $\frac{6}{5}$ **d.** $1\frac{2}{5}$ **d.** $1\frac{1}{3}$ **d.** $1\frac{1}{10}$

246

Add. Write the sum in simplest form.

1. $\frac{3}{8} + \frac{2}{8}$ 2. $\frac{2}{5} + \frac{1}{5}$ 3. $\frac{4}{7} + \frac{2}{7}$ 4. $\frac{7}{10} + \frac{2}{10}$ 5. $\frac{2}{9} + \frac{5}{9}$

6. $\frac{1}{4} + \frac{1}{4}$ 7. $\frac{3}{6} + \frac{1}{6}$ 8. $\frac{3}{12} + \frac{5}{12}$ 9. $\frac{2}{9} + \frac{1}{9}$ 10. $\frac{5}{15} + \frac{5}{15}$

11. $\frac{6}{8} + \frac{5}{8}$ 12. $\frac{2}{7} + \frac{5}{7}$ 13. $\frac{5}{9} + \frac{8}{9}$ 14. $\frac{2}{3} + \frac{1}{3}$ 15. $\frac{3}{6} + \frac{4}{6}$

16. $\frac{7}{12} + \frac{9}{12}$ 17. $\frac{7}{8} + \frac{7}{8}$ 18. $\frac{7}{10} + \frac{5}{10}$ 19. $\frac{4}{16} + \frac{12}{16}$ 20. $\frac{6}{12} + \frac{5}{12}$ 21. $\frac{2}{4} + \frac{3}{4}$

★22. $\frac{1}{4} + \frac{2}{4} + \frac{3}{4}$ ★23. $\frac{2}{7} + \frac{3}{7} + \frac{1}{7}$ ★24. $\frac{3}{6} + \frac{5}{6} + \frac{6}{6}$ ★25. $\frac{7}{9} + \frac{5}{9} + \frac{6}{9}$

Solve.

26. Jean spends $\frac{2}{6}$ hour caring for Bon Bon on Monday. She spends $\frac{4}{6}$ hour with Bon Bon on Tuesday. How long does Jean spend with Bon Bon on Monday and Tuesday?

27. A special feed machine records how much each calf eats. Nell, a young calf, eats $\frac{2}{3}$ pound of feed in the morning and $\frac{2}{3}$ pound of feed in the afternoon. How much does Nell eat?

★28. Jean's family drinks the following amounts of milk in three days: $\frac{3}{4}$ gallon, $\frac{5}{4}$ gallon, $\frac{2}{4}$ gallon. How much milk does Jean's family drink in all?

ANOTHER LOOK

Complete.

1. 2 cm = ▨ mm   2. 16 mm = ▨ cm

3. 8.5 mm = ▨ cm   4. 4 km = ▨ m

5. 250 mm = ▨ km   6. 8.75 km = ▨ m

7. 375 mg = ▨ g   8. 6 g = ▨ mg

9. 1.3 g = ▨ mg

More Practice, page 435   **247**

Estimating Fraction Sums

A. You can estimate the value of a fraction by comparing the numerator and the denominator. If the numerator is much smaller than the denominator, you can round the fraction to 0.

$\frac{2}{9}$, $\frac{1}{10}$, $\frac{2}{15}$, and $\frac{8}{93}$ are all close to 0.

If the denominator is about twice as large as the numerator, you can round the fraction to $\frac{1}{2}$.

$\frac{5}{8}$, $\frac{7}{15}$, $\frac{13}{24}$, and $\frac{45}{93}$ are all close to $\frac{1}{2}$.

If the numerator is nearly the same as the denominator, you can round the fraction to 1.

$\frac{9}{10}$, $\frac{6}{7}$, $\frac{7}{9}$, and $\frac{91}{93}$ are all close to 1.

B. You can use what you know about the value of a fraction to estimate the sum of fractions.

Estimate $\frac{4}{5} + \frac{5}{8}$.

Round each fraction to 0, $\frac{1}{2}$, or 1.

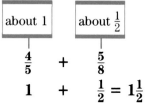

about 1	about $\frac{1}{2}$
$\frac{4}{5}$ +	$\frac{5}{8}$
1 +	$\frac{1}{2}$ = $1\frac{1}{2}$

The estimated sum is about $1\frac{1}{2}$.

Another example:

Estimate $\frac{7}{9} + \frac{2}{15} + \frac{3}{7}$.

about 1	about 0	about $\frac{1}{2}$
$\frac{7}{9}$ +	$\frac{2}{15}$ +	$\frac{3}{7}$
1 +	0 +	$\frac{1}{2}$ = $1\frac{1}{2}$

Write 0, $\frac{1}{2}$, or 1 to complete.

1. $\frac{6}{7}$ is close to ▨.

2. $\frac{5}{11}$ is close to ▨.

3. $\frac{3}{23}$ is close to ▨.

4. $\frac{1}{5}$ is close to ▨.

5. $\frac{8}{9}$ is close to ▨.

6. $\frac{4}{9}$ is close to ▨.

Estimate.

7. $\frac{1}{8} + \frac{13}{16}$

8. $\frac{2}{5} + \frac{9}{10}$

9. $\frac{7}{9} + \frac{1}{5}$

10. $\frac{2}{5} + \frac{7}{8}$

11. $\frac{4}{19} + \frac{3}{8}$

12. $\frac{1}{12} + \frac{7}{8}$

13. $\frac{8}{9} + \frac{5}{6}$

14. $\frac{5}{12} + \frac{1}{15}$

15. $\frac{3}{5} + \frac{4}{9}$

16. $\frac{4}{9} + \frac{5}{6}$

17. $\frac{6}{7} + \frac{1}{15} + \frac{8}{9}$

18. $\frac{12}{13} + \frac{4}{7} + \frac{19}{20}$

19. $\frac{2}{13} + \frac{5}{11} + \frac{5}{9}$

20. $\frac{7}{15} + \frac{5}{12} + \frac{6}{11}$

21. $\frac{8}{9} + \frac{3}{5} + \frac{7}{15}$

22. $\frac{1}{5} + \frac{7}{8} + \frac{9}{10}$

Solve.

23. Flor is raising a baby duck. She feeds it $\frac{5}{8}$ cup oats and $\frac{1}{3}$ cup sunflower seeds. Estimate the total amount of food Flor feeds her duckling.

24. Flor feeds her goose $\frac{2}{3}$ cup corn mixed with $\frac{3}{4}$ cup barley. Can she fit the mixture into a bowl that can hold 2 cups?

25. Flor's friends José and Sam visit Flor to see her duckling. They each give Flor a package that holds about $\frac{7}{8}$ cup sunflower seeds. If Flor already had a package with about $\frac{4}{5}$ cup sunflower seeds, estimate how many cups of the seed she has now.

★26. José and Sam start an ant farm. They begin with $\frac{4}{5}$ cup sand. Later, they add $\frac{3}{8}$ cup more. If their ant farm can hold 2 cups sand, do they have room for another $\frac{1}{2}$ cup?

FOCUS: ESTIMATION

These fractions are missing either a numerator or a denominator. Complete each fraction.

Write a fraction close to 0.

1. $\frac{▨}{12}$　**2.** $\frac{▨}{9}$　**3.** $\frac{▨}{24}$　**4.** $\frac{▨}{18}$　**5.** $\frac{3}{▨}$　**6.** $\frac{8}{▨}$

Write a fraction close to $\frac{1}{2}$.

7. $\frac{▨}{9}$　**8.** $\frac{▨}{15}$　**9.** $\frac{▨}{21}$　**10.** $\frac{5}{▨}$　**11.** $\frac{9}{▨}$

Adding Unlike Fractions

A. Pecans are one of the South's most important crops. Georgia grows the most pecans and produces $\frac{1}{5}$ of the nation's total. Texas is second with $\frac{1}{10}$ of the crop. What part of the total pecan crop is grown in Texas and Georgia?

Add $\frac{1}{5} + \frac{1}{10}$.

To add fractions with unlike denominators, you need to write equivalent fractions that have a common denominator.

$$\frac{1}{5} + \frac{1}{10}$$
$$\downarrow \qquad \downarrow$$
$$\frac{1}{5} = \frac{1 \times 2}{5 \times 2} = \frac{2}{10} \qquad \frac{2}{10} + \frac{1}{10} = \frac{3}{10}$$

Georgia and Texas produce $\frac{3}{10}$ of the nation's total pecan crop.

B. Be sure to write the sum in simplest form.

$$\begin{array}{r} \frac{7}{12} = \frac{7}{12} \\ + \frac{1}{6} = \frac{2}{12} \\ \hline \frac{9}{12} = \frac{3}{4} \end{array} \qquad \begin{array}{r} \frac{3}{4} = \frac{9}{12} \\ + \frac{3}{6} = \frac{6}{12} \\ \hline \frac{15}{12} = 1\frac{3}{12} = 1\frac{1}{4} \end{array}$$

Checkpoint   Write the letter of the correct answer.

Add. The answer must be in simplest form.

1. $\frac{1}{4} + \frac{3}{8}$ **2.** $\frac{2}{3} + \frac{5}{6}$ **3** $\frac{3}{5} + \frac{1}{2}$ **4.** $\frac{1}{4} + \frac{5}{6}$

a. $\frac{4}{32}$ **a.** $\frac{10}{18}$ **a.** $\frac{3}{10}$ **a.** $\frac{5}{24}$

b. $\frac{4}{12}$ **b.** $\frac{7}{9}$ **b.** $\frac{4}{7}$ **b.** $\frac{6}{10}$

c. $\frac{4}{8}$ **c.** $\frac{9}{6}$ **c.** $1\frac{1}{10}$ **c.** $1\frac{1}{12}$

d. $\frac{5}{8}$ **d.** $1\frac{1}{2}$ **d.** $\frac{11}{10}$ **d.** $\frac{26}{24}$

Add. Write the sum in simplest form.

1. $\frac{2}{4} + \frac{1}{2}$

2. $\frac{4}{9} + \frac{1}{3}$

3. $\frac{5}{10} + \frac{2}{5}$

4. $\frac{3}{8} + \frac{2}{4}$

5. $\frac{2}{7} + \frac{5}{14}$

6. $\frac{3}{4} + \frac{1}{2}$

7. $\frac{2}{3} + \frac{5}{7}$

8. $\frac{4}{11} + \frac{2}{3}$

9. $\frac{7}{10} + \frac{3}{8}$

10. $\frac{3}{7} + \frac{2}{3}$

11. $\frac{3}{5} + \frac{4}{9}$

12. $\frac{5}{6} + \frac{4}{12}$

13. $\frac{3}{5} + \frac{9}{10}$

14. $\frac{7}{18} + \frac{3}{8}$

15. $\frac{2}{9} + \frac{1}{2}$

16. $\frac{3}{4} + \frac{5}{12}$

17. $\begin{array}{r} \frac{5}{12} \\ + \frac{5}{6} \\ \hline \end{array}$

18. $\begin{array}{r} \frac{3}{7} \\ + \frac{8}{9} \\ \hline \end{array}$

19. $\begin{array}{r} \frac{5}{12} \\ + \frac{9}{13} \\ \hline \end{array}$

20. $\begin{array}{r} \frac{1}{9} \\ + \frac{2}{3} \\ \hline \end{array}$

21. $\begin{array}{r} \frac{1}{2} \\ + \frac{2}{3} \\ \hline \end{array}$

22. $\begin{array}{r} \frac{1}{5} \\ + \frac{4}{15} \\ \hline \end{array}$

★23. $\frac{3}{9} + \frac{5}{6} + \frac{1}{2}$

★24. $\frac{1}{12} + \frac{5}{16} + \frac{1}{4}$

★25. $\frac{7}{10} + \frac{3}{5} + \frac{1}{4}$

★26. $\frac{4}{7} + \frac{3}{4} + \frac{1}{2}$

Solve.

27. Dori is making nut butter. Her recipe calls for $\frac{1}{2}$ pound of peanuts and $\frac{1}{3}$ pound of cashews. How many pounds of nuts does Dori use?

28. John buys $\frac{2}{3}$ pound of trail mix for $3.00. He buys $\frac{1}{2}$ pound of granola for $1.25. How much does he spend?

29. When Millie makes her nut mix, she uses $\frac{1}{3}$ walnuts, $\frac{2}{9}$ macadamia nuts, and $\frac{4}{9}$ almonds. How much of Millie's mix is walnuts and almonds?

ANOTHER LOOK

Choose the appropriate unit: *mm, cm,* or *km.*

1. the length of a notebook

2. the length of an ant

3. the length of a river

4. the length of an eyelash

Subtracting Like and Unlike Fractions

A. The chief grain crops grown in the United States are wheat, rice, corn, and barley. Wheat is $\frac{4}{10}$ of grain production. Barley is $\frac{1}{10}$ of grain production. How much more of the grain production is wheat than barley?

Find $\frac{4}{10} - \frac{1}{10}$.

To subtract fractions with like denominators, subtract the numerators. Write the difference over the denominator.

$$\frac{4}{10} - \frac{1}{10} = \frac{4-1}{10} = \frac{3}{10}$$

Of the grain production, $\frac{3}{10}$ more is wheat.

B. Sometimes you need to write fractions as equivalent fractions that have a common denominator in order to subtract.

$$\frac{7}{10} - \frac{3}{5} \qquad \boxed{\frac{3 \times 2}{5 \times 2} = \frac{6}{10}}$$

$$\frac{7}{10} - \frac{6}{10} = \frac{7-6}{10} = \frac{1}{10}$$

$$\frac{5}{6} = \frac{20}{24}$$
$$-\frac{6}{8} = \frac{18}{24}$$
$$\overline{\frac{2}{24}} = \frac{1}{12}$$

Checkpoint Write the letter of the correct answer.

Subtract. The answer must be in simplest form.

1. $\frac{9}{10} - \frac{2}{10}$

a. $\frac{7}{0}$

b. $\frac{7}{10}$

c. $1\frac{1}{10}$

d. $\frac{11}{10}$

2. $\frac{7}{10} - \frac{1}{10}$

a. $\frac{6}{10}$

b. $\frac{3}{5}$

c. $\frac{4}{5}$

d. $\frac{8}{10}$

3. $\frac{8}{9} - \frac{2}{3}$

a. $\frac{2}{9}$

b. $\frac{1}{2}$

c. $\frac{5}{6}$

d. $\frac{6}{6}$

4. $\frac{4}{5} - \frac{2}{4}$

a. $\frac{3}{10}$

b. $\frac{6}{20}$

c. $\frac{6}{40}$

d. $\frac{2}{1}$

Subtract. Write the difference in simplest form.

1. $\frac{5}{9} - \frac{2}{9}$

2. $\frac{9}{12} - \frac{5}{12}$

3. $\frac{5}{6} - \frac{1}{6}$

4. $\frac{2}{3} - \frac{1}{3}$

5. $\frac{13}{16} - \frac{5}{16}$

6. $\frac{5}{6} - \frac{3}{18}$

7. $\frac{2}{3} - \frac{1}{6}$

8. $\frac{9}{14} - \frac{2}{7}$

9. $\frac{4}{9} - \frac{5}{27}$

10. $\frac{3}{8} - \frac{1}{4}$

11. $\frac{3}{4} - \frac{3}{5}$

12. $\frac{5}{7} - \frac{1}{3}$

13. $\frac{10}{15} - \frac{2}{10}$

14. $\frac{7}{12} - \frac{4}{8}$

15. $\frac{7}{8} - \frac{3}{4}$

16. $\begin{array}{r} \frac{8}{9} \\ - \frac{2}{3} \\ \hline \end{array}$

17. $\begin{array}{r} \frac{9}{14} \\ - \frac{3}{7} \\ \hline \end{array}$

18. $\begin{array}{r} \frac{3}{5} \\ - \frac{3}{7} \\ \hline \end{array}$

19. $\begin{array}{r} \frac{1}{2} \\ - \frac{2}{9} \\ \hline \end{array}$

20. $\begin{array}{r} \frac{6}{7} \\ - \frac{2}{7} \\ \hline \end{array}$

21. $\begin{array}{r} \frac{5}{6} \\ - \frac{1}{3} \\ \hline \end{array}$

★22. $\frac{7}{10} - \frac{n}{10} = \frac{3}{10}$

★23. $\frac{n}{8} - \frac{6}{8} = \frac{3}{8}$

★24. $\frac{n}{15} - \frac{4}{15} = \frac{11}{15}$

★25. $\frac{5}{18} - \frac{n}{18} = \frac{4}{18}$

Solve.

26. Workers harvest $\frac{2}{3}$ of the wheat crop in the morning. After lunch, the remaining $\frac{1}{3}$ of the crop is harvested. How much more was harvested in the morning?

27. Jody spends $\frac{1}{2}$ hour cleaning the tractor. Lon spends $\frac{1}{4}$ hour cleaning the wagon. How much time do both Jody and Lon spend cleaning?

28. Dan's crew harvested $\frac{9}{10}$ acre of wheat in one hour. Tom's crew harvested $\frac{1}{2}$ acre of wheat in one hour. How much more acreage did Dan's crew harvest than Tom's?

★29. Of the harvested wheat, $\frac{9}{10}$ will be sold. The rest will be used on the farm. What fraction of the wheat harvested will be used on the farm?

CHALLENGE

Find the pattern. Copy and complete.

1. $\frac{2}{5}, \frac{4}{5}, \frac{6}{5}, \frac{8}{5}, \frac{10}{5}, \frac{\blacksquare}{\blacksquare}, \frac{\blacksquare}{\blacksquare}, \frac{\blacksquare}{\blacksquare}$

2. $\frac{1}{3}, \frac{1}{9}, \frac{1}{27}, \frac{1}{81}, \frac{\blacksquare}{\blacksquare}, \frac{\blacksquare}{\blacksquare}, \frac{\blacksquare}{\blacksquare}$

More Practice, page 435

Adding Mixed Numbers

A. Sandy lives in California. In the summer, she has a fruit-juice stand. On her first day, Sandy sells $2\frac{3}{8}$ gallons of apple juice. She also sells $3\frac{1}{8}$ gallons of orange juice. How many gallons of juice does Sandy sell?

Add $2\frac{3}{8} + 3\frac{1}{8}$.

Add the fractions.	Add the whole numbers.	Write the sum in simplest form.
$2\frac{3}{8}$	$2\frac{3}{8}$	$2\frac{3}{8}$
$+\,3\frac{1}{8}$	$+\,3\frac{1}{8}$	$+\,3\frac{1}{8}$
$\frac{4}{8}$	$5\frac{4}{8}$	$5\frac{4}{8} = 5\frac{1}{2}$

Sandy sells $5\frac{1}{2}$ gallons of juice.

B. Add $5\frac{1}{2} + 3\frac{1}{4}$.

Find equivalent fractions with a common denominator.		Add the fractions.	Add the whole numbers.
$5\frac{1}{2} =$	$5\frac{2}{4}$	$5\frac{1}{2} = 5\frac{2}{4}$	$5\frac{1}{2} = 5\frac{2}{4}$
$+\,3\frac{1}{4} =$	$3\frac{1}{4}$	$+\,3\frac{1}{4} = 3\frac{1}{4}$	$+\,3\frac{1}{4} = 3\frac{1}{4}$
		$\frac{3}{4}$	$8\frac{3}{4}$

Checkpoint   Write the letter of the correct answer.

Add. The answer must be in simplest form.

1. $2\frac{1}{3} + 3\frac{1}{3}$

a. $5\frac{1}{9}$
b. $5\frac{1}{6}$
c. $5\frac{2}{3}$
d. $23\frac{2}{3}$

2. $3\frac{3}{4} + 4$

a. $3\frac{7}{4}$
b. $7\frac{3}{4}$
c. $12\frac{3}{4}$
d. $34\frac{3}{4}$

3. $4\frac{1}{5} + 1\frac{4}{10}$

a. $4\frac{3}{5}$
b. $5\frac{1}{3}$
c. $5\frac{3}{5}$
d. $5\frac{6}{10}$

4. $7\frac{3}{4} + 6\frac{1}{5}$

a. $13\frac{3}{20}$
b. $13\frac{4}{9}$
c. $13\frac{19}{20}$
d. $13\frac{38}{40}$

Add. Write the sum in simplest form.

1. $3\frac{4}{7}$
  $+ 5\frac{2}{7}$

2. $8\frac{9}{22}$
  $+ 12\frac{11}{22}$

3. $17\frac{1}{4}$
  $+ 4\frac{1}{4}$

4. $2\frac{1}{8}$
  $+ 8\frac{5}{8}$

5. $9\frac{3}{10}$
  $+ 4\frac{3}{10}$

6. $3\frac{1}{3}$
  $+ 9\frac{1}{3}$

7. $2\frac{13}{23}$
  $+ 1\frac{6}{23}$

8. $12\frac{2}{7}$
  $+ 11\frac{7}{10}$

9. $5\frac{3}{5}$
  $+ 5\frac{1}{5}$

10. $8\frac{7}{10}$
  $+ 13\frac{1}{10}$

11. $4\frac{1}{2}$
  $+ 4\frac{1}{4}$

12. $9\frac{2}{5}$
  $+ 2\frac{7}{15}$

13. $13\frac{4}{9}$
  $+ 15\frac{7}{18}$

14. $11\frac{1}{8}$
  $+ 3\frac{13}{16}$

15. $5\frac{1}{6}$
  $+ 1\frac{3}{4}$

16. $14\frac{1}{8} + 11\frac{1}{4}$

17. $7\frac{3}{14} + 2\frac{3}{7}$

18. $6\frac{1}{6} + 4\frac{8}{18}$

19. $14\frac{2}{3} + 1\frac{1}{3}$

20. $6\frac{2}{5} + 11\frac{2}{15}$

21. $3\frac{5}{6} + 8\frac{1}{6}$

22. $7\frac{7}{10} + 21\frac{5}{20}$

23. $11\frac{6}{7} + 9\frac{1}{21}$

24. $3\frac{4}{9} + 16\frac{1}{9}$

25. $4 + 1\frac{13}{14}$

26. $2\frac{11}{16} + 6\frac{3}{16}$

27. $15\frac{1}{2} + 7\frac{3}{17}$

★28. $12\frac{1}{3} + 1\frac{1}{2} + 2\frac{1}{6}$

★29. $4\frac{1}{8} + 7\frac{1}{24} + 11$

★30. $5\frac{1}{4} + 3\frac{1}{6} + 7\frac{3}{8}$

Solve.

31. Sandy plans to make fruit juice. She picks $1\frac{1}{3}$ bushels of grapes. She picks another $2\frac{1}{4}$ bushels of apricots. How many bushels of fruit does Sandy pick?

★32. Sandy mixed $1\frac{1}{4}$ quarts cranberry juice with $1\frac{1}{3}$ quarts apple juice. She added another $1\frac{1}{4}$ quarts strawberry juice. How much fruit punch did she make?

CHALLENGE

A group of farmers are taking their prize pigs to a county fair. On their way, they encounter a broad, deep river. Two boys are on the far bank with a small boat. The boat can only hold one farmer and one pig, or the two boys. How might the farmers get across?

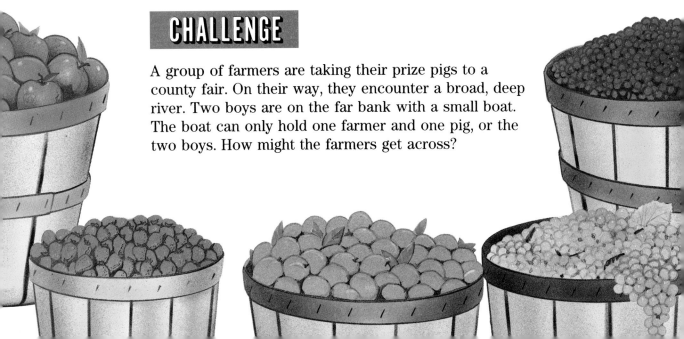

Adding Mixed Numbers with Renaming

A. Regina shops at the local market. She bought $5\frac{5}{16}$ pounds of ripe bananas and $4\frac{7}{8}$ pounds of green bananas. Did Regina buy more than 10 pounds of bananas?

You can estimate to answer this question.

Estimate $5\frac{5}{16} + 4\frac{7}{8}$.

Add the whole numbers.

$5\frac{5}{16} + 4\frac{7}{8}$
$\downarrow \qquad \downarrow$
$5 \quad + 4 \ = 9$

Estimate the sum of the fractions.

$5\frac{5}{16} + 4\frac{7}{8}$
$\downarrow \qquad \downarrow$
$\frac{1}{2} + 1 = 1\frac{1}{2}$

Add the sum of the whole numbers and the estimated sum of the fractions.

$9 + 1\frac{1}{2} = 10\frac{1}{2}$

So, $5\frac{5}{16} + 4\frac{7}{8} \approx 10\frac{1}{2}$.

Regina bought more than 10 pounds of bananas.

B. Regina's mother wanted to know how many pounds of bananas were bought.

Because an exact answer is needed, you need to add $5\frac{5}{16} + 4\frac{7}{8}$.

Find fractions with a common denominator.

$5\frac{5}{16} = 5\frac{5}{16}$
$+ 4\frac{7}{8} = 4\frac{14}{16}$

Add.

$5\frac{5}{16} = 5\frac{5}{16}$
$+ 4\frac{7}{8} = 4\frac{14}{16}$
$\overline{\qquad\qquad 9\frac{19}{16}}$

Write the sum in simplest form.

$9\frac{19}{16} = 9 + 1\frac{3}{16}$
$\qquad = 10\frac{3}{16}$

Regina bought $10\frac{3}{16}$ pounds of bananas.

Checkpoint Write the letter of the correct answer.

Add. The answer must be in simplest form.

1. $4\frac{6}{7} + 2\frac{1}{4}$

a. $6\frac{3}{28}$

b. $6\frac{31}{28}$

c. $7\frac{3}{28}$

d. $7\frac{31}{28}$

2. $3\frac{3}{9} + 5\frac{5}{6}$

a. $8\frac{21}{18}$

b. $9\frac{1}{6}$

c. $9\frac{3}{18}$

d. $9\frac{21}{18}$

3. $11\frac{11}{36} + 1\frac{2}{3}$

a. $11\frac{1}{36}$

b. $12\frac{1}{36}$

c. $12\frac{13}{36}$

d. $12\frac{35}{36}$

Write the letter of the best estimate.

1. $1\frac{2}{5} + 2\frac{3}{7}$

a. about 3
b. about 4
c. about 5

2. $9\frac{5}{8} + 3\frac{3}{4}$

a. about 12
b. about 14
c. about 15

3. $8\frac{1}{3} + 1\frac{7}{8} + \frac{4}{5}$

a. about 9
b. about 10
c. about 11

4. $2\frac{3}{5} + 1\frac{7}{8} + 3\frac{5}{6}$

a. about 6
b. about 9
c. about 10

Add. Write the answer in simplest form.

5. $7\frac{1}{2}$
  $+ 5\frac{3}{4}$

6. $6\frac{5}{8}$
  $+ 1\frac{2}{3}$

7. $9\frac{3}{4}$
  $+ 2\frac{15}{16}$

8. $3\frac{4}{5}$
  $+ 4\frac{7}{10}$

9. $1\frac{5}{8}$
  $+ 6\frac{1}{2}$

10. $3\frac{4}{5} + 2\frac{2}{3}$

11. $5\frac{3}{7} + 8\frac{3}{4}$

12. $1\frac{3}{5} + 5\frac{5}{8}$

13. $4\frac{1}{2} + 11\frac{4}{7}$

14. $2\frac{6}{7} + 4\frac{1}{3}$

15. $5\frac{1}{2} + 3\frac{3}{5}$

16. $6\frac{1}{8} + 7\frac{15}{16}$

17. $9\frac{2}{3} + 3\frac{5}{8}$

18. $4\frac{3}{5} + 6\frac{4}{5}$

★19. $(9\frac{3}{5} + 6\frac{1}{20}) + 5\frac{1}{8}$

★20. $(2\frac{1}{8} + 3\frac{3}{4}) + 4\frac{1}{2}$

★21. $(7\frac{2}{3} + 5\frac{1}{5}) + 6\frac{3}{4}$

Solve.

22. Mr. Manti sells sausage. One morning, he sells $7\frac{1}{2}$ pounds of sausage to Regina and $5\frac{3}{4}$ pounds to Regina's cousin, Sophia. How many pounds of sausage does Mr. Manti sell to Regina and Sophia?

★23. Claudia brought 13 loaves of her delicious bread to sell at the market. She traded $1\frac{1}{2}$ loaves for some tomatoes. She sold $9\frac{1}{2}$ loaves. How many loaves did she have left?

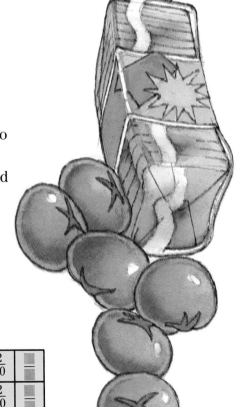

CHALLENGE

Add to complete the fraction squares. Write each fraction in simplest form.

1.

2.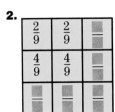

3.

$\frac{1}{10}$	$\frac{2}{10}$	
$\frac{3}{10}$	$\frac{2}{10}$	

More Practice, page 436

Subtracting Mixed Numbers

A. The Pickens family is having a barn raising. They need planks that are $4\frac{1}{4}$ feet long. Ned Class brings $5\frac{3}{4}$-foot planks. How much should Ned trim from the planks?

Find $5\frac{3}{4} - 4\frac{1}{4}$.

Subtract the fractions.	Subtract the whole numbers.	Write the difference in simplest form.
$5\frac{3}{4}$ $-\ 4\frac{1}{4}$ $\overline{\frac{2}{4}}$	$5\frac{3}{4}$ $-\ 4\frac{1}{4}$ $\overline{1\frac{2}{4}}$	$5\frac{3}{4}$ $-\ 4\frac{1}{4}$ $\overline{1\frac{2}{4},\ \text{or}\ 1\frac{1}{2}.}$

Ned should trim $1\frac{1}{2}$ feet from the planks.

B. Find $8\frac{1}{2} - 4\frac{1}{4}$.

Find equivalent fractions with a common denominator.	Compare fractions. Rename if necessary.	Subtract.
$8\frac{1}{2} = 8\frac{2}{4}$ $-\ 4\frac{1}{4} = 4\frac{1}{4}$	$8\frac{1}{2} = 8\frac{2}{4}$ $-\ 4\frac{1}{4} = 4\frac{1}{4}$	$8\frac{1}{2} = 8\frac{2}{4}$ $-\ 4\frac{1}{4} = 4\frac{1}{4}$ $\overline{4\frac{1}{4}}$

Other examples:

$$5\frac{3}{5}$$
$$-\ 2$$
$$\overline{3\frac{3}{5}}$$

$$4\frac{2}{3} = 4\frac{8}{12}$$
$$-\ 1\frac{1}{4} = 1\frac{3}{12}$$
$$\overline{3\frac{5}{12}}$$

Checkpoint Write the letter of the correct answer.

Subtract. The answer must be in simplest form.

1. $5\frac{3}{5} - 3$

a. $\frac{8}{5}$

b. 2

c. $2\frac{3}{5}$

d. $8\frac{3}{5}$

2. $6\frac{5}{8} - 4\frac{3}{8}$

a. $\frac{2}{8}$

b. $2\frac{1}{10}$

c. $2\frac{1}{4}$

d. $10\frac{1}{4}$

3. $7\frac{1}{2} - 2\frac{1}{8}$

a. $5\frac{1}{4}$

b. $5\frac{3}{8}$

c. $5\frac{5}{8}$

d. $9\frac{3}{8}$

Subtract. Write the difference in simplest form.

1. $2\frac{1}{2}$
   $-\ 1$

2. $6\frac{2}{3}$
   $-\ 3$

3. $3\frac{3}{4}$
   $-\ 2\frac{1}{4}$

4. $8\frac{5}{8}$
   $-\ 2\frac{1}{8}$

5. $9\frac{4}{7}$
   $-\ 8\frac{2}{7}$

6. $4\frac{1}{2}$
   $-\ 2\frac{1}{4}$

7. $9\frac{2}{3}$
   $-\ 3\frac{1}{6}$

8. $7\frac{1}{4}$
   $-\ 3\frac{1}{8}$

9. $9\frac{5}{8}$
   $-\ 2\frac{1}{2}$

10. $11\frac{1}{5}$
    $-\ \ 9\frac{1}{10}$

11. $5\frac{1}{3} - 5\frac{1}{4}$

12. $8\frac{3}{4} - 5\frac{1}{5}$

13. $9\frac{1}{4} - 6\frac{1}{6}$

14. $4\frac{2}{3} - 1\frac{2}{5}$

15. $12\frac{6}{7} - 3\frac{1}{4}$

16. $9\frac{1}{2} - 3\frac{1}{4}$

17. $3\frac{2}{3} - 1\frac{1}{4}$

18. $6\frac{1}{2} - 5$

19. $8\frac{5}{8} - 5\frac{1}{8}$

20. $8\frac{3}{5} - 2\frac{3}{10}$

21. $6\frac{1}{2} - 2\frac{1}{4}$

22. $7\frac{6}{7} - 1\frac{11}{14}$

23. $5\frac{2}{3} - 3\frac{4}{9}$

24. $4\frac{3}{4} - 1\frac{1}{12}$

25. $5\frac{7}{10} - 3\frac{1}{5}$

26. $7\frac{4}{5} - 5\frac{1}{2}$

Solve.

27. The Pickens family makes lunch for their barn-raising neighbors. They make $8\frac{1}{2}$ pounds of three-bean salad. Only $3\frac{1}{4}$ pounds are eaten. How much three-bean salad is left?

★28. Jamie Plunkett brings 25 pounds of flour for baking. Mr. Pickens uses $10\frac{1}{2}$ pounds of flour for rolls and $11\frac{1}{2}$ pounds for bread. How much flour does he have left?

CHALLENGE

The sum of each pie slice is printed in the pie's center. Copy each drawing. Then subtract to fill in each slice.

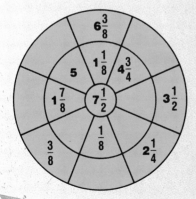

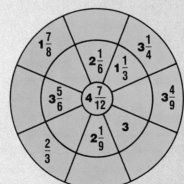

More Practice, page 436

Subtracting Mixed Numbers with Renaming

A. Potter and Maggie opened a roadside vegetable stand for the summer. On opening day, they started with $4\frac{1}{4}$ bushels of corn. They sold $3\frac{3}{4}$ bushels. How much corn was left?

To find how much corn was left, you subtract. Find $4\frac{1}{4} - 3\frac{3}{4}$.

Find fractions with a common denominator.	Compare fractions. Rename if necessary.	Subtract.
$\begin{array}{r} 4\frac{1}{4} \\ -3\frac{3}{4} \\ \hline \end{array}$	$\frac{1}{4} < \frac{3}{4}$ $4\frac{1}{4} = 3 + 1\frac{1}{4}$ $= 3\frac{5}{4}$	$\begin{array}{r} 4\frac{1}{4} = 3\frac{5}{4} \\ -3\frac{3}{4} = 3\frac{3}{4} \\ \hline \frac{2}{4} = \frac{1}{2} \end{array}$

Of the corn, $\frac{1}{2}$ bushel was left.

B. Find $4\frac{1}{4} - 2\frac{1}{3}$.

Find fractions with a common denominator.	Compare fractions. Rename if necessary.	Subtract.
$\begin{array}{r} 4\frac{1}{4} = 4\frac{3}{12} \\ -2\frac{1}{3} = 2\frac{4}{12} \\ \hline \end{array}$	$\frac{3}{12} < \frac{4}{12}$ $4\frac{3}{12} = 3 + 1\frac{3}{12}$ $= 3\frac{15}{12}$	$\begin{array}{r} 4\frac{3}{12} = 3\frac{15}{12} \\ -2\frac{4}{12} = 2\frac{4}{12} \\ \hline 1\frac{11}{12} \end{array}$

Another example:

$$\begin{array}{r} 7 = 6\frac{2}{2} \\ -3\frac{1}{2} = 3\frac{1}{2} \\ \hline 3\frac{1}{2} \end{array} \qquad \boxed{7 = 6 + \frac{2}{2} = 6\frac{2}{2}}$$

Checkpoint Write the letter of the correct answer.

Subtract. The answer must be in simplest form.

1. $2\frac{1}{5} - 1\frac{4}{5}$ **a.** $\frac{2}{5}$ **b.** $1\frac{2}{5}$ **c.** $1\frac{5}{5}$ **d.** $3\frac{2}{5}$

2. $7\frac{2}{3} - 1\frac{3}{4}$ **a.** $5\frac{9}{12}$ **b.** $5\frac{11}{12}$ **c.** $6\frac{11}{12}$ **d.** $7\frac{11}{12}$

3. $5 - 3\frac{2}{3}$ **a.** $1\frac{1}{3}$ **b.** $1\frac{2}{3}$ **c.** $2\frac{1}{3}$ **d.** $8\frac{2}{3}$

Subtract. Write the difference in simplest form.

1. $5\frac{1}{4}$
$-2\frac{3}{4}$

2. $3\frac{3}{10}$
$-1\frac{7}{10}$

3. $7\frac{1}{6}$
$-1\frac{1}{6}$

4. $4\frac{7}{16}$
$-2\frac{13}{16}$

5. $8\frac{1}{4}$
$-4\frac{1}{4}$

6. $6\frac{1}{5}$
$-6\frac{1}{20}$

7. $18\frac{1}{3}$
$-8\frac{5}{6}$

8. $3\frac{2}{5}$
$-2\frac{13}{20}$

9. $12\frac{1}{6}$
$-5\frac{5}{12}$

10. $14\frac{2}{3}$
$-6\frac{11}{12}$

11. $7\frac{4}{9} - 5\frac{7}{9}$

12. $19\frac{7}{17} - 18\frac{11}{17}$

13. $11 - 2\frac{5}{7}$

14. $7 - 1\frac{9}{10}$

15. $13\frac{3}{6} - 3\frac{15}{18}$

16. $8\frac{3}{5} - 1\frac{4}{5}$

17. $14 - 6\frac{12}{21}$

18. $5\frac{1}{3} - 3\frac{2}{3}$

★19. $(4\frac{1}{4} - \frac{3}{4}) - \frac{1}{4}$

★20. $(1\frac{7}{11} - \frac{9}{11}) - \frac{3}{22}$

★21. $(19 - 7\frac{3}{7}) - \frac{3}{14}$

Solve.

22. Maggie received a delivery of 14 dozen eggs. Of these, $5\frac{5}{12}$ dozen were broken and had to be sent back. How many dozen did Maggie keep?

23. Maggie picked $5\frac{1}{2}$ pounds of plums in the morning. By the end of the day, Potter had sold $1\frac{2}{3}$ pounds of them. The rest were given to a hospital. How many pounds were given to the hospital?

★24. Potter and Maggie kept a chart of their first day's sales so that they would know what to stock the second day. Copy and complete the chart to decide what they should stock the second day.

VEGETABLES SOLD ON THE FIRST DAY

Vegetables	Stocked (pounds)	Sold (pounds)	Not sold
green peppers	8	$7\frac{3}{8}$	▨
mushrooms	5	$2\frac{3}{4}$	▨
peas	$7\frac{1}{2}$	3	▨
radishes	$8\frac{1}{4}$	$2\frac{3}{4}$	▨

PROBLEM SOLVING
Interpreting the Quotient and the Remainder

Some problems require division. Pay special attention to the answer you get when you divide. You may need to

1. drop the remainder,
2. round the answer to the next greater whole number, or
3. use only the remainder.

Alice is putting up a fence in her nursery. The section of shrubbery she wants to fence measures 122 feet around. She is building the fence with wood posts and wire mesh which is sold in 16 foot rolls.

$$\begin{array}{r} 7\ \mathrm{R}10 \\ 16\overline{)122} \\ 112 \\ \hline 10 \end{array}$$

Divide

Read each question. Think about how the answers differ for each question.

Question	Action	Answer
1. How many whole rolls of wire mesh will be used?	Drop the remainder.	7 whole rolls will be used.
2. How many rolls does she need to buy?	Round the quotient to the next greater whole number.	8 rolls are needed. (7 full rolls will be used. 1 roll will be cut.)
3. How many feet of the last roll will she use?	Use only the remainder.	10 feet will be used.

Write the letter of the correct answer.

1. Alice orders 250 bags of fertilizer to sell at her nursery. She sends a truck to pick them up. The truck holds 38 bags at a time. How many round trips will the truck make?
$$250 \div 38 = 6 \text{ R22}$$

   a. 6 round trips
   b. 22 round trips
   c. 7 round trips

2. Alice is planting tulip bulbs. She can plant up to 16 bulbs per pot. She has 5,238 bulbs. How many pots will be completely filled with bulbs?
$$5,258 \div 16 = 328 \text{ R10}$$

   a. 328 pots
   b. 10 pots
   c. 329 pots

Solve.

3. Alice is planting saplings. Each sapling needs 5 square feet of space. How many saplings can she plant in 117 square feet of earth?

4. Alice's nursery has 210 decorative hedges. A landscaper buys 25 equal batches of hedges in one month. How many hedges are left in the nursery?

5. A customer buys 150 flower pots. The car he has holds 42 pots at a time. How many carloads will it take to transport the pots?

6. Alice is building a greenhouse for african violets. She has a 48-square-foot pane of glass. How many 5-square-foot windows can she cut?

7. Alice runs a sale on fragrant plants. Customers can buy a dozen for a reduced price. The dozen plants are packed into a box that has just enough room for them. If Alice sells 3,084 plants during the sale, how many boxes are used?

8. This year Alice sold 1,072 evergreens. She has to place her order for next year's stock. She expects to sell at least as many evergreens next year as she did this year. She has to order evergreens in shipments of 55. How many shipments should she order?

9. Alice is setting up her seed display. She can put 14 packs of seeds into each slot. How many slots will she need to display 1,138 packages of seeds?

★10. A greenhouse at Alice's nursery has 340 planters. Each planter holds 10 plants. Alice has 1,092 plants in the greenhouse. How many planters are completely empty?

CALCULATOR

Clear the calculator.

Press these numbers and commands on your calculator. Watch the display. What does it show?

The display should show 12, 24, 36.

Each time you press the $=$, the calculator adds 12 more to the total. This gives multiples of 12: 12×1, 12×2, 12×3, and so on. Find the first ten multiples of 12: 12, 24, 36, 48, 60, 72, 84, 96, 108, 120

You can use the calculator to help find the least common multiple of two different numbers. Find the first ten multiples of 15: Clear the calculator.

Press:

Display: 15, 30, 45, 60, 75, 90, 105, 120, 135, 150

You can see that 60 and 120 are in both sets of multiples. So, 60 is the LCM of 12 and 15.

Find the least common multiple of these sets of numbers on your calculator.

Numbers	Ten Multiples	Common Multiples	LCM
1. 28 7			
2. 50 60			
3. 22 33			
4. 3 5			
5. 14 6			

GROUP PROJECT

What Makes Your Garden Grow?

The problem: Jenny and Josh both had champion squash plants at the fair. "It was the right amount of sunlight," said Jenny. Josh believed it was the right amount of water. Can you find out the effect of sun and water on a plant's growth?

To see the effect of sunlight, follow these steps.
- Take two containers, and fill them with good soil.
- Plant several kidney, pinto, or lima beans in each pot.
- Mark one pot with a picture of the sun.
- Place both pots in a warm, light spot, and water them.
- After they sprout, allow both pots only 4 hours of sun. Then cover them with a cardboard box.
- Give the pot with the sun 10 more minutes of sun every day, and the other pot 10 minutes less.
- At the end of 24 days, measure each plant. What is the difference?

Design your own experiment to show the difference between enough water and too little water on plant growth.

CHAPTER TEST

Find the least common multiple. (page 227)

1. 6, 8

Find the greatest common factor. (page 229)

2. 16, 24, 64

Write *prime* or *composite* to describe the number. (page 231)

3. 21   **4.** 43

Write the fraction for the part that is shaded. (page 233)

5.   **6.** △ △ △ ○ ○ ▢ ▢ ▢ ▢

Complete. (page 235)

7. $\frac{2}{5} = \frac{\blacksquare}{10}$ **8.** $\frac{5}{8} = \frac{10}{\blacksquare}$

Write in simplest form. (page 239)

9. $\frac{9}{27}$ **10.** $\frac{8}{40}$

Write as a whole number or as a mixed number. (page 241)

11. $\frac{9}{4}$ **12.** $\frac{16}{3}$ **13.** $\frac{49}{7}$

Write as a fraction. (page 241)

14. $4\frac{2}{3}$ **15.** $6\frac{3}{8}$ **16.** $4\frac{1}{2}$

Compare. Write >, <, or = for ●. (page 243)

17. $\frac{2}{5}$ ● $\frac{4}{9}$ **18.** $1\frac{3}{8}$ ● $1\frac{4}{12}$

Write in order from the least to the greatest. (page 243)

19. $\frac{1}{5}, \frac{1}{9}, \frac{1}{4}$ **20.** $\frac{1}{2}, \frac{5}{8}, \frac{2}{3}$

Estimate. (page 253)

21. $\frac{3}{8} + \frac{1}{16}$ **22.** $5\frac{1}{7} + 2\frac{3}{8}$

Add. Write the sum in simplest form. (pages 247, 249, and 253)

23. $\frac{3}{12} + \frac{11}{12}$ **24.** $\frac{2}{7} + \frac{3}{4}$ **25.** $\frac{5}{6} + \frac{3}{5}$

26. $\begin{array}{r} 6\frac{1}{3} \\ + 2\frac{1}{3} \\ \hline \end{array}$ **27.** $\begin{array}{r} 5\frac{7}{8} \\ + 3\frac{1}{8} \\ \hline \end{array}$ **28.** $\begin{array}{r} 4\frac{2}{3} \\ + 2\frac{3}{8} \\ \hline \end{array}$ **29.** $\begin{array}{r} 19\frac{7}{9} \\ + 8\frac{1}{6} \\ \hline \end{array}$

Subtract. Write the difference in simplest form.
(pages 251, 259, and 261)

30. $\frac{5}{6} - \frac{2}{6}$

31. $\frac{3}{4} - \frac{2}{9}$

32. $\frac{4}{5} - \frac{1}{6}$

33. $\begin{array}{r} 6 \\ - 2\frac{1}{3} \\ \hline \end{array}$

34. $\begin{array}{r} 3\frac{5}{8} \\ - 2\frac{7}{8} \\ \hline \end{array}$

35. $\begin{array}{r} 5\frac{1}{7} \\ - 1\frac{3}{28} \\ \hline \end{array}$

36. $\begin{array}{r} 4\frac{2}{3} \\ - 2\frac{3}{8} \\ \hline \end{array}$

Solve. Write the letter of the operation you would use
to solve the problem. (pages 234–235)

37. Bob worked 38 hours on a dairy
farm. Herb worked twice as many
hours. How long did Herb work on
the dairy farm?

    a. multiplication   **b.** division

38. The 90 students in the agriculture
course are placed in classes of no
more than 20 students. How many
classes are there?

    a. multiplication   **b.** division

Write the letter of the correct answer. (pages 262–263)

39. Mr. Meyers harvested enough millet
to fill all his grain bins, with 1,697
pounds left. He rented extra bins,
each to hold 96 pounds of grain.
How many bins did Mr. Meyers
have to rent?

    a. 17   **b.** 18   **c.** 17 R64   **d.** $\frac{2}{3}$

40. Mr. Meyers's soy crop totaled
10,214 pounds. He stored the crop
in bins holding 92 pounds each. He
sold all the soy except for that in
the partly filled bin. How much soy
did Mr. Meyers keep?

    a. 111   **b.** 111 R2   **c.** R2   **d.** $\frac{1}{46}$

BONUS

In each square below, choose the three fractions or
mixed numbers that, when added, give the greatest
sum. Then choose the three fractions or mixed
numbers that, when added, give the least sum.

1.

$\frac{4}{5}$	$\frac{6}{7}$ $\frac{2}{3}$	$\frac{7}{10}$
	$\frac{8}{12}$ $\frac{5}{8}$	$\frac{3}{5}$
$\frac{4}{16}$	$\frac{9}{11}$	$\frac{1}{8}$

2.

$2\frac{3}{8}$	$3\frac{3}{8}$	$2\frac{5}{6}$
$4\frac{3}{10}$	$2\frac{2}{3}$	$3\frac{7}{8}$
$3\frac{15}{20}$	$2\frac{20}{25}$	$4\frac{4}{5}$

RETEACHING

Before you can subtract mixed numbers, you may have to rename.

Find $7\frac{1}{5} - 4\frac{1}{3}$.

Find fractions that have like denominators.

$$7\frac{1}{5} = 7\frac{3}{15}$$
$$-4\frac{1}{3} = 4\frac{5}{15}$$

Compare fractions. Rename if necessary.

$$\frac{3}{15} < \frac{5}{15}$$
$$7\frac{3}{15} = 6 + 1\frac{3}{15}$$
$$= 6\frac{18}{15}$$

Subtract. Write the answer in simplest form.

$$7\frac{3}{15} = 6\frac{18}{15}$$
$$-4\frac{5}{15} = 4\frac{5}{15}$$
$$\overline{\phantom{-4\frac{5}{15} =\ } 2\frac{13}{15}}$$

Another example:

$$9 = 8\frac{5}{5}$$
$$-4\frac{1}{5} = 4\frac{1}{5}$$
$$\overline{\phantom{-4\frac{1}{5} =\ } 4\frac{4}{5}}$$

$$\boxed{9 = 8 + \frac{5}{5} = 8\frac{5}{5}}$$

Subtract. Write the answer in simplest form.

1. $17\frac{1}{5}$ $-15\frac{2}{5}$

2. $9\frac{3}{10}$ $-7\frac{9}{10}$

3. $14\frac{1}{6}$ $-12\frac{5}{6}$

4. $16\frac{2}{5}$ $-7\frac{4}{5}$

5. $14\frac{3}{11}$ $-2\frac{9}{11}$

6. $13\frac{1}{5}$ $-3\frac{3}{4}$

7. $21\frac{5}{6}$ $-9\frac{9}{10}$

8. $5\frac{3}{11}$ $-3\frac{1}{3}$

9. $18\frac{1}{9}$ $-8\frac{5}{6}$

10. $7\frac{3}{7}$ $-4\frac{3}{5}$

11. $13\frac{2}{5}$ $-7\frac{4}{25}$

12. $3\frac{1}{5}$ $-1\frac{1}{3}$

13. $17\frac{5}{9}$ $-12\frac{1}{4}$

14. $4\frac{3}{7}$ $-2\frac{1}{4}$

15. $23\frac{5}{6}$ $-11\frac{2}{9}$

16. $9\frac{2}{5} - 4\frac{1}{2}$

17. $4 - 2\frac{1}{3}$

18. $2\frac{3}{8} - 1\frac{7}{8}$

19. $14\frac{2}{9} - 8\frac{2}{3}$

20. $12 - 3\frac{1}{5}$

21. $20\frac{2}{7} - 3\frac{1}{3}$

22. $8\frac{3}{10} - 5\frac{7}{10}$

23. $13\frac{2}{5} - 12\frac{3}{4}$

ENRICHMENT

Cross Products

To find the cross products of two fractions, multiply each numerator by the denominator in the other fraction. If the cross products are equal, the fractions are equivalent.

Are $\frac{4}{8}$ and $\frac{1}{2}$ equivalent fractions?

Cross multiply.

$$4 \times 2 \quad \bullet \quad 1 \times 8$$
$$8 = 8$$

The cross products are equal; so, the fractions are equivalent.

Write $=$ or $\neq$. Use the cross products.

1. $\frac{3}{4} \bullet \frac{6}{8}$ **2.** $\frac{5}{7} \bullet \frac{6}{9}$ **3.** $\frac{4}{23} \bullet \frac{7}{29}$ **4.** $\frac{2}{9} \bullet \frac{3}{8}$

5. $\frac{5}{15} \bullet \frac{3}{9}$ **6.** $\frac{2}{3} \bullet \frac{9}{12}$ **7.** $\frac{8}{18} \bullet \frac{4}{9}$ **8.** $\frac{5}{6} \bullet \frac{7}{9}$

9. $\frac{4}{10} \bullet \frac{6}{15}$ **10.** $\frac{5}{7} \bullet \frac{10}{14}$ **11.** $\frac{3}{8} \bullet \frac{4}{9}$ **12.** $\frac{3}{10} \bullet \frac{5}{15}$

13. $\frac{13}{19} \bullet \frac{7}{12}$ **14.** $\frac{3}{5} \bullet \frac{9}{15}$ **15.** $\frac{7}{8} \bullet \frac{9}{11}$ **16.** $\frac{10}{25} \bullet \frac{2}{5}$

17. $\frac{12}{36} \bullet \frac{1}{3}$ **18.** $\frac{9}{20} \bullet \frac{11}{25}$ **19.** $\frac{8}{16} \bullet \frac{1}{2}$ **20.** $\frac{5}{6} \bullet \frac{10}{12}$

Solve. Use cross products.

21. A mountaineering club makes two expeditions, one in May and one in July. In May, $\frac{4}{5}$ of the club goes on the expedition. In July, $\frac{2}{3}$ of the club goes on the expedition. Does the same fraction of the club go on each expedition?

22. Dan the milkman delivers $\frac{6}{10}$ of his weekly milk supply on Monday. On Tuesday, he delivers $\frac{3}{5}$ of his milk supply. Does Dan deliver the same fraction of his milk supply on each day?

CUMULATIVE REVIEW

Write the letter of the correct answer.

1. $742.7 \div 10$

   a. 7.427
   b. 74.27
   c. 7,427
   d. not given

2. A 10,000-meter run is how many kilometers long?

   a. 0.001 km
   b. 1 km
   c. 10 km
   d. not given

3. $96.15 \div 6$

   a. 16.02
   b. 16.025
   c. 16.25
   d. not given

4. Which unit would you use to measure the length of a pencil point?

   a. millimeters
   b. centimeters
   c. meters
   d. not given

5. $4.395 \div 15$

   a. 0.233
   b. 0.293
   c. 0.333
   d. not given

6. $6,372 + 3,498 + 65 + 412$

   a. 10,006
   b. 11,446
   c. 12,400
   d. not given

7. 0.7×0.25

   a. 0.175
   b. 0.555
   c. 0.715
   d. not given

8. Order from the least to the greatest: 7.53, 7.503, 7.35, 5.

   a. 5, 7.53, 7.503, 7.35
   b. 7.35, 7.53, 5, 7.503
   c. 5, 7.35, 7.503, 7.53
   d. not given

9. Estimate $3751 \div 5$.

   a. 7
   b. 70
   c. 700
   d. not given

10. Fred's car cost $8,799.00. Al's car cost $11,851.72. Herb's car cost $7,327.51 more than the combined cost of Fred's and Al's cars. What was the cost of Herb's car?

   a. $13,323.21
   b. $19,179.23
   c. $27,978.23
   d. not given

11. One year, Hank's Sporting Goods sold 327,523 baseballs, and Sal's Sporting Goods sold 299,858 baseballs. How many more baseballs did Hank's sell than Sal's?

   a. 27,665
   b. 38,775
   c. 627,387
   d. not given

Suppose your class is holding a carnival to raise money. What games would you organize? How much would you charge for each? How much money could you earn?

9 MULTIPLYING AND DIVIDING FRACTIONS

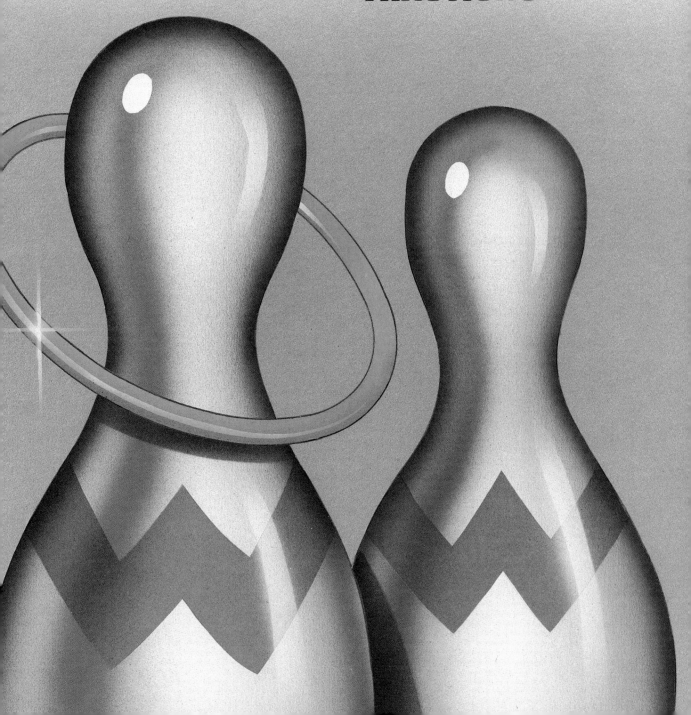

Multiplying Fractions

When the actors met for their first rehearsal, they discovered that boxes of props covered $\frac{1}{2}$ of the stage area. The stage crew quickly cleared $\frac{1}{3}$ of that area so that the rehearsal could begin. What part of the stage did the crew clear?

You can draw a picture to show $\frac{1}{3}$ of $\frac{1}{2}$.

Here is $\frac{1}{2}$ of the stage.

Here is $\frac{1}{3}$ of $\frac{1}{2}$.

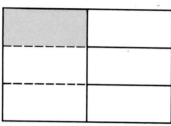

$\frac{1}{3}$ of $\frac{1}{2}$ is the same as $\frac{1}{6}$ of 1 whole.
The crew cleared $\frac{1}{6}$ of the stage.

You can use multiplication to find $\frac{1}{3}$ of $\frac{1}{2}$.
Multiply $\frac{1}{3} \times \frac{1}{2}$.

Multiply the numerators.
Multiply the denominators. $\frac{1}{3} \times \frac{1}{2} = \frac{1 \times 1}{3 \times 2} = \frac{1}{6}$

If necessary, write the product in simplest form.

Multiply $\frac{2}{3} \times \frac{3}{4}$. $\frac{2}{3} \times \frac{3}{4} = \frac{2 \times 3}{3 \times 4} = \frac{6}{12} = \frac{1}{2}$

Checkpoint   Write the letter of the correct answer.

Multiply. The answer should be in simplest form.

1. $\frac{1}{5} \times \frac{3}{5}$

a. $\frac{3}{25}$

b. $\frac{4}{25}$

c. $\frac{4}{10}$

d. $\frac{4}{5}$

2. $\frac{2}{3} \times \frac{3}{10}$

a. $\frac{1}{5}$

b. $\frac{5}{13}$

c. $\frac{5}{30}$

d. $\frac{6}{13}$

3. $\frac{5}{6} \times \frac{3}{4}$

a. $\frac{1}{3}$

b. $\frac{4}{5}$

c. $\frac{5}{8}$

d. $\frac{3}{2}$

Multiply. Write the answer in simplest form.

1. $\frac{1}{7} \times \frac{1}{5}$ 2. $\frac{1}{3} \times \frac{1}{9}$ 3. $\frac{1}{4} \times \frac{1}{2}$ 4. $\frac{1}{7} \times \frac{1}{7}$ 5. $\frac{1}{5} \times \frac{1}{8}$

6. $\frac{4}{9} \times \frac{2}{5}$ 7. $\frac{7}{8} \times \frac{3}{4}$ 8. $\frac{2}{3} \times \frac{5}{7}$ 9. $\frac{4}{7} \times \frac{2}{9}$ 10. $\frac{3}{8} \times \frac{3}{4}$

11. $\frac{2}{4} \times \frac{1}{2}$ 12. $\frac{1}{8} \times \frac{4}{5}$ 13. $\frac{1}{3} \times \frac{3}{4}$ 14. $\frac{5}{9} \times \frac{1}{5}$ 15. $\frac{1}{2} \times \frac{4}{5}$

16. $\frac{5}{6} \times \frac{2}{5}$ 17. $\frac{3}{10} \times \frac{5}{6}$ 18. $\frac{3}{4} \times \frac{2}{5}$ 19. $\frac{3}{4} \times \frac{5}{6}$ 20. $\frac{5}{8} \times \frac{2}{5}$

21. $\frac{4}{9} \times \frac{3}{4}$ ★22. $\frac{1}{2} \times \frac{1}{3} \times \frac{1}{4}$ ★23. $\frac{3}{7} \times \frac{1}{7} \times \frac{1}{3}$ ★24. $\frac{1}{4} \times \frac{3}{5} \times \frac{2}{3}$ ★25. $\frac{2}{5} \times \frac{2}{3} \times \frac{5}{8}$

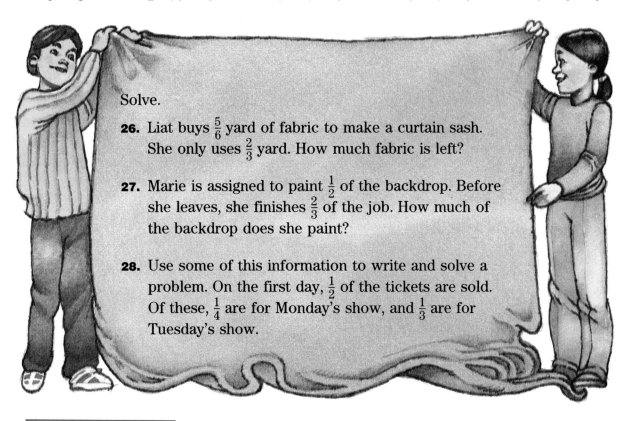

Solve.

26. Liat buys $\frac{5}{6}$ yard of fabric to make a curtain sash. She only uses $\frac{2}{3}$ yard. How much fabric is left?

27. Marie is assigned to paint $\frac{1}{2}$ of the backdrop. Before she leaves, she finishes $\frac{2}{3}$ of the job. How much of the backdrop does she paint?

28. Use some of this information to write and solve a problem. On the first day, $\frac{1}{2}$ of the tickets are sold. Of these, $\frac{1}{4}$ are for Monday's show, and $\frac{1}{3}$ are for Tuesday's show.

CHALLENGE

Take a shortcut when you multiply fractions. First divide by a common factor.

2 is a common factor of 2 and 4. Divide by 2; then multiply.

$$\frac{3}{\overset{}{\underset{2}{4}}} \times \frac{\overset{1}{2}}{7} = \frac{3 \times 1}{2 \times 7} = \frac{3}{14}$$

Use the shortcut to multiply.

1. $\frac{1}{3} \times \frac{6}{7}$ 2. $\frac{1}{2} \times \frac{4}{9}$ 3. $\frac{3}{5} \times \frac{5}{8}$ ★4. $\frac{2}{3} \times \frac{3}{10}$ ★5. $\frac{3}{4} \times \frac{8}{15}$

Multiplying Fractions and Whole Numbers

A. There are 12 students in the stage crew. Of these, $\frac{1}{3}$ are responsible for painting scenery. How many students paint scenery?

Find $\frac{1}{3}$ of 12. $\frac{1}{3} \times 12$

Write the whole number as a fraction. $\frac{1}{3} \times \frac{12}{1}$

Multiply the fractions. $\frac{1}{3} \times \frac{12}{1} = \frac{1 \times 12}{3 \times 1} = \frac{12}{3}$

Write the product as a whole number. $\frac{12}{3} = 4$

4 students paint scenery.

B. Sometimes the answer is a fraction or a mixed number.

Find $\frac{1}{10} \times 5$. $\frac{1}{10} \times \frac{5}{1} = \frac{5}{10} \text{ or } \frac{1}{2}$

Find $6 \times \frac{3}{4}$. $\frac{6}{1} \times \frac{3}{4} = \frac{18}{4} = 4\frac{2}{4} \text{ or } 4\frac{1}{2}$

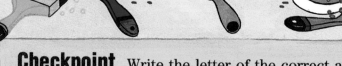

Checkpoint   Write the letter of the correct answer.

Multiply. The answer should be in simplest form.

1. $\frac{2}{5} \times 5$

    a. $\frac{10}{25}$

    b. $\frac{2}{25}$

    c. 2

    d. $1\frac{2}{5}$

2. $2 \times \frac{5}{12}$

    a. $\frac{5}{6}$

    b. $\frac{7}{12}$

    c. $\frac{5}{24}$

    d. $\frac{10}{24}$

3. $\frac{5}{6} \times 4$

    a. $\frac{20}{24}$

    b. $\frac{5}{24}$

    c. $1\frac{1}{2}$

    d. $3\frac{1}{3}$

Multiply. Write the answer in simplest form.

1. $\frac{1}{2} \times 8$ **2.** $4 \times \frac{3}{4}$ **3.** $10 \times \frac{2}{5}$ **4.** $\frac{3}{4} \times 8$ **5.** $6 \times \frac{2}{3}$

6. $2 \times \frac{1}{3}$ **7.** $\frac{2}{9} \times 4$ **8.** $\frac{3}{10} \times 3$ **9.** $5 \times \frac{1}{8}$ **10.** $\frac{2}{5} \times 2$

11. $3 \times \frac{2}{9}$ **12.** $\frac{5}{12} \times 2$ **13.** $2 \times \frac{3}{10}$ **14.** $\frac{3}{16} \times 4$ **15.** $2 \times \frac{3}{8}$

16. $\frac{5}{8} \times 10$ **17.** $6 \times \frac{3}{4}$ **18.** $\frac{5}{6} \times 8$ **19.** $12 \times \frac{7}{10}$ **20.** $14 \times \frac{3}{8}$

21. $\frac{5}{9} \times 6$ **22.** $6 \times \frac{3}{8}$ **23.** $64 \times \frac{1}{8}$ **24.** $6 \times \frac{7}{10}$ **25.** $\frac{1}{3} \times 25$

★**26.** $7 \times \frac{1}{8} \times \frac{1}{2}$ ★**27.** $\frac{1}{4} \times 5 \times \frac{2}{5}$ ★**28.** $\frac{2}{3} \times \frac{1}{6} \times 8$

Solve.

29. Margot draws a lamppost that is $\frac{2}{3}$ the height of the backdrop. If the backdrop is 9 feet high, how high is the lamppost?

30. Maya spends 2 hours gathering materials needed to create special effects. Then Ralph takes over for $\frac{1}{2}$ hour. How much time is spent on gathering the materials?

31. A gallon of paint usually costs $20. Students who shop at Myer's Paint Shop receive a discount equal to $\frac{1}{5}$ of the regular price. What is the amount of the discount?

32. Jeffrey has designed 40 posters to advertise the play. If he gives the publicity committee $\frac{3}{4}$ of them, how many posters does he have left?

CALCULATOR

Use your calculator to find the product of $\frac{7}{8} \times \frac{6}{7}$. You know that $\frac{7}{8}$ is the same as $8\overline{)7}$, or $7 \div 8$. So, $\frac{7}{8} \times \frac{6}{7}$ is the same as $(7 \div 8) \times (6 \div 7)$. You can use your calculator to find the product.

 0.75, or $\frac{75}{100}$, or $\frac{3}{4}$

Use your calculator to solve.

1. $\frac{3}{4} \times \frac{2}{3}$ **2.** $\frac{3}{10} \times \frac{1}{3}$ **3.** $\frac{3}{8} \times \frac{4}{5}$ **4.** $\frac{3}{4} \times \frac{5}{6}$

5. $\frac{3}{8} \times \frac{2}{3}$ **6.** $\frac{9}{10} \times \frac{5}{12}$ **7.** $\frac{7}{8} \times \frac{64}{8}$ **8.** $\frac{9}{15} \times \frac{36}{27}$

Multiplying Fractions and Mixed Numbers

A. This year's musical is going to include a rainbow of colors and costumes. For the boys' shirts, $\frac{1}{2}$ bolt of red satin is needed. $5\frac{3}{4}$ times that is needed for the girls' skirts. How much red satin is needed for the skirts?

Find $5\frac{3}{4} \times \frac{1}{2}$.

Rename the mixed number.	Multiply the fractions.	Write the product as a mixed number.
$5\frac{3}{4} = \frac{23}{4}$	$\frac{23}{4} \times \frac{1}{2} = \frac{23}{8}$	$\frac{23}{8} = 2\frac{7}{8}$

For the girls' skirts, $2\frac{7}{8}$ bolts of satin are needed.

B. In some examples, you can write fractions for both factors before multiplying.

Find $6 \times 2\frac{1}{5}$. $\frac{6}{1} \times \frac{11}{5} = \frac{66}{5} = 13\frac{1}{5}$

Find $3\frac{1}{9} \times 2\frac{1}{2}$. $\frac{28}{9} \times \frac{5}{2} = \frac{140}{18} = 7\frac{7}{9}$

Checkpoint   Write the letter of the correct answer.

Multiply. The answer should be in simplest form.

1. $\frac{2}{3} \times 1\frac{7}{8}$

2. $2 \times 4\frac{1}{8}$

3. $3\frac{1}{3} \times 5\frac{1}{4}$

a. $1\frac{7}{12}$

b. $1\frac{5}{12}$

c. $1\frac{1}{3}$

d. $1\frac{1}{4}$

a. $\frac{1}{16}$

b. $3\frac{1}{4}$

c. $8\frac{1}{4}$

d. $8\frac{1}{8}$

a. $\frac{210}{12}$

b. $15\frac{1}{2}$

c. $15\frac{3}{4}$

d. $17\frac{1}{2}$

Multiply. Write each product in simplest form.

1. $\frac{1}{2} \times 6\frac{2}{3}$ **2.** $\frac{2}{5} \times 5\frac{1}{2}$ **3.** $9\frac{1}{3} \times \frac{1}{6}$ **4.** $4\frac{1}{2} \times \frac{2}{3}$

5. $7 \times 3\frac{3}{4}$ **6.** $2\frac{2}{5} \times 5$ **7.** $4 \times 3\frac{1}{8}$ **8.** $2\frac{1}{2} \times 6$

9. $3\frac{2}{5} \times 4\frac{1}{4}$ **10.** $2\frac{1}{4} \times 1\frac{1}{6}$ **11.** $8\frac{2}{5} \times 3\frac{1}{8}$ **12.** $2\frac{1}{2} \times 3\frac{1}{3}$

13. $2\frac{1}{3} \times 1\frac{3}{5}$ **14.** $\frac{5}{6} \times 5\frac{1}{13}$ **15.** $\frac{2}{3} \times 8$ **16.** $3 \times 1\frac{5}{6}$

17. $1\frac{3}{5} \times 4\frac{3}{4}$ **18.** $\frac{1}{2} \times 3\frac{3}{5}$ **19.** $\frac{1}{3} \times 1\frac{1}{8}$ **20.** $2 \times 2\frac{1}{9}$

Solve.

21. The sparkling silver fabric used to create the nighttime sky weighs $10\frac{1}{2}$ pounds. The red velvet for the sunset weighs $\frac{2}{3}$ as much. How much does the red velvet weigh?

22. Use the advertisement to make up your own problem.

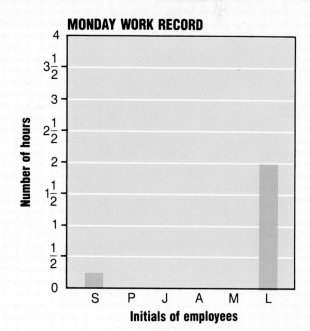

SALE
at Buttons and Bows
Get ¼ yard free for every
2½ yards of fabric you buy.

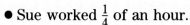

CHALLENGE

Each day the head of the costume department keeps track of how many hours each person works. Copy the graph, and use the data below to complete it. Who worked the most hours?

- Sue worked $\frac{1}{4}$ of an hour.
- Pete worked 2 times as much as Sue.
- José worked $1\frac{1}{2}$ times as much as Pete.
- Anne worked $4\frac{2}{3}$ times as much as José.
- Max worked $\frac{3}{7}$ of what Anne worked.
- Lily worked $1\frac{1}{3}$ times as much as Max.

MONDAY WORK RECORD

Number of hours vs. Initials of employees

PROBLEM SOLVING
Estimation

You can solve problems by using estimated amounts. Often when you estimate, you have to decide to what place you should round the numbers. You can estimate more closely by rounding numbers to different places.

The after-school film-club members rate the latest movies. The movies are rated as follows:

Points	Rating	
240 and under	*Tree House*	⌐
241–280	*Standing Pat*	★
281–320	*Greetings*	★★
321–360	*Broken Trail*	★★★
361–400	*Rain Go Away*	★★★★

Liza, a club member, is adding the points for 2 different films. She wants to quickly compare their ratings. She decides to estimate. Liza first rounds to the nearest ten and finds that the sums are the same. So, she rounds to the nearest whole number.

	A Winner Never Quits	*Forever*
Acting	86.685----90----87	85.75----90----86
Cinematography	83.66-----80----84	84.75----80----85
Script	85.25-----90----85	89.50----90----90
Effects, costumes, etc.	71.25-----70----71	74.25----70----74
	330 327	330 335

When you are comparing amounts that are close, you need a more exact estimate. If the more exact estimates are close, you can round to the nearest tenth or even hundredth. By rounding to the nearest whole number, Liza found that *Forever* received a slightly higher rating than *A Winner Never Quits*.

Estimate to solve.

1. The film club shows films every Wednesday from 3:30 to 5:30 P.M. The club members must leave no later than 5:45 P.M. This Wednesday they are showing

   A Trail of Leaves—$22\frac{1}{2}$ min

   The Last Game —$36\frac{1}{2}$ min

   Clyde, the Pig —27 min

   Math Class —$15\frac{1}{4}$ min

   Will they be finished viewing by 5:30—5:35 P.M.?

2. One group of film-club members is videotaping a drama. So far they have spent

   Videotape——————————$9.39
   Duplicating the tape ————$12.95
   Equipment rental—————$49.98
   ($24.99/day)
   Props, Costumes ————$27.88
   They have a budget of $105.00.
   Do they have enough money to rent equipment for another day?

3. What rating should the following film receive? (Use the ratings on page 278.) *Greetings:* 84.575, 84.575, 78.66, 66.66.

4. The video club needs a camera and lights. A camera costs $279.85 to purchase. Lights cost $89.95 to buy. Rented equipment costs $24.99 a day for 12 days. Is it better over the year to buy or rent the equipment?

5. Ralph, the film-club president, counts 24 students who have paid the $0.90 admission to the showing. He also knows that last week the club earned $28.75. Has the club earned the $49.98 needed to pay for the rental of video equipment?

★6. Use the list below to make two film-club programs that take up at least 100 min but not more than 125 min each.

   Scarlet Rose $33\frac{1}{4}$ min

   Tree House $32\frac{1}{4}$ min

   Standing Pat $45\frac{1}{2}$ min

   Parachute $16\frac{2}{3}$ min

   Broken Trail $38\frac{1}{2}$ min

   Rain Go Away $75\frac{1}{2}$ min

★7. The group wants to purchase videos for their video library. The list on the right shows the videos they want and their prices. They have $164.41. Which videos could they purchase?

Together	$19.96
Raise the Flag	$21.75
Patty Malone	$29.00
Shiny Days	$18.75
Viola	$25.98
School Days	$36.98
Vegetable Soup	$29.99

Dividing Whole Numbers by Fractions

Daniela does the makeup for the actors in the play. She uses sticks of greasepaint. She needs $\frac{1}{2}$ stick to make up one actor. How many actors can she make up with 4 sticks of greasepaint?

You need to find how many $\frac{1}{2}$'s there are in 4 sticks.

To find how many, you divide.
You can use a picture to help you.

There are 8 halves in 4. **$4 \div \frac{1}{2} = 8$**

Check by multiplying. $8 \times \frac{1}{2} = 4$

Daniela can make up 8 actors.

Another example: $5 \div \frac{1}{3} = $ ■

There are 15 thirds in 5. **$5 \div \frac{1}{3} = 15$**

Check by multiplying. $15 \times \frac{1}{3} = 5$

Checkpoint Write the letter of the correct answer.

Divide.

1.

$3 \div \frac{1}{3} = $

a. 1

b. $3\frac{1}{3}$

c. 3

d. 9

2.

$6 \div \frac{1}{4} = $

a. $1\frac{1}{2}$

b. 6

c. 24

d. $6\frac{1}{4}$

3.

$3 \div \frac{1}{2} = $

a. $3\frac{1}{2}$

b. 5

c. 6

d. 7

Divide.

1. $3 \div \frac{1}{4}$

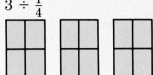

2. $5 \div \frac{1}{4}$

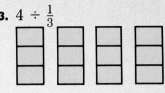

3. $4 \div \frac{1}{3}$

4. $6 \div \frac{1}{3}$

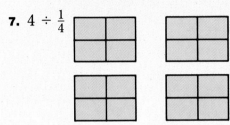

5. $9 \div \frac{1}{4}$

6. $4 \div \frac{1}{6}$

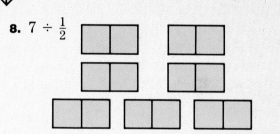

7. $4 \div \frac{1}{4}$

8. $7 \div \frac{1}{2}$

Solve. Draw a picture to help you.

9. The refreshment committee has 5 bottles of cider at the opening-night party. If $\frac{1}{4}$ bottle fills 1 cup, how many cups of cider are there?

10. The decoration committee has 15 yards of crepe paper to decorate the room. They use $\frac{2}{3}$ of it. How many yards do they use?

11. There are 6 cantaloupes to serve 48 people at the party. If each person eats $\frac{1}{4}$ melon, will there be enough melon for all? If not, how many more melons are needed?

12. Use this information to write and solve a problem of your own. Leonard has 6 boxes of napkins. Every table in the party room needs $\frac{1}{8}$ of a box of napkins.

MIDCHAPTER REVIEW

Multiply or divide. The answer should be in simplest form.

1. $\frac{7}{8} \times \frac{3}{4}$

2. $\frac{2}{3} \times \frac{3}{7}$

3. $5 \times \frac{1}{2}$

4. $\frac{9}{10} \times 11$

5. $2\frac{1}{4} \times 4\frac{4}{7}$

6. $2\frac{1}{3} \times 1\frac{3}{5}$

7. $6 \div \frac{1}{5}$

8. $12 \div \frac{1}{7}$

PROBLEM SOLVING
Making a Table To Find a Pattern

Sometimes you can make a table to find a pattern that will help you solve a problem.

> The River City High School is planning its yearly musical. In order to play certain music, the orchestra has to have a certain number of kinds of instruments. If there are 5 stringed instruments, there must be 3 woodwind instruments. If there are 10 strings, there must be 6 woodwinds. If there are 15 strings, there must be 9 woodwinds. How many woodwinds will be needed if 30 strings are used?

You can make a table to solve this problem by finding a pattern.

ORCHESTRA

Strings	5	10	15	■
Woodwinds	3	6	9	■

You can see that there is a pattern in the way the number of strings increases. The number increases by 5 each time. So, the next number in the table will be 20. The increase in woodwinds also shows a pattern. That number increases each time by 3. The next number in the table will be 12. Once you have figured out the pattern, you can use it to solve the problem by filling in the table until you reach 30 stringed instruments.

ORCHESTRA

Strings	5	10	15	20	25	30
Woodwinds	3	6	9	12	15	18

If 30 stringed instruments are used, the orchestra will need 18 woodwinds.

Copy and complete the table to solve.

1. Day 1's rehearsal for the musical ran for 30 minutes. Day 2's rehearsal was 4 minutes longer. Day 3's rehearsal was 5 minutes longer than the previous day's. Rehearsal on Day 4 was 6 minutes longer than the one on Day 3. If the pattern continues, on which day would the rehearsal run for 60 minutes?

Day	1	2	3						
Rehearsal time									

2. After how many days would the rehearsal last 90 minutes?

Solve. Make a table if needed.

3. The school auditorium has two levels: the main level and the balcony. When tickets for the musical went on sale, 4 balcony tickets were sold for every 3 main-level tickets. How many balcony tickets were sold if 45 tickets for the main level were sold?

4. If 33 tickets for the main level were sold, how many tickets for the balcony were sold?

5. One piece of scenery for the musical is attached to a pulley. When the stage manager uses a crank to turn the wheel of the pulley $1\frac{1}{2}$ times, the scenery is lowered 1 foot. How many times does the manager need to turn the pulley wheel in order to lower the scenery 12 feet?

6. How far is the scenery lowered after $13\frac{1}{2}$ turns?

7. During intermission, refreshments were sold. Jan sold rice cakes with cheese. For every 4 cakes she sold, she gave away 1. How many cakes did she give away if she sold 48?

Time

A. You can read time two ways.

9:45

Read 9:45 as 45 minutes after 9, or 15 minutes before 10.

A.M. is the time between 12:00 midnight and 12:00 noon.

P.M. is the time between 12:00 noon and 12:00 midnight.

B. Marcy is setting up a booth at her school fair. She began at 9:45 A.M. She finished at 1:00 P.M. How much time did Marcy need to set up her booth?

To find **elapsed time,** count the minutes and then the hours.

9:45 A.M. to 10:00 A.M. is 15 minutes.   10:00 A.M. to 1:00 P.M. is 3 hours.

Marcy needs 3 hours 15 minutes to set up her booth.

C. These units are used to measure time.

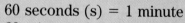

60 seconds (s) = 1 minute	7 days (d) = 1 week
60 minutes (min) = 1 hour	52 weeks (wk) = 1 year (y)
24 hours (h) = 1 day	12 months (mo) = 1 year

365 days = 1 year

To rename larger units with smaller units, you can multiply.

4 wk = ■ d   1 wk = 7 d

4 × 7 = 28

4 wk = 28 d

To rename smaller units with larger units, you can divide.

48 h = ■ d   24 h = 1 d

48 ÷ 24 = 2

48 h = 2 d

Write *A.M.* or *P.M.*

1. The sun rises.　　**2.** School begins.　　**3.** Bedtime.

How much time has elapsed?

4. from  to

5. from  to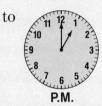

6. from [clock showing 3:20] P.M. to [clock showing 5:45] P.M.

7. from [clock showing 6:30] A.M. to [clock showing 12:07] P.M.

Complete.

8. 4 d = ■ h　　　**9.** 6 min = ■ s　　**10.** 4 wk = ■ d　　**11.** 3 h = ■ min

12. 72 h = ■ d　　**13.** 3 y = ■ d　　**14.** $\frac{1}{2}$ h = ■ min　　**15.** $\frac{1}{5}$ y = ■ d

Solve.

16. One class began to put up a used-toy booth at 8:45 A.M. They finished the booth at 10:15 A.M. How long did it take to build the booth?

17. Roberto began to work in the antique-clothing booth at 10:45 A.M. He worked steadily until 12:00 noon. For how long did Roberto work?

18. Copy and complete the chart.

Student	Begin	End	Time worked
John	2:15 P.M.	4:30 P.M.	■
Marco	10:50 A.M.	■	4 h 25 min
Kate	1:25 P.M.	5:20 P.M.	■
Maggie	■	4:10 P.M.	4 h 15 min

ANOTHER LOOK

Multiply.

1.　203
　　　× 21

2.　304
　　　× 34

3.　752
　　　× 57

4.　197
　　　× 68

5.　850
　　　× 73

6. 48 × 700　　**7.** 62 × 348　　**8.** 86 × 427　　**9.** 87 × 402　　**10.** 58 × 329

Adding and Subtracting Time

A. Basketball practice lasted for 1 h 40 min on Monday, and 2 h 35 min on Tuesday. For how long did the team practice in the two days?

Add 1 h 40 min + 2 h 35 min.

Add the minutes.

$$\begin{array}{r} 1\text{ h }40\text{ min} \\ +\ 2\text{ h }35\text{ min} \\ \hline 75\text{ min} \end{array}$$

Add the hours.
Rename if necessary.

$$\begin{array}{r} 1\text{ h }40\text{ min} \\ +\ 2\text{ h }35\text{ min} \\ \hline 3\text{ h }75\text{ min} \end{array}$$

> **Think:**
> 3 h 75 min
> = 3 h + 1 h + 15 min
> = 4 h 15 min

The team practiced for 4 h 15 min in two days.

B. In two days, the team practiced for 4 h 15 min. John practiced for 2 h 55 min. For how much more time did the rest of the team practice?

Subtract 4 h 15 min − 2 h 55 min.

Subtract the minutes.
Rename if necessary.

$$\begin{array}{r} \overset{3}{\cancel{4}}\text{ h }\overset{75}{\cancel{15}}\text{ min} \\ -\ 2\text{ h }55\text{ min} \\ \hline 20\text{ min} \end{array}$$

Subtract the hours.

$$\begin{array}{r} \overset{3}{\cancel{4}}\text{ h }\overset{75}{\cancel{15}}\text{ min} \\ -\ 2\text{ h }55\text{ min} \\ \hline 1\text{ h }20\text{ min} \end{array}$$

The rest of the team practiced for 1 h 20 min more than John.

Checkpoint Write the letter of the correct answer.

Add or subtract.

1.
$$\begin{array}{r} 5\text{ h }35\text{ min} \\ +\ 4\text{ h }42\text{ min} \end{array}$$

a. 10 h 27 min
b. 9 h 77 min
c. 10 h 7 min
d. 10 h 17 min

2.
$$\begin{array}{r} 4\text{ min }26\text{ s} \\ -\ 3\text{ min }12\text{ s} \end{array}$$

a. 0 min 14 s
b. 1 min 14 s
c. 1 min 38 s
d. 7 min 38 s

3.
$$\begin{array}{r} 3\text{ h }48\text{ min} \\ -\ 2\text{ h }57\text{ min} \end{array}$$

a. 1 min
b. 51 min
c. 1 h 51 min
d. 91 min

Add or subtract.

1. 4 h 16 min
    + 3 h 37 min

2. 2 h 28 min
    + 2 h 13 min

3. 7 min 19 s
    + 4 min 24 s

4. 5 min 32 s
    + 3 min 9 s

5. 1 min 25 s
    + 3 min 40 s

6. 4 h 18 min
    + 1 h 51 min

7. 7 min 52 s
    + 2 min 43 s

8. 2 h 27 min
    + 3 h 44 min

9. 7 h 52 min
    − 4 h 17 min

10. 8 min 42 s
    − 7 min 26 s

11. 6 h 35 min
    − 4 h 27 min

12. 9 min 36 s
    − 2 min 14 s

13. 5 min 13 s
    − 3 min 22 s

14. 7 h 28 min
    − 2 h 34 min

15. 11 h 13 min
    − 9 h 41 min

16. 3 min 2 s
    − 2 min 15 s

★17. 3 h 52 min 37 s
     + 2 h 21 min 43 s

★18. 7 h 19 min 22 s
     − 6 h 13 min 34 s

★19. 8 h 14 min 31 s
     − 5 h 25 min 37 s

Solve.

20. José learned his team's new play in only 1 h 20 min of practice on Tuesday and 1 h 45 min on Thursday. How long did it take José to learn the new play?

21. At 1:25 P.M. on the day of the big game, Steve realizes he forgot his sneakers. He has to be on the court by 3:15 P.M. How much time does he have to get his sneakers?

FOCUS: ESTIMATION

You can round to estimate numbers of hours.

54 h 47 min ≈ ■ h

54 h 47 min ≈ 60 h

Think: less than 30 minutes, round down.

more than 30 minutes, round up.

| 47 > 30 |
| 47 ≈ 1 h |

Round to estimate the number of hours.

1. 3 h 22 min

2. 10 h 49 min

3. 6 h 31 min

4. 9 h 18 min

5. 1 h 11 min

6. 57 min

7. 8 h

8. 4 h 1 min

More Practice, page 438

PROBLEM SOLVING
Using a Schedule

When you plan to take a bus, a train, or a plane from one place to another, you will find that a schedule is helpful. A bus schedule, for example, shows you where and when each bus stops along its route.

The regional softball play-off will take place at Allenwood School. Many students from neighboring towns will go to the game by bus.

Here is the information you should look for on a schedule.
- The heading tells you the endpoints of the bus route.
- The column headings show the towns and the streets where each bus stops.
- Some towns have two bus stops located at different streets.
- A dash (——) means that the bus does not stop.
- Since all the times shown are in the morning, A.M. appears under the headings for each column.

Use the schedule to answer this question. Gregory plans to take the bus that leaves from River Road in Newton at 7:50 A.M. At what time will he arrive in Allenwood?

Look at the schedule. Locate the entry that shows 7:50 A.M. under the Newton–River Road column. Follow the entry across to the Allenwood–Western Avenue column. The bus that leaves River Road at 7:50 A.M. will arrive in Allenwood at 8:48 A.M.

Here is the bus schedule they will use.

NEWTON TO ALLENWOOD							
Newton River Road	**Newton Broad Street**	**Mayville Elm Street**	**Davis Gold Avenue**	**Hudson Pine Street**	**Gladstone Wyoming Avenue**	**Gladstone Main Street**	**Allenwood Western Avenue**
A.M.	A.M.	A.M.	A.M.	A.M.	A.M.	A.M.	A.M.
5:50	6:05	6:11	6:17	6:24	6:29	6:33	6:43
——	——	6:21	6:27	6:34	6:39	6:43	6:53
6:05	6:20	6:26	——	6:37	——	6:44	6:54
7:00	7:16	7:23	7:30	7:37	7:42	7:47	7:57
7:10	7:26	7:33	7:40	7:47	7:52	7:57	8:07
7:25	——	——	——	7:52	——	——	8:12
——	——	7:53	8:00	8:07	8:12	8:17	8:27
7:40	7:56	8:03	8:10	8:17	8:22	8:27	8:37
7:50	8:07	8:14	8:20	8:27	8:33	8:38	8:48
——	——	8:24	8:30	8:37	8:43	8:48	8:58
8:10	8:27	8:34	8:40	8:47	8:53	8:58	9:08

Use the bus schedule to verify each statement. Write *true* or *false*.

1. The 7:50 A.M. bus from River Road arrives in Allenwood before the 8:14 A.M. bus from Mayville.

2. The 7:40 A.M. bus from River Road arrives in Allenwood 10 minutes earlier than the 7:50 A.M. bus.

3. It takes less time to ride from Davis to Hudson than from Mayville to Hudson.

4. The earliest bus to leave Mayville arrives in Gladstone before 6:30 A.M.

Use the bus schedule to solve each problem.

5. Jean and Mike want to arrive in Allenwood at 8:27 A.M. Jean lives in Hudson. Mike lives in Davis. They want to meet on the bus. At what time should each of them board the bus?

6. Maria lives near Broad Street in Newton. She wants to arrive in Allenwood shortly before 9:00 A.M. Which bus should she take?

7. Perry wants to go to Main Street in Gladstone. He boards the 7:33 A.M. bus in Mayville. If he rode for 19 minutes, would he be at his destination?

8. Stella boarded the 8:00 A.M. bus in Davis. She left the bus at Wyoming Avenue in Gladstone. Assuming the bus arrived on time, for how long did she ride?

9. The 6:20 A.M. bus from Broad Street in Newton is delayed 4 minutes. At what time will it arrive at Main Street in Gladstone?

10. How much longer does the trip from River Road in Newton to Allenwood take on the 7:50 A.M. bus than on the 5:50 A.M. bus? Why do you think so?

★11. Rick boards a bus at Broad Street in Newton at 7:26 A.M. Clara boards a bus in Hudson at 7:37 A.M. Who arrives in Allenwood first? How many minutes later does the second person arrive?

12. Paul lives in Davis. He wants to be in Allenwood 45 minutes before tickets for the big game go on sale. Tickets go on sale at 9:00 A.M. It will take about 5 minutes to walk from the Allenwood bus stop to the ticket booth. Which bus should Paul take?

Length: Inches

A. The members of the Carpenter's Club are building birdhouses. They need nails that are at least $1\frac{1}{2}$ inches long. Is this nail long enough?

1 inch

An **inch (in.)** is a customary unit of length. It can be used to measure small objects such as a paper clip or a safety pin.

The nail measures

2 inch to the nearest inch.

$1\frac{1}{2}$ in. to the nearest $\frac{1}{2}$ in.

$1\frac{2}{4}$ in. to the nearest $\frac{1}{4}$ in.

$1\frac{5}{8}$ in. to the nearest $\frac{1}{8}$ in.

$1\frac{9}{16}$ in. to the nearest $\frac{1}{16}$ in.

The nail measures $1\frac{9}{16}$ in. long. It is long enough.

B. You can find the distance around an object by measuring its sides and finding the sum of its measurements. Measure the distance around this shape.

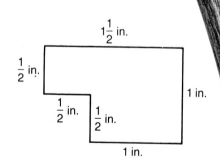

$1\frac{1}{2}$ in.

$\frac{1}{2}$ in.

$\frac{1}{2}$ in.

$\frac{1}{2}$ in.

1 in.

1 in.

$1\frac{1}{2} + \frac{1}{2} + \frac{1}{2} + \frac{1}{2} + 1 + 1 = 5$

The distance is 5 inches.

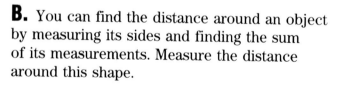

Measure this pencil to the nearest

1. inch. **2.** $\frac{1}{2}$ in. **3.** $\frac{1}{4}$ in. **4.** $\frac{1}{8}$ in. **5.** $\frac{1}{16}$ in.

Use a ruler to draw a line that is

6. $\frac{3}{4}$ in. **7.** $2\frac{3}{8}$ in. **8.** $3\frac{5}{16}$ in. **9.** $1\frac{7}{8}$ in.

Measure the distance around each.

10. **11.** **12.**

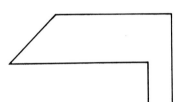

Solve.

13. Joe's birdhouse will be shaped like a triangle. A perch will go around all three sides. How many inches around will the perch be?

14. Ann uses two screwdrivers. One is $7\frac{3}{4}$ in. long. The other is $1\frac{1}{2}$ in. shorter. How long is the second screwdriver?

15. Ronnie is using nails of three different sizes. Each size is $\frac{1}{4}$ in. shorter than the next. The shortest nail is $1\frac{5}{8}$ in. What are the lengths of the other two nails?

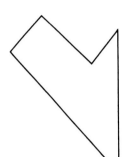

14 in.

14 in.

$10\frac{1}{2}$ in.

CHALLENGE

How long are these items? First estimate to the nearest $\frac{1}{2}$ in. Then measure. Copy and complete the table.

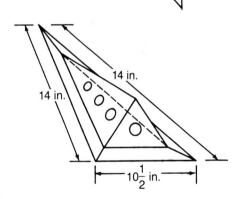

Object	this book	your thumb	your hand	your foot	a pencil
Estimate	▦	▦	▦	▦	▦
Actual length	▦	▦	▦	▦	▦

Feet, Yards, and Miles

If you have ever attended a track meet, you may have seen a 50-yard dash or a 6-mile marathon. Distances in track meets are often measured in yards and miles.

Here are several customary units of length that you should know.

$$12 \text{ inches (in.)} = 1 \text{ foot (ft)}$$
$$3 \text{ ft} = 1 \text{ yard (yd)}$$
$$1{,}760 \text{ yd} = 1 \text{ mile (mi)}$$
$$5{,}280 \text{ ft} = 1 \text{ mi}$$

A **foot** can be used to measure objects. An egg carton is about a foot.

A **yard** can be used to measure objects. A baseball bat is about a yard.

A **mile** can be used to measure length. The distances between cities are measured in miles.

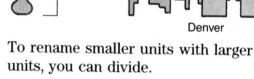

Portland

1 ft

1 yd

miles

Denver

To rename larger units with smaller units, you can multiply.

4 ft = ▨ in.
$\boxed{1 \text{ ft} = 12 \text{ in.}}$
4 × 12 = 48
4 ft = 48 in.

To rename smaller units with larger units, you can divide.

78 ft = ▨ yd
$\boxed{3 \text{ ft} = 1 \text{ yd}}$
78 ÷ 3 = 26
78 ft = 26 yd

Other examples:

5 ft 7 in. = ▨ in.
$\boxed{5 \times 12 = 60}$
60 in. + 7 in. = 67 in.
5 ft 7 in. = 67 in.

2 mi = ▨ yd
$\boxed{1{,}760 \text{ yd} = 1 \text{ mi}}$
2 × 1,760 = 3,520
2 mi = 3,520 yd

Checkpoint Write the letter of the correct answer.

Complete.

1. 3 ft = ▨ in.

a. $\frac{1}{4}$
b. 24
c. 32
d. 36

2. 54 ft = ▨ yd

a. $4\frac{1}{2}$
b. 18
c. 19
d. 648

3. 3 mi = ▨ yd

a. 880
b. 1,760
c. 5,280
d. 5,290

Write *inches*, *feet*, or *yards*.

1. A desk may be 28 ▦ high.

2. An adult may be $5\frac{3}{4}$ ▦ tall.

3. A doorway may be 30 ▦ wide.

4. A person's arm may be 2 ▦ long.

Choose the appropriate unit of measure. Write *yd* or *mi*.

5. the distance a car travels in an hour

6. the distance you throw a ball

7. the length of a football field

8. the length of a piece of fabric

9. the distance to the sun

10. the length of shadows

11. the distance you can hit a golf ball

12. the distance you can walk in a day

Complete.

13. 5 ft = ▦ in.

14. 36 in. = ▦ ft

15. 120 in. = ▦ ft

16. 2 ft = ▦ in.

17. 5 ft 3 in. = ▦ in.

18. $1\frac{1}{2}$ ft = ▦ in.

19. 4 ft 7 in. = ▦ in.

20. 4 mi = ▦ yd

21. 5 yd = ▦ ft

22. 18 ft = ▦ yd

23. 1,760 yd = ▦ mi

24. 5,280 yd = ▦ mi

25. 440 yd = ▦ mi

26. $2\frac{1}{3}$ yd = ▦ in.

27. 2,640 ft = ▦ in.

Solve.

The results of Friday's track meet were posted on Monday morning. Copy and complete the chart to answer the questions.

28. Elva thought she ran the same distance as Frank. Was she correct?

29. Compare Bob's distance with Sam's and Lisa's. List them in order from who ran the greatest distance to who ran the least distance.

30. Sylvia ran this course around the school grounds. Find her total distance, and list it on the chart. Which runners ran farther than Sylvia? How many yards farther did they run?

TOTAL DISTANCE RUN

Runner	In feet	In yards	In miles
Elva	▦	1,320	▦
Sam	▦	880	▦
Frank	▦	▦	$\frac{3}{4}$
Lisa	▦	▦	$\frac{5}{8}$
Bob	▦	352	▦
Sylvia	▦	▦	▦

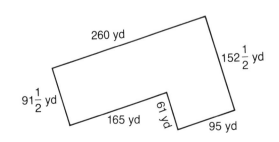

260 yd

$152\frac{1}{2}$ yd

$91\frac{1}{2}$ yd

165 yd

61 yd

95 yd

Cups, Pints, Quarts, and Gallons

Cups (c), pints (pt), quarts (qt), and gallons (gal) are customary units of volume.

2 c = 1 pt
2 pt = 1 qt
4 qt = 1 gal

1 cup (c)

1 pint (pt)

1 quart (qt)

1 gallon (gal)

To rename larger units with smaller units, you can multiply.

3 pt = ▧ c $\boxed{1 \text{ pt} = 2 \text{ c}}$
2 × 3 = 6
3 pt = 6 c

To rename smaller units with larger units, you can divide.

14 c = ▧ pt $\boxed{2 \text{ c} = 1 \text{ pt}}$
14 ÷ 2 = 7
14 c = 7 pt

Write *cups, pints, quarts,* or *gallons.*

1. A full tank of gas may hold 15 ▧.

2. A juice glass may hold 2 ▧.

3. A swimming pool's capacity may be measured in ▧.

Complete.

4. 2 gal = ▧ qt   **5.** 3 qt = ▧ pt   **6.** 1 pt = ▧ c   **7.** 3 gal = ▧ qt

8. 4 c = ▧ pt   **9.** 8 qt = ▧ gal   **10.** 6 c = ▧ pt   **11.** 4 pt = ▧ qt

12. $2\frac{1}{4}$ gal = ▧ qt   **13.** 3 c = ▧ pt   **★14.** 3 gal 2 qt = ▧ qt

Solve.

15. Maria bakes 14 loaves of bread for the picnic. She needs 1 pt milk for each loaf. How many quarts of milk does she need?

16. Ralph needs $1\frac{1}{2}$ gal water to cook vegetables for the picnic. He has a 2-c measuring cup. How many times must he fill the cup?

17. Use the recipe to write and solve a word problem.

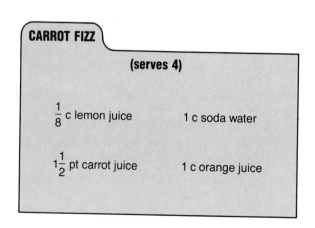

CARROT FIZZ

(serves 4)

$\frac{1}{8}$ c lemon juice   1 c soda water

$1\frac{1}{2}$ pt carrot juice   1 c orange juice

Ounces, Pounds, and Tons

Ounces (oz), pounds (lb), and **tons (T)** are customary units of weight.

16 oz = 1 lb
2,000 lb = 1 T

1 ounce (oz)   **1 pound (lb)**   **1 ton (T)**

To rename larger units with smaller units, you can multiply.

2 T = ▨ lb $\boxed{1\ T = 2{,}000\ lb}$
2 × 2,000 = 4,000
2 T = 4,000 lb

To rename smaller units with larger units, you can divide.

192 oz = ▨ lb $\boxed{16\ oz = 1\ lb}$
192 ÷ 16 = 12
192 oz = 12 lb

Write *ounces, pounds,* or *tons.*

1. An elephant may weigh 4 ▨.

2. A glass of milk may weigh 10 ▨.

3. A greeting card may weigh 1 ▨.

4. A full suitcase may weigh 25 ▨.

Complete.

5. 2 lb = ▨ oz

6. 4 lb = ▨ oz

7. 10 T = ▨ lb

8. 5 lb = ▨ oz

9. 112 oz = ▨ lb

10. 160 oz = ▨ lb

11. 10,000 lb = ▨ T

12. 500 lb = ▨ T

13. $1\frac{1}{4}$ lb = ▨ oz

14. $2\frac{3}{4}$ lb = ▨ oz

15. 224 oz = ▨ lb

16. 80 oz = ▨ lb

★**17.** 2 lb 6 oz = ▨ oz

★**18.** 5 lb 4 oz = ▨ oz

★**19.** 2 T 4 lb = ▨ lb

Solve.

20. The Green Thumb Club buys a 1-lb bag of seeds at Jack's Seed Store. The bag contains 8 packets. How many ounces does each packet weigh?

21. The club buys 35 oz of bluebell seeds for $8.10, and Jack adds a bonus of 10 oz. Including the bonus, how much did the club pay for each ounce of seed?

Fahrenheit Temperature

A. Temperature can be measured by using **degrees Fahrenheit (°F).** Look at the thermometer. The temperature is 30 degrees above zero, or 30°F. If the thermometer reads 30 degrees below zero, you write ⁻30°F.

B. If the temperature falls from 30°F to ⁻30°F, by how many degrees has the temperature changed?

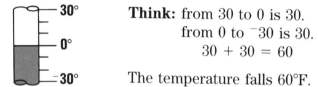

Think: from 30 to 0 is 30.
from 0 to ⁻30 is 30.
$30 + 30 = 60$

The temperature falls 60°F.

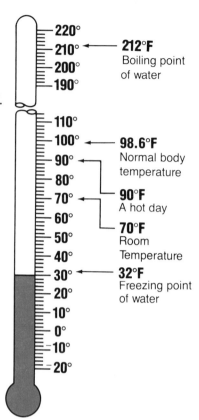

212°F	Boiling point of water
98.6°F	Normal body temperature
90°F	A hot day
70°F	Room Temperature
32°F	Freezing point of water

Write the temperature.

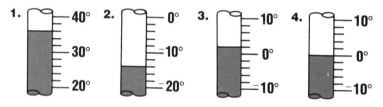

1. 40° 30° 20° 2. 0° ⁻10° ⁻20° 3. 10° 0° ⁻10° 4. 10° 0° ⁻10°

Copy and complete the chart.

	5.	6.	7.	8.
Starting temperature	0°F	⁻10°F	41°F	
Temperature change	rose 15°			rose 7°
Final temperature		3°F	⁻11°F	⁻9°F

Choose the more suitable temperature for

9. a swim in a lake.   104°F or 45°F

10. a warm shower.   70°F or 115°F

Solve. For Problem 12, use the Infobank.

11. Mona records a temperature of 3°F at 8 A.M. in Springfield. Her noon recording is 18°F. By how many degrees has the temperature risen?

12. Use the information on page 419 to write and solve your own word problem.

Celsius Temperature

A. Temperature can also be measured by using **degrees Celsius (°C).** Look at the thermometer. It shows a normal room temperature of 20° above zero or 20°C. Below-zero Celsius temperatures can also be recorded. If the thermometer reads 20 degrees below zero, you write ⁻20°C.

B. If the temperature rises from ⁻20°C to 20°C, by how many degrees has the temperature changed?

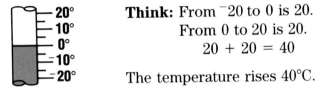

Think: From ⁻20 to 0 is 20.
From 0 to 20 is 20.
20 + 20 = 40

The temperature rises 40°C.

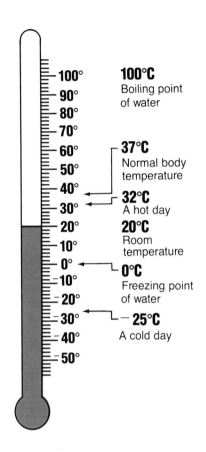

100°	**100°C** Boiling point of water
90°	
80°	
70°	
60°	**37°C** Normal body temperature
50°	
40°	**32°C** A hot day
30°	
20°	**20°C** Room temperature
10°	
0°	**0°C** Freezing point of water
⁻10°	
⁻20°	
⁻30°	**⁻25°C** A cold day
⁻40°	
⁻50°	

Write the temperature.

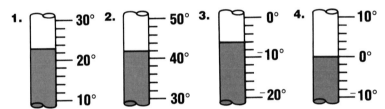

1. 30° 20° 10° **2.** 50° 40° 30° **3.** 0° ⁻10° ⁻20° **4.** 10° 0° ⁻10°

Copy and complete the chart.

	5.	6.	7.	8.
Starting temperature	12°C	⁻2°C	51°C	
Temperature change	fell 12°	rose 10°		fell 21°
Final temperature			39°C	⁻5°C

Choose the more suitable temperature for

9. building a snowman. ⁻2°C 35°C

10. a bowl of chicken soup. 49°C 77°C

Solve.

11. Dan uses an almanac to find world temperatures. The page is smudged, and he cannot tell whether the temperature at the North Pole is 40°C or ⁻40°C. Which is the more likely temperature?

297

PROBLEM SOLVING
Using a Time-Zone Map

To find the difference in time between two places, you can use a time-zone map.

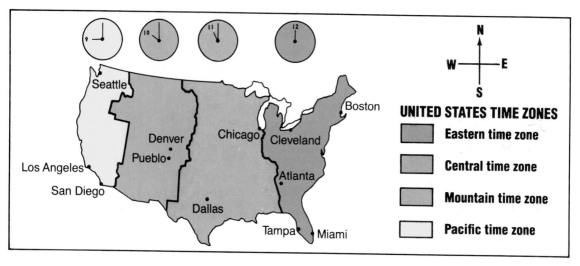

Chris lives in Cleveland. She wants to telephone her friend Pablo in San Diego at 10:00 A.M., Pacific time. At what time should she place the call?

Use the information on the time-zone map to help you solve this problem. This is how it is arranged.

- The headings tell you the names of the time zones.

- The colors and the heavy black lines show the areas of the time zones.

- The clocks show what time it is in each time zone when it is 12:00 noon in the Eastern time zone.

To find out when Chris should place her call:

- Locate the two places. Cleveland is in the Eastern time zone. San Diego is in the Pacific time zone.

- Look at the clocks. The time in San Diego is 3 hours earlier than it is in Cleveland. So, Eastern time is 3 hours later than Pacific time.

Chris should place her call at 1:00 P.M., Eastern time.

Write *true* or *false* for each statement.

1. Dallas and Tampa are in the same time zone.

2. It is 1 hour earlier in Boston than it is in Denver.

3. The Central time zone and the Pacific time zone are 2 hours apart.

Solve.

4. Debbie lives in Atlanta. Sue lives in Seattle. When Debbie wakes up at 7:30 A.M., what time is it where Sue lives?

5. Sue goes to bed in Seattle at 11:00 P.M. What time is it where Debbie lives?

6. At noon, Paul telephones his brother in Atlanta. Paul lives in Tampa. At what time does Paul's brother receive the call?

7. Gina drives from Cleveland to Chicago. Should she set her watch forward or backward? by how many hours?

8. Dan lives in Los Angeles. He left home at 10:00 A.M. He returned $11\frac{1}{2}$ hours later. At what time did he return home?

9. At 8:00 A.M., Dan's father left Denver to drive to Pueblo. He returned to Denver at 6:15 P.M. For how long had he been gone?

10. Kim lives on the West Coast, and Jan lives on the East Coast. Each begins dinner at 6:30 P.M. Both finish 1 hour later. At what time do both of them finish dinner?

11. It is 4:25 P.M. in Dallas. In another city, it is 1 hour later. In which time zone is that other city?

12. The National Guitar Company has offices in San Diego, Denver, Chicago, and Miami. All four offices are open daily from 8:30 A.M. to 5:00 P.M. At 3:30 P.M., Mountain time, which of the company's offices are open and which are closed?

★13. Suppose you fly from San Diego, California, to Boston, Massachusetts. Your plane takes off at 3:30 P.M., Pacific time. It lands at 11:40 P.M., Eastern time. How long did the flight take?

CALCULATOR

You can use a calculator to help you solve problems that involve many computations or computations with large numbers.

Units of Time
60 seconds = 1 minute
60 minutes = 1 hour
24 hours = 1 day
7 days = 1 week
12 months = 1 year
365 days = 1 year

Solve. Use the table if needed.

1. If your heart beats 75 times a minute, how many times does your heart beat in an hour?

2. How many times does your heart beat in a day?

3. How many times does your heart beat in a week?

4. How many times does your heart beat in a year?

5. Time your own heartbeat for 1 minute. Then answer problems 1–4 using that number.

6. If you started to count one number each second 24 hours each day, how long would it take you to count to 1 million? Write your answer as __ days, __ hours, __ minutes, __ seconds.

7. Could you count to 1 billion in your lifetime?

8. About how many years would it take?

9. Choose which will give the greatest amount.

   a. You are given 1 penny per second for 1 year.
   b. You are given $1 per minute for 3 weeks.
   c. You are given $2 a day for 20 years.

GROUP PROJECT

Having a Field Day

The problem: You are in charge of organizing the sports events for you school's field day. Decide how to schedule the events. Make a copy of the chart to help you.

Key Facts

- There are 100 students participating.
- You now have 8 adult volunteers to help organize and referee the events.
- You must have enough events to keep everyone involved for 4 hours.
- The athletic field is 100 yards long and 50 yards wide (the size of a football field).
- You want to award three ribbons for each event.

Key Questions

- What events should you include?
- Do you have enough space for more than one event per hour?
- How many people should participate in each event?
- How will students sign up for the events?
- Do you have enough adult help?
- Who will judge the events?
- Will you award ribbons after each event or at the end of the day?
- Will you invite parents to observe?

	Events	Players	Referee
First Hour			
Second Hour			
Third Hour			
Fourth Hour			

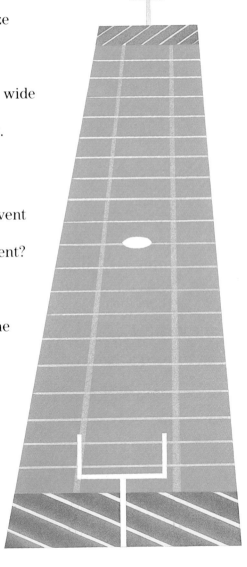

CHAPTER TEST

Multiply. Write the product in simplest form. (pages 272, 274, and 276)

1. $\frac{1}{9} \times \frac{1}{6}$

2. $\frac{2}{3} \times \frac{5}{8}$

3. $\frac{4}{9} \times \frac{7}{8}$

4. $5 \times \frac{2}{3}$

5. $\frac{9}{10} \times 5$

6. $13 \times \frac{4}{7}$

7. $6\frac{1}{8} \times 9\frac{7}{10}$

8. $12\frac{5}{6} \times \frac{7}{9}$

Divide. (page 280)

9. $9 \div \frac{1}{8}$

10. $12 \div \frac{1}{6}$

11. $2 \div \frac{1}{4}$

12. $6 \div \frac{1}{3}$

How much time has passed? (page 284)

Complete. (page 286)

13. from to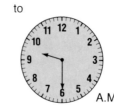

14. $3 \text{ d} = \blacksquare \text{ h}$

15. $8 \text{ min} = \blacksquare \text{ s}$

Add or subtract. (page 286)

16.   5 h 26 min
      + 3 h 45 min

17.   7 h 11 min
      − 5 h 25 min

Measure the line to the nearest (page 290) ————————

18. inch

19. $\frac{1}{4}$ inch

20. $\frac{1}{8}$ inch

21. $\frac{1}{2}$ inch

Choose the appropriate unit. (page 294)
Write *in., ft, yd,* or *mi.*

22. the length of a highway

23. the height of a volleyball net

24. the length of a key

Choose the appropriate unit. (pages 292, 294, and 296)
Write *c, pt, qt,* or *gal.*

25. a bath tub

26. a milk bottle

27. a tea pot

Write *ounces, pounds,* or *tons.*

28. a kitten

29. a blue whale

30. a gorilla

Complete. (page 294)

31. $3 \text{ gal} = \blacksquare \text{ qt} = \blacksquare \text{ pt} = \blacksquare \text{ c}$

32. $64 \text{ oz} = \blacksquare \text{ lb}$

33. $2 \text{ mi} = \blacksquare \text{ yd} = \blacksquare \text{ ft} = \blacksquare \text{ in.}$

Copy and complete the chart. (pages 296 and 297)

	34.	35.	36.
Starting temperature	▓	217°C	17°C
Temperature change	rose 13°	▓	fell 25°
Final temperature	19°F	185°C	▓

Solve. Make a table if needed. (page 283)

37. The principal in Judy's school hands out a new-activities list to all the students in the school. There are only 4 copies for every 9 students. If there are 486 students in the school, how many copies does he hand out?

38. One day, all the students in Jack's school went to the auditorium for a presentation. There were 27 students for every 3 teachers in the auditorium. If there were 351 students in the auditorium, how many teachers were there?

Use the time zone map to solve the problems. (page 299)

39. If it is 9:00 P.M. in New York, what time is it in San Francisco?

40. What time is it in Chicago if it is 7:00 A.M. in Denver?

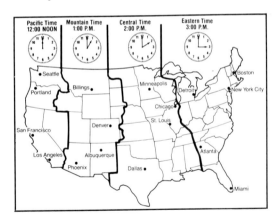

BONUS

Multiply the fractions in each row. Write the answers in simplest form.

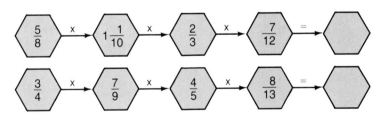

RETEACHING

A. Before you can multiply fractions and mixed numbers, you need to rename the mixed number as a fraction.

When renaming mixed numbers, remember these steps.

1. Multiply the whole numbers by the denominator.
$2\frac{3}{4} \rightarrow 4 \times 2 = 8$

2. Add the product to the numerator. $8 + 3 = 11$

3. Write the sum over the denominator. $2\frac{3}{4} \rightarrow \frac{11}{4}$

Find $3\frac{7}{9} \times \frac{1}{3}$.

Rename the mixed number.	Multiply the fractions.	Show the answer as a mixed number.
$3\frac{7}{9} = \frac{34}{9}$	$\frac{34}{9} \times \frac{1}{3} = \frac{34}{27}$	$\frac{34}{27} = 1\frac{7}{27}$

B. Sometimes you need to write fractions for both factors before multiplying.

Find $4 \times 6\frac{1}{9}$. $\frac{4}{1} \times \frac{55}{9} = \frac{220}{9} = 24\frac{4}{9}$

Another example:

Find $7\frac{5}{6} \times 3\frac{3}{4}$. $\frac{47}{6} \times \frac{15}{4} = \frac{705}{24} = 29\frac{3}{8}$

Multiply. Write the answer in simplest form.

1. $3\frac{1}{6} \times \frac{2}{3}$

2. $10\frac{1}{2} \times \frac{1}{7}$

3. $\frac{1}{2} \times 2\frac{8}{9}$

4. $\frac{1}{10} \times 3\frac{3}{4}$

5. $\frac{6}{7} \times 7\frac{1}{6}$

6. $5 \times 3\frac{2}{5}$

7. $2\frac{4}{21} \times 9$

8. $1\frac{5}{7} \times \frac{1}{2}$

9. $7\frac{7}{9} \times 2\frac{3}{5}$

10. $2\frac{3}{4} \times 2\frac{5}{6}$

11. $2\frac{7}{10} \times 8\frac{7}{9}$

12. $6\frac{2}{5} \times 7\frac{3}{8}$

13. $8\frac{5}{12} \times \frac{3}{5}$

14. $6\frac{4}{7} \times 1\frac{1}{4}$

15. $4\frac{2}{3} \times 2\frac{3}{8}$

16. $3 \times 2\frac{3}{4}$

17. $\frac{3}{5} \times 9\frac{5}{6}$

18. $1\frac{9}{10} \times 7$

19. $2\frac{1}{2} \times 10\frac{8}{9}$

20. $2\frac{1}{3} \times \frac{4}{9}$

21. $4 \times \frac{6}{7}$

22. $7\frac{1}{2} \times \frac{5}{6}$

23. $5\frac{11}{15} \times 6\frac{2}{3}$

24. $5\frac{1}{6} \times 5\frac{1}{7}$

ENRICHMENT

Reading a Road Map

A **road map** is like a satellite picture of the major streets and highways in an area. It shows the routes and distances between cities and towns. The *key* tells you what each symbol means.

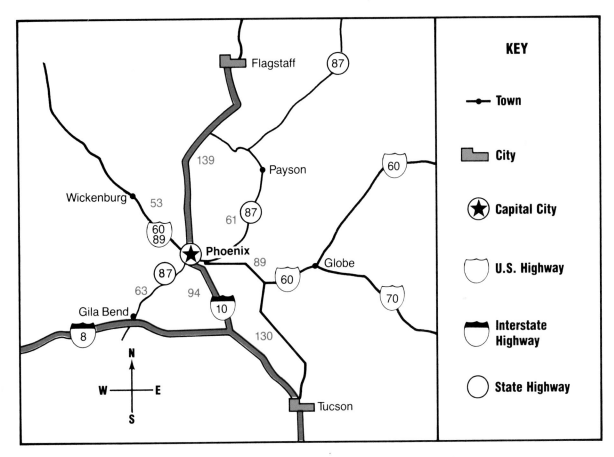

Note: The red numbers along the highways show the distances in miles between cities and towns.

Find the mileage from

1. Phoenix to Flagstaff.   **2.** Phoenix to Globe.   **3.** Gila Bend to Payson.

Find the fastest route between

4. Tucson and Phoenix.   **5.** Phoenix and Gila Bend.   **6.** Globe and Wickenburg.

7. Tucson and Payson.   **8.** Gila Bend and Globe.   **9.** Flagstaff and Tucson.

CUMULATIVE REVIEW

Write the letter of the correct answer.

1. A fraction for the part that is shaded.

    a. $\frac{4}{9}$ **b.** $\frac{5}{9}$

    c. $\frac{4}{5}$ **d.** not given

2. Write in order from the greatest to the least: $\frac{19}{25}, \frac{5}{20}, \frac{7}{10}, \frac{4}{5}$.

    a. $\frac{7}{10}, \frac{4}{5}, \frac{5}{20}, \frac{19}{25}$ **b.** $\frac{4}{5}, \frac{5}{20}, \frac{7}{10}, \frac{19}{25}$

    c. $\frac{4}{5}, \frac{19}{25}, \frac{7}{10}, \frac{5}{20}$ **d.** not given

3. Write the answer in simplest form: $1\frac{5}{8} + 3\frac{3}{6}$.

    a. $4\frac{1}{8}$ **b.** $5\frac{1}{8}$

    c. $5\frac{3}{8}$ **d.** not given

4. Write the answer in simplest form: $\frac{2}{5} - \frac{3}{10}$.

    a. $\frac{1}{10}$ **b.** $\frac{2}{5}$

    c. $\frac{1}{2}$ **d.** not given

5. Write the answer in simplest form: $4 - 3\frac{2}{5}$.

    a. $\frac{3}{5}$ **b.** $1\frac{3}{5}$

    c. $2\frac{2}{5}$ **d.** not given

6. Write the answer in simplest form: $5\frac{1}{6} - 2\frac{3}{4}$.

    a. $2\frac{5}{12}$ **b.** $3\frac{5}{12}$

    c. $3\frac{11}{12}$ **d.** not given

7. $1.4 \text{ kg} = \blacksquare \text{ mg}$

    a. 0.0014 mg   **b.** 1,400 mg

    c. 140 mg   **d.** not given

8. $0.26 + 3.997 + 0.05$

    a. 4.2475   **b.** 4.317

    c. 3.275   **d.** not given

9. 3.67×0.9

    a. 0.313   **b.** 3.303

    c. 4.303   **d.** not given

10. $28,496 \div 9$

    a. 3,166 R2   **b.** 3,165

    c. 3,174 R4   **d.** not given

11. There are 1,375 people on the train to Tulsa. If each car on the train holds 65 people, how many cars are there?

    a. 21   **b.** 22

    c. 23   **d.** not given

12. The dining car on the train has 3 sittings for each meal. They serve 135 people at each breakfast sitting. A total of 150 more people are served at lunch. How many people are served at lunch?

    a. 345   **b.** 455

    c. 555   **d.** not given

Your class is planning a sight-seeing trip to the state capital. How will you travel there? How will you spend your time once you arrive? What will your costs be? Find a way to show the percent of the total cost that will be spent on each item.

10 RATIO AND PERCENT

Ratios

A. Every day at rush hour, there is a traffic jam at the Grove Street intersection in Rockville City. The city planners made a study of the area. In 1 hour, 15 cars and 4 buses stopped traffic. They compared the number of cars to the number of buses and decided to create a separate bus lane. What is the ratio of cars to buses?

We use **ratios** to compare two numbers.

The ratio of cars to buses is 15 to 4.

B. When you compare numbers in a ratio, be careful to write them in the correct order. You can write ratios in several ways.

Compare the number of buses to the number of cars.

$$4 \text{ to } 15 \quad 4 \text{ out of } 15 \quad 4:15 \quad \frac{4}{15}$$

Compare the number of buses to the total number of vehicles.

$$4 \text{ to } 19 \quad 4 \text{ out of } 19 \quad 4:19 \quad \frac{4}{19}$$

Compare the total number of vehicles to the number of cars.

$$19 \text{ to } 15 \quad 19:15 \quad \frac{19}{15}$$

Checkpoint Write the letter of the correct answer.

Choose the ratio that matches.

1. 8:11

a. 11:19
b. 8 to 11
c. $\frac{11}{8}$
d. 8:3

2. 9 out of 10

a. 1:9
b. $\frac{10}{9}$
c. 9:10
d. 10:9

3. $\frac{22}{15}$

a. 15 to 22
b. 22:15
c. 15 to 7
d. 22:7

Write the ratio.

1. bicycles to tricycles

2. unicycles to bicycles

3. green cycles to yellow cycles

4. tricycles to unicycles

5. bicycles to all cycles

6. all cycles to green cycles

7. buses to cars

8. trucks to buses

9. red vehicles to blue vehicles

10. cars to buses

11. cars to all vehicles

12. all vehicles to red vehicles

Write each ratio as a fraction.

13. 1 out of 2

14. 7 to 8

15. 4:5

16. 100 to 350

17. 19 out of 20

18. 60:70

19. 21 out of 28

20. 3 to 4

Solve.

21. A study shows that, in one city, there are 5 bus riders to every 7 train riders. Write this ratio three ways.

22. City planners set aside 1 bicycle lane for every 6 automobile lanes. What is the ratio of bicycle lanes to automobile lanes?

CHALLENGE

Take a survey of how your classmates travel to school each morning. Copy the table, and record your results. Write a ratio for

1. the number of students who ride bicycles to the number who walk.

2. the number who ride the school bus to the number who travel by car.

MEANS OF TRANSPORTATION TO SCHOOL

Bus	
Car	
Walk	
Bicycle	

Equal Ratios

A. Mr. Chen wants to save fuel and to lessen traffic by starting a car pool at his company. He knows that he needs 1 car for every 5 people. How many cars will he need if 20 people join the car pool?

You can use **equal ratios** to find how many cars are needed. Two ratios are equal if they can be written as equivalent fractions.

Write the ratios comparing the number of cars to the number of people.

$$\frac{1}{5} \xleftarrow{\text{cars}} \xrightarrow{} \frac{n}{20}$$

Find equivalent fractions.

| Think: $5 \times 4 = 20$. | $\frac{1}{5} = \frac{1 \times 4}{5 \times 4} = \frac{4}{20}$ |

Mr. Chen needs 4 cars for 20 people.

B. Mr. Chen makes a table to keep track of how many cars and people there are in the car pool. How many people can ride if he has 6 cars?

Complete.

Cars	1	2	3	4	5	6
People	5	10	15	20	25	▪

$\frac{1}{5} = \frac{6}{n}$ Think: $1 \times 6 = 6.$ $\frac{1}{5} = \frac{1 \times 6}{5 \times 6} = \frac{6}{30}$

So, 30 people can ride in the car pool.

C. You can tell whether two ratios are equal by checking to see if they are equivalent fractions.

Are $\frac{3}{5}$ and $\frac{18}{35}$ equal ratios?

| Think: $3 \times 6 = 18$. | $\frac{3}{5} = \frac{3 \times 6}{5 \times 6} = \frac{18}{30}$ |

$$\frac{18}{30} \neq \frac{18}{35}$$

So, $\frac{3}{5} \neq \frac{18}{35}$.

Find the two equal ratios.

1. $\frac{1}{3} = \frac{2}{6} = \frac{}{} = \frac{}{}$

2. $\frac{2}{5} = \frac{4}{10} = \frac{}{} = \frac{}{}$

3. $\frac{3}{8} = \frac{6}{16} = \frac{}{} = \frac{}{}$

4. $\frac{2}{9} = \frac{4}{18} = \frac{}{} = \frac{}{}$

5. $\frac{3}{2} = \frac{9}{6} = \frac{}{} = \frac{}{}$

6. $\frac{5}{3} = \frac{10}{6} = \frac{}{} = \frac{}{}$

Find the missing number.

7. $\frac{3}{7} = \frac{15}{n}$

8. $\frac{6}{9} = \frac{48}{n}$

9. $\frac{5}{8} = \frac{n}{56}$

10. $\frac{11}{5} = \frac{n}{35}$

11. $\frac{3}{27} = \frac{n}{9}$

12. $\frac{12}{5} = \frac{60}{n}$

13. $\frac{10}{13} = \frac{n}{39}$

14. $\frac{8}{5} = \frac{n}{35}$

Copy and complete each ratio table.

15.

Number of cars	2	4	▨	8
Number of headlights	4	▨	12	▨

16.

Number of cars	▨	5	▨	10	12	▨
Number of doors	12	▨	32	40	▨	60

Write = or ≠ for ●.

17. $\frac{4}{50} \; ● \; \frac{20}{250}$

18. $\frac{3}{12} \; ● \; \frac{21}{86}$

19. $\frac{9}{15} \; ● \; \frac{56}{90}$

20. $\frac{18}{3} \; ● \; \frac{72}{12}$

21. $\frac{3}{11} \; ● \; \frac{18}{65}$

22. $\frac{8}{9} \; ● \; \frac{32}{36}$

23. $\frac{9}{13} \; ● \; \frac{36}{56}$

24. $\frac{5}{12} \; ● \; \frac{50}{122}$

Solve. For Problem 26, use the Infobank.

25. Mr. Chen is driving from New York to Chicago on business. He drives at a rate of 94 miles every 2 hours. How far does he drive in 4 hours?

26. Use the information on page 419 to solve. How far will the intercity bus travel in 2 hours?

ANOTHER LOOK

Multiply.

1.
$$\begin{array}{r} 2{,}354 \\ \times \quad 45 \\ \hline \end{array}$$

2.
$$\begin{array}{r} 7{,}841 \\ \times \quad 73 \\ \hline \end{array}$$

3.
$$\begin{array}{r} 6{,}003 \\ \times \quad 11 \\ \hline \end{array}$$

4.
$$\begin{array}{r} 2{,}206 \\ \times \quad 61 \\ \hline \end{array}$$

5.
$$\begin{array}{r} 7{,}011 \\ \times \quad 52 \\ \hline \end{array}$$

6.
$$\begin{array}{r} 3{,}498 \\ \times \quad 39 \\ \hline \end{array}$$

PROBLEM SOLVING
Writing a Simpler Problem

Sometimes decimals, fractions, and large numbers make a problem seem difficult. It may be easier to use simpler numbers than those in the problem. Once you see how to get the answer, you can use the more difficult numbers to find the exact answer.

SCHOOL BUS

> Rick owns the Go-to-School Bus Company. Each day, 4,368 students ride his buses to go to school. Each bus holds 42 students. How many busloads of students ride Rick's buses during a 5-day week?

Substitute simpler numbers for the numbers in the problem.

$$4,368 \longrightarrow 4,000$$
$$42 \longrightarrow 40$$

You multiply to find the total number of students who ride the bus in 5 days.

$$4,000 \times 5 = 20,000$$

Then you divide to find the number of busloads.

$$20,000 \div 40 = 500$$

Now you multiply and then divide, using the actual numbers in the problem.

$$4,368 \times 5 = 21,840 \qquad 21,840 \div 42 = 520$$

During a 5-day week, Rick's buses carry 520 busloads of students to school.

Write the letter of the better plan for simplifying each problem.

1. John has $52,000 dollars that he wants to use to buy vans. Each van costs $12,942. Does he have enough money to buy 4 vans?

   a. Step 1: $12 × 4 = $48
       Step 2: $52 > $48

   b. Step 1: $52 − $12 = $40
       Step 2: $40 ÷ 4 = $10

2. A bus driver travels $13\frac{3}{10}$ miles from his garage to the first stop on his route. There are 24 stops along this route. The stops are $1\frac{1}{4}$ miles apart. How far does the bus travel from its garage to its last stop?

   a. Step 1: 24 × 1 = 24
       Step 2: 24 − 13 = 11

   b. Step 1: 24 × 1 = 24
       Step 2: 24 + 13 = 37

Solve. Use simpler numbers if needed.

3. A bridge is being built, which will shorten Lee's ride to school by $3\frac{7}{10}$ miles. She used to ride $4\frac{1}{4}$ miles to the old bridge and $5\frac{1}{2}$ miles from there to school. How long will Lee's ride be when she uses the new bridge?

4. Last year, John spent $48,789 to maintain his buses. He expects to spend twice as much this year. He has already spent $72,642. How much more money does he expect to pay for maintenance this year?

5. Many students ride bicycles to Hillsboro School. Seth rides $1\frac{3}{4}$ miles. Paul rides $1\frac{3}{10}$ miles farther than Seth. Myra rides $2\frac{1}{2}$ miles. How much farther does Paul ride than Myra?

6. John hired 12 new bus drivers. Each will earn $19,342 per year. The budget allows $240,000 to pay new drivers. Has he budgeted enough money to pay all the new drivers he has hired?

★7. On Monday, 1,302 students rode buses to school, and 504 students walked. Because it rained on Tuesday, $\frac{1}{2}$ of the students who walked on Monday rode buses. Each bus holds 42 students. How many busloads of students rode to school on Tuesday?

★8. Last year, 1,542 students went on school trips. It cost $7.89 per student. This year, there are 327 fewer students going on trips. But, it cost $9.45 per student. What is the difference in the total cost in the last two years?

Scale Drawings

A. A **scale drawing** is used as a model for an object. Designers draw "to scale" by reducing or enlarging all parts of the object.

Look at this scale drawing. The measurements of the Wright house have been drawn to the scale of 1 in.:10 ft. This means that 1 in. in the drawing is equivalent to 10 ft in the actual house.

How long is the back wall of the actual house?

You can use equal ratios to find the actual length of the wall.

Write a ratio for the scale.

1 in.:10 ft, or $\frac{1}{10}$.

Write a ratio for the length of the wall.

length in drawing $\longrightarrow$ $\frac{5 \text{ in.}}{n \text{ ft}}$
actual length $\longrightarrow$

Use equal ratios to find the unknown length.

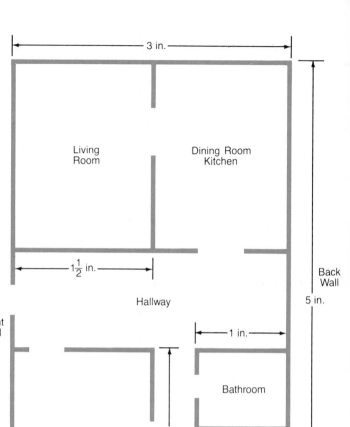

Scale: 1 in.: 10 feet

length in drawing (in.) $\longrightarrow$ $\dfrac{1}{10} = \dfrac{5}{n}$ $\boxed{\text{Think:} \\ 1 \times 5 = 5.}$ $\dfrac{1}{10} = \dfrac{1 \times 5}{10 \times 5} = \dfrac{5}{50}$
actual length (ft) $\longrightarrow$

The actual length of the back wall is 50 ft.

B. How long will a 20-foot-long bookcase appear to be in the scale drawing?

length in drawing (in.) $\longrightarrow$ $\dfrac{1}{10} = \dfrac{n}{20}$ $\boxed{\text{Think:} \\ 10 \times 2 = 20.}$ $\dfrac{1}{10} = \dfrac{1 \times 2}{10 \times 2} = \dfrac{2}{20}$
actual length (ft) $\longrightarrow$

The bookcase will appear to be 2 inches long in the drawing.

314

Use the scale drawing at the right to find the length and width of each room.

1. Living room
   $l = $ ▨; $w = $ ▨

2. Dining room
   $l = $ ▨; $w = $ ▨

3. Kitchen
   $l = $ ▨; $w = $ ▨

4. Study
   $l = $ ▨; $w = $ ▨

5. Pantry
   $l = $ ▨; $w = $ ▨

6. Bathroom
   $l = $ ▨; $w = $ ▨

Solve. Use Scale Drawing A.

7. What is the length of the wall that separates the dining room from the living room?

★8. Mrs. Wright is extending her living room and adding new furniture. The room will be 9 m long and 5 m wide. Use a scale of 1 cm : 0.5 m to draw the room. Then draw each piece of furniture from the chart in your scale drawing.

Scale Drawing A

Scale: 1 cm: 3 m

Furniture	Actual length	Actual width
Sofa	2 m	1 m
Cabinets	3 m	0.5 m
Coffee table	1 m	0.5 m

MIDCHAPTER REVIEW

Write each ratio as a fraction.

1. 6 to 9   2. 17 out of 20   3. 14 to 30

4. 7 : 12

Write = or ≠ for ●.

5. $\frac{17}{34}$ ● $\frac{35}{68}$ 6. $\frac{4}{9}$ ● $\frac{36}{81}$ 7. $\frac{24}{64}$ ● $\frac{3}{8}$

Use Scale Drawing B to find the length and the width of each room.

8. Kitchen
   $l = $ ▨; $w = $ ▨

9. Bathroom
   $l = $ ▨; $w = $ ▨

10. Living room
    $l = $ ▨; $w = $ ▨

11. Bedroom
    $l = $ ▨; $w = $ ▨

Scale Drawing B

Scale: 1 in.: 12 ft

PROBLEM SOLVING
Using a Recipe

A recipe tells you the ingredients that are required, the steps to follow in preparing the dish, and the number of people it will serve. You may want to change the recipe to increase or decrease the number of servings.

MIGHTY MEAT LOAF

Makes 6 servings

3 eggs, beaten $\frac{3}{4}$ cup minced onion

4 pounds ground beef $\frac{1}{4}$ cup milk

$2\frac{1}{2}$ cups bread crumbs $\frac{1}{4}$ cup ketchup

Blend eggs and meat. Add other ingredients. Mix well. Shape into an oval loaf. Bake in a shallow pan for about 1 hour at 350°F.

Bill wants to make Mighty Meat Loaf for a family car trip. He wants to make enough for 12 servings. How many eggs should he use?

You can write equal ratios to decide how many eggs Bill will need.

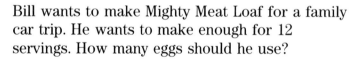

number of eggs   number of eggs

in recipe $\longrightarrow$ $\dfrac{3}{6} = \dfrac{\blacksquare}{12}$ $\longleftarrow$ needed

number of servings $\rightarrow$ $\longleftarrow$ number of servings

in recipe   wanted

Find equivalent fractions.

$$6 \times 2 = 12 \qquad \frac{3}{6} = \frac{3 \times 2}{6 \times 2} = \frac{6}{12}$$

Bill will need 6 eggs for 12 servings.

Solve. Write the letter of the best answer.

1. Bill wants to make 18 servings of the meat loaf. The recipe yields 9 servings and calls for $\frac{1}{4}$ cup of milk. How much milk will Bill use?

   a. $\frac{1}{3}$ cup   **b.** $\frac{1}{2}$ cup
   c. 3 cups   **d.** 4 cups

2. The meat-loaf recipe calls for $1\frac{1}{4}$ cup of bread crumbs. This amount is used to make 6 servings. How many cups of bread crumbs will Bill need for 24 servings?

   a. $\frac{2}{24}$ cup   **b.** $8\frac{1}{4}$ cup
   c. 5 cups   **d.** 36 cups

Solve.

3. One soup recipe calls for 2 cups of broth. The recipe yields 5 servings. How much broth would you need for 15 servings?

4. A casserole recipe calls for $1\frac{1}{2}$ cups grated cheese. The casserole serves 8 people. How much cheese would you need for 24 servings?

5. To cook 3 cups of rice, you need 6 cups of water. What is the ratio of rice to water? How much water would you need for 6 cups of rice? for 9 cups of rice?

6. Larry is making date-nut bread. The recipe yields 1 loaf. Larry plans to bake 3 loaves. The recipe calls for $\frac{3}{4}$ cup chopped dates. How many cups of dates should Larry use?

★7. A chow-mein recipe calls for one 5–ounce can of water chestnuts. The recipe serves 6. Clara wants to serve 24 people. The store stocks only 10-ounce cans. How many 10-ounce cans should Clara buy?

★8. Sam's egg-salad recipe calls for 1 tablespoon of mustard. The recipe serves 9. Sam wants to serve 6. How much mustard should Sam use?
(HINT: 3 teaspoons = 1 tablespoon)

Ratio and Percent

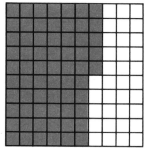

100 squares

A. You can think of a **percent (%)** as the ratio of a number to 100. *Percent* means "per hundred."

The figure is made up of 100 squares; 65 are shaded.

The ratio of shaded squares to all the squares is 65 to 100. This is 65 "per hundred," or 65%.

What percent of the squares are not shaded?

$\frac{35}{100} = 35\%$

35% of the squares are not shaded.

B. A poll of airline passengers was taken at one of America's busiest airports. The poll showed that 73 out of the 100 passengers polled were traveling on business. What percent of the passengers polled were traveling on business?

73 out of 100 = 73%

Other examples:

Write as a percent.

3 out of 100	83:100	27 to 100	$\frac{14}{100}$
$\frac{3}{100} = 3\%$	$\frac{83}{100} = 83\%$	$\frac{27}{100} = 27\%$	$\frac{14}{100} = 14\%$

Write the percent that describes the shaded part of the figure.

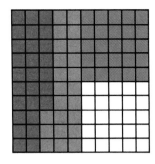

1. red   **2.** purple

3. gray   **4.** unshaded

Write the percent.

5. $\frac{5}{100}$ **6.** $\frac{42}{100}$ **7.** $\frac{25}{100}$ **8.** $\frac{68}{100}$ **9.** $\frac{6}{100}$

10. $\frac{99}{100}$ **11.** $\frac{22}{100}$ **12.** $\frac{36}{100}$ **13.** $\frac{18}{100}$ **14.** $\frac{56}{100}$

15. $3:100$ **16.** $10:100$ **17.** $90:100$ **18.** $62:100$

19. 26 to 100   **20.** 10 to 100   **21.** 99 to 100   **22.** 17 to 100

23. 11 out of 100   **24.** 8 out of 100   **25.** 32 out of 100   **26.** 43 out of 100

27. 4 per hundred   **28.** $2:100$ **29.** a ratio of 87 to 100   **★30.** $\frac{165}{100}$

Solve.

31. For a flight from Dallas, Texas, to Boston, Massachusetts, 19 out of the 100 tickets have been sold. What percent of the tickets have been sold?

32. A total of 631 passengers fly Wildfire Airways in three days. If 120 people fly on day one and 274 fly on day three, how many people fly on day two?

33. Use the advertisement to write and solve your own problem.
FLY SWIFTWAY AIRLINES
99 times out of 100, we get there faster! 89 out of 100 of our pilots are more experienced!

★34. A ticket agent sells 125 seats for a flight that has only 100 seats. Write the percent for the number of seats the agent sold. By what percent has the agent oversold the flight?

CHALLENGE

Jane is older than Susie. Susie is older than William. Lily is older than William but younger than Jane. Betsy is older than Susie but younger than Lily. Is Betsy older than William?

Percents and Decimals

A. The table lists the seven busiest air routes in the United States. There are about 16,580 flights on these routes each month. Of these flights, 11% fly from Honolulu to Kahului, over the Pacific Ocean. The remaining 89% fly over the continental United States.

You read percents every day. In order to compute with percents, you can write the percents as decimals.

Route	% of flights
Dallas–Houston	17%
Los Angeles–San Francisco	17%
New York–Chicago	16%
New York–Washington, D.C.	15%
New York–Boston	13%
Honolulu–Kahului	11%
New York–Miami	11%

Write 89% as a decimal.
Think: $89\% = \frac{89}{100}$, or 0.89.

Write 2% as a decimal.
Think: $2\% = \frac{2}{100}$, or 0.02.

B. You can write a decimal as a percent.

Write 0.61 as a percent.
Think: $0.61 = \frac{61}{100}$, or 61%.

Write 0.7 as a percent.
Think: $0.7 = 0.70 = \frac{70}{100}$, or 70%.

Write 0.03 as a percent.
Think: $0.03 = \frac{3}{100}$, or 3%.

Checkpoint Write the letter of the correct answer.

Choose the corresponding decimal or percent.

1. 0.25
a. 0.0025%
b. 0.25%
c. 4%
d. 25%

2. 5%
a. 0.0005
b. 0.05
c. 0.50
d. 5

3. 0.8
a. 0.8%
b. 8%
c. 80%
d. 800%

4. 55%
a. 0.0055
b. 0.55
c. 5.5
d. 55

Write as a percent.

1. 0.22
2. 0.15
3. 0.84
4. 0.63
5. 0.50
6. 0.01
7. 0.04
8. 0.09
9. 0.72
10. 0.9
11. 0.05
12. 0.9
13. 0.6
★14. 3.00
★15. 2.50

Write as a decimal.

16. 34%
17. 97%
18. 11%
19. 19%
20. 49%
21. 6%
22. 69%
23. 80%
24. 30%
25. 3%
26. 71%
27. 1%
28. 9%
★29. 200%
★30. 130%

Solve.

31. Several airlines fly passengers from New York to Boston. A survey of passengers shows that 85% are regular commuters, 10% are vacationing, and 5% are traveling for other reasons. Write each percent as a decimal.

32. Since there are fewer passengers on Saturdays, one airline schedules 0.8 of the number of flights that it schedules for each weekday. What percent of the number of weekday flights is flown on Saturdays?

33. Passengers are allowed to check 60 lb of baggage. How many pounds can be checked by 7 people?

★34. A 100-seat jet has 31 empty seats. What percent of the jet's seats are full?

CHALLENGE

Trace the diagram. See whether you can draw four lines that divide the figure into eleven regions, each of which contains only one dot.

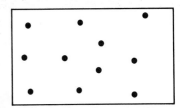

Percents and Fractions

A. The United States is the world's largest supplier of farm machinery. Cargo ships transport farm equipment from the United States to countries around the world. On one ship, 75% of the cargo consists of tractors.

Some facts can be expressed as percents or as fractions. You can write a fraction to describe the part of the cargo that consists of tractors.

Write 75% as a fraction.

percent	fraction	Write in simplest form.

$$75\% \longrightarrow \frac{75}{100} \longrightarrow \frac{3}{4}$$

Of the ship's cargo, $\frac{3}{4}$ consists of tractors.

B. On another ship, $\frac{4}{5}$ of the cargo is refrigerators. What percent of the total cargo is this?

Write $\frac{4}{5}$ as a percent.

Find an equivalent fraction with a denominator of 100.

$$\frac{4}{5} = \frac{4 \times 20}{5 \times 20} = \frac{80}{100}$$

Write the fraction as a percent.

$$\frac{80}{100} = 80\%$$

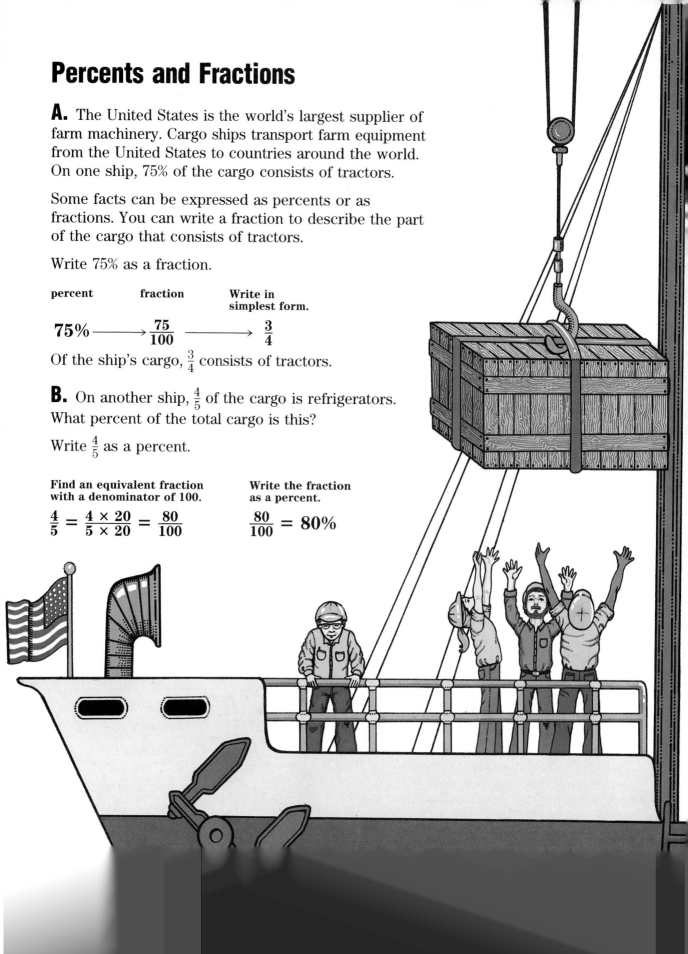

Write each percent as a fraction in simplest form.

1. 6%   **2.** 10%   **3.** 14%   **4.** 20%   **5.** 35%

6. 55%   **7.** 43%   **★8.** 132%   **★9.** 184%   **★10.** 167%

Write each fraction as a percent.

11. $\frac{2}{50}$ **12.** $\frac{1}{10}$ **13.** $\frac{4}{5}$ **14.** $\frac{1}{4}$ **15.** $\frac{2}{5}$

16. $\frac{6}{20}$ **17.** $\frac{19}{50}$ **★18.** $\frac{21}{10}$ **★19.** $\frac{20}{5}$ **★20.** $\frac{36}{25}$

Copy and complete the chart. Write each fraction in simplest form.

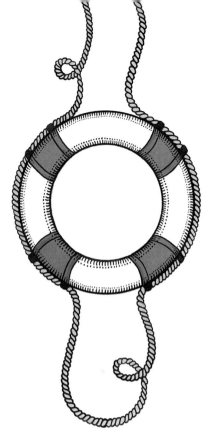

	21.	22.	23.	24.	25.	26.	27.	28.	29.
Fraction	▦	▦	▦	$\frac{1}{4}$	▦	▦	$\frac{24}{25}$	▦	$\frac{1}{5}$
Percent	▦	50%	▦	▦	22%	▦	▦	33%	▦
Decimal	0.99	▦	0.6	▦	▦	0.04	▦	▦	▦

Solve.

30. A cargo ship carries 120 containers of airplane parts. Each container weighs 2,300 kg. What is the total weight of the containers?

★31. About 20% of the work on a new ship is completed. What fraction of the work has been completed on the ship?

CALCULATOR

You can use a calculator to find a percent that is equivalent to a fraction.

Find the percent that is equivalent to $\frac{5}{8}$.

Divide.

$$\frac{5}{8} = 5 \div 8 = 0.625$$

Round to the nearest hundredth.

$$0.625 \longrightarrow 0.63$$

Write the decimal as a percent.

$$0.63 \longrightarrow \frac{63}{100} \longrightarrow 63\%$$

The percent equivalent of $\frac{5}{8}$ is 63%.

Write as a percent. Use a calculator.

1. $\frac{3}{8}$ **2.** $\frac{7}{16}$ **3.** $\frac{4}{7}$ **4.** $\frac{7}{50}$ **5.** $\frac{1}{4}$ **6.** $\frac{19}{26}$

More Practice, page 440

Percent of a Number

In the United States, travel is one of the most popular vacation activities. Each year, millions of people travel to places throughout the United States and to other countries.

A travel agency arranges vacation trips for 1,500 people. Of these people, 65% travel by airplane. How many people travel by airplane?

Find 65% of 1,500.

Write 65% as a decimal.

$$65\% \longrightarrow \frac{65}{100} \longrightarrow 0.65$$

Multiply.

$$
\begin{array}{r}
1{,}5\,0\,0 \\
\times\ \ \ 0.6\,5 \\
\hline
7\,5\,0\,0 \\
9\,0\,0\,0\,0 \\
\hline
9\,7\,5.0\,0
\end{array}
$$

Of the 1,500 people, 975 people travel by airplane.

Another example:

$$
\begin{aligned}
4\% \text{ of } 525 &= n \\
0.04 \times 525 &= n \\
21 &= n \\
4\% \text{ of } 525 &= 21
\end{aligned}
$$

Checkpoint Write the letter of the correct answer.

Find the percent of each number

1. 35% of 400

a. 32
b. 140
c. 140%
d. 14,000

2. 50% of 2,500

a. $\frac{1}{2}$
b. 1,250
c. 11,500
d. 125,000

3. 9% of 7,200

a. 638
b. 648
c. 800
d. 6,480

4. 81% of 100

a. 0.81
b. 8.1
c. 81
d. 100

Find the percent of each number.

1. 10% of 100
2. 75% of 20
3. 30% of 210
4. 90% of 30

5. 20% of 60
6. 80% of 640
7. 8% of 800
8. 60% of 590

9. 5% of 710
10. 90% of 660
11. 55% of 480
12. 25% of 212

13. 40% of 80
14. 70% of 450
15. 8% of 720
16. 30% of 350

17. 65% of 120
18. 50% of 44
19. 78% of 1,100
20. 4% of 200

21. 70% of 700
22. 25% of 4
★23. 125% of 16
★24. 225% of 100

Solve.

25. Ellen's class planned a bike trip during summer vacation. Only 40% of Ellen's class went on the trip. If there were 30 people in her class, how many people went biking?

26. Ellen's friends are planning a 9-day hike of 108 miles. How many miles should they hike each day if they want to hike the same number of miles each day?

27. Fritz has $300.00 to spend for his vacation. He budgets 60% of his money for transportation. How much money does he budget for transportation?

★28. The twenty-five members of the bicycle club want to fly to Italy in order to bike through the countryside. If a 400-seat plane is 90% full, how many seats are left?

FOCUS: MENTAL MATH

Sometimes it is easier to find the percent of a number if you think of the percent as a fraction.

$$50\% = \frac{1}{2} \quad 25\% = \frac{1}{4} \quad 10\% = \frac{1}{10}$$

Use fractions to find the percent of the number.

50% of 120	25% of 360	10% of 780
$= \frac{1}{2} \times 120$	$= \frac{1}{4} \times 360$	$= \frac{1}{10} \times 780$
$= \frac{120}{2}$	$= \frac{360}{4}$	$= \frac{780}{10}$
$= 60$	$= 90$	$= 78$

Find the percent of each number.

1. 10% of 200
2. 25% of 180
3. 50% of 66

4. 50% of 998
5. 10% of 510
6. 25% of 328

So welcome to Fiji. Enjoy our shopping and our wonderful resorts. Enjoy the magic of a world untouched by the anger and violence that has affected so many destinations.

Finding Percents

A. The FunTime Toy Company owns 20 trucks. The owners plan to paint 5 of them red. What percent of the trucks will be painted red?

You need to find what percent of 20 is 5. Compare the part to the whole.

Write a fraction.	Find an equivalent fraction with a denominator of 100.	Write the fraction as a percent.
part $\longrightarrow$ $\dfrac{5}{20}$ whole $\longrightarrow$	$\dfrac{5 \times 5}{20 \times 5} = \dfrac{25}{100}$	**25%**

Of all FunTime's trucks, 25% will be painted red.

B. Sometimes you have to simplify to find an equivalent fraction with a denominator of 100.

8 is what percent of 40?

Write a fraction.	Simplify.	Find an equivalent fraction that has a denominator of 100.	Write the fraction as a percent.
part $\longrightarrow$ $\dfrac{8}{40}$ whole $\longrightarrow$	$\dfrac{8}{40} \div \dfrac{8}{8} = \dfrac{1}{5}$	$\dfrac{1 \times 20}{5 \times 20} = \dfrac{20}{100}$	**20%**

8 is 20% of 40.

Checkpoint   Write the letter of the correct answer.

Solve.

1. What percent of 50 is 5?

a. 10
b. 20
c. 55
d. 250

2. What percent of 25 is 2?

a. 4
b. 8
c. $12\frac{1}{2}$
d. 50

3. 6 is what percent of 10?

a. 6
b. 10
c. 60
d. 600

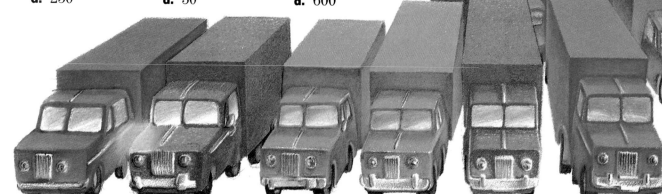

Copy and complete the chart.

	1.	**2.**	**3.**	**4.**	**5.**	**6.**	**7.**	**8.**
Part	15	18	32	7	20	18	8	12
Whole	75	60	80	35	25	25	32	50
Percent	▪	▪	▪	▪	▪	▪	▪	▪

Compute.

9. What percent of 25 is 15?

10. 18 is what percent of 50?

11. What percent of 45 is 9?

12. 24 is what percent of 60?

13. 32 is what percent of 80?

14. What percent of 90 is 27?

15. 3 is what percent of 10?

16. What percent of 50 is 13?

Find the percent that each part is of the whole.

17. 9 calico cats
10 cats
▪% are calico.

18. 1 goldfish
5 fish
▪% are goldfish.

19. 19 black umbrellas
20 umbrellas
▪% are black.

★20. 2 blue bicycles
8 yellow bicycles
10 white bicycles
▪% are yellow.

★21. 9 pink flowers
6 red flowers
10 orange flowers
▪% are pink.

★22. 12 brown dogs
36 spotted dogs
2 tan dogs
▪% are tan.

Solve.

23. The FunTime Toy Company uses 12 out of their 20 trucks to deliver electronic games. What percent of the trucks deliver electronic games?

★24. On one day, 17 of FunTime's trucks were delivering. The other 3 trucks remained in the warehouse. What percent of the trucks remained in the warehouse?

ANOTHER LOOK

Multiply.

1. 0.07
× 0.8

2. 0.04
× 0.9

3. 0.05
× 0.3

4. 0.09
× 0.7

5. 0.03
× 0.4

6. 1.07
× 0.7

7. 7.04
× 5.3

8. 9.07
× 0.6

9. 7.02
× 9.2

10. 3.02
× 5.9

More Practice, page 440

PROBLEM SOLVING
Using a Circle Graph

You can use a circle graph to show data given in percents and as fractional parts of a whole. The circle is divided into parts. A larger fraction or percent is shown by a larger part of the graph.

**NUMBER OF CARS OWNED
BY GREENTOWN FAMILIES**

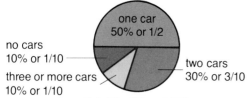

one car
50% or 1/2

no cars
10% or 1/10

three or more cars
10% or 1/10

two cars
30% or 3/10

Total families: 1,180

- The title tells you that this graph shows the number of cars owned by families in Greentown.
- Each part shows what percent or fraction of the families own a certain number of cars.

  The percents must add up to 100%. The fractions must add up to 1.

- The size of each part shows the size of its percent or fractional part.

  How many families own two cars?
  First, find the part labeled *two cars*. What percent of the families own two cars? What fractional part of the families own two cars?

  The graph tells you that 30%, or $\frac{3}{10}$, of the families in Greentown own two cars.

  Then, find 30% of 1,180. $\quad$ 1,180
  Write 30% as a decimal. $\quad \underline{\times \quad 0.3}$
  $\qquad 30\% = 0.3 \qquad\qquad$ 354

Of the Greentown families, 354 own two cars.

Write *true* or *false*.

1. The sum of the percents shown in a circle graph may be greater than 100%.

2. If a circle graph were divided into 4 equal parts, each part would be $\frac{1}{4}$ of the whole.

328

Use the circle graph to answer Exercises 3–8.

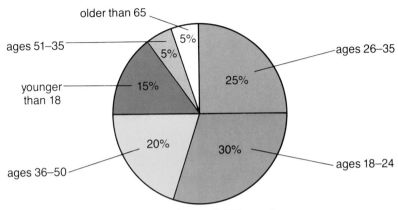

SUMMER ATTENDANCE AT TRANSPORTATION MUSEUM BY AGE GROUP

older than 65 — 5%
ages 51–35 — 5%
younger than 18 — 15%
ages 26–35 — 25%
ages 36–50 — 20%
ages 18–24 — 30%

Total attendance: 9,900 people

3. What percent of the people who visited the museum were in the 26–35 age group?

4. The greatest percent of people who visited the museum were in which age group?

5. How many people were in the younger-than-18 age group?

6. How many people in the 26–35 age group attended the museum?

Circle graphs can be used to compare information. Use the three circle graphs below to solve each problem.

APPROXIMATE NUMBER OF MOTOR VEHICLES IN USE

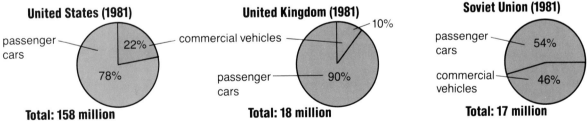

United States (1981)
passenger cars — 78%
commercial vehicles — 22%
Total: 158 million

United Kingdom (1981)
commercial vehicles — 10%
passenger cars — 90%
Total: 18 million

Soviet Union (1981)
passenger cars — 54%
commercial vehicles — 46%
Total: 17 million

9. About how many passenger cars were in use in the United Kingdom in 1981?

10. Which of the countries had the greatest percent of passenger cars in use?

★11. Which country had the greatest number of passenger cars in use?

★12. To the nearest ten thousand, about how many more passenger cars than commercial vehicles are in use in the Soviet Union?

READING MATH

Scientists traveled from all over the world to see Dr. Magwich's plans for a new kind of flying machine made of superlight, superstrong clay. When they arrived, Dr. Magwich found that her list of supplies had been damaged by water and was unreadable. This is what Dr. Magwich had.

3/4 GALLONS
31 1/2 POUNDS
7 YARDS

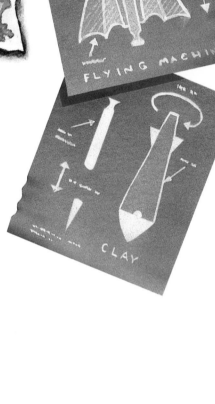

Dr. Magwich knew she needed clay, some nylon cloth for the wings, and some special liquid fuel. How much of each did she need?

Math has its own vocabulary. Such terms as *gallons*, *pounds*, and *yards* stand for units of measure. Dr. Magwich knew that *gallons* measure liquid, *pounds* measure solid weight, and *yards* measure length. Tell which item she put with each unit of measurement and write in the missing word. Choose from these terms.

pints	gallons	prime number
feet	inches	simplest form
pounds	miles	denominator
yards	numerator	mixed number

1. $31\frac{1}{2}$ is a kind of fraction called a _____.

2. If Dr. Magwich wrote $\frac{3}{6}$ as $\frac{1}{2}$, she would have written the fraction in its _____.

3. In $\frac{3}{4}$, 4 is the _____.

4. To measure enough fuel to fit in 2 cups, Dr. Magwich would use _____.

5. To show how far the flying machine would have to fly to reach the next city, Dr. Magwich would measure by _____.

6. A number that can be divided only by itself and 1 is called a _____.

GROUP PROJECT

Champ's Used Skateboards

The problem: Leroy is called "Champ" because he has been the skateboard champion in the neighborhood for so long. Everyone comes to him to find out about the latest safety equipment and for help fixing their boards. So Champ has decided to open a shop called Champ's Used Skateboards. What should he charge for his goods and services? Can you help? Discuss the Key Facts with your classmates; then make a price list.

Champs Boards

new ___$60___   now _____

The Professionals

new ___$120–$150___   now _____

Retired Reactor Gear

Helmet   new ___$32___   now _____

Gloves with fingers   new ___$15___   now _____

Gloves without fingers   new ___$14___   now _____

Knee Guards   new ___$26___   now _____

Elbow Guards   new ___$22___   now _____

Last Month's Magazines   new ___$1.50___   now _____

Key Facts

- Repairing skateboards can require a great deal of labor.
- Supplies needed are screws, paint, sandpaper, and glue.
- If the wheels need to be replaced, they cost $40 if they are new.
- Champ can take $5 to $10 off the price if an old skateboard is traded in when a new one is purchased.

CHAPTER TEST

Write each ratio as a fraction. (page 308)

1. $2:5$ **2.** 5 out of 7   **3.** $10:3$ **4.** 6 to 9

Write each fraction as a ratio. (page 308)

5. $\frac{3}{8}$ **6.** $\frac{12}{7}$ **7.** $\frac{8}{9}$ **8.** $\frac{14}{25}$

Find the missing number. (page 310)

9. $\frac{2}{3} = \frac{8}{n}$ **10.** $\frac{6}{11} = \frac{42}{n}$ **11.** $\frac{4}{5} = \frac{n}{25}$ **12.** $\frac{16}{7} = \frac{n}{28}$

Use the scale drawing at the right to
find the length and width of each room.
(page 314)

13. Living room   **14.** Study

15. Parlor   **16.** Balcony

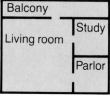

SCALE DRAWING

Balcony

Living room   Study

Parlor

Scale: 1 cm : 4m

Write the percent that describes the
shaded part of the figure. (page 318)

17. red   **18.** blue

19. yellow   **20.** white

Write as a decimal. (page 320)

21. 18%   **22.** 52%   **23.** 9%   **24.** 27%

Write each percent as a fraction in simplest form.
(page 322)

25. 11%   **26.** 32%   **27.** 49%   **28.** 83%

Find the percent of each number. (page 325)

29. 20% of 100   **30.** 40% of 600   **31.** 80% of 35   **32.** 6% of 950

Compute. (page 327)

33. What percent
of 16 is 4?   **34.** 9 is what
percent of 25?   **35.** What percent
of 55 is 11?   **36.** 36 is what
percent of 90?

Solve. Use simpler numbers if needed. (pages 312–313)

37. There are 25 students in Sam's class. Of that number, 40% ride the bus to school. In John's class, there are 35 students, and 60% of the students in his class ride the bus to school. How many students ride the bus to school in both John's and Sam's classes?

38. Jane's family drives 1,311 miles in 3 days to visit their grandmother. After the first day, they had driven $\frac{1}{3}$ of the way to their grandmother's. How many miles did they drive the first day?

Use the circle graph to solve. (pages 328–329)

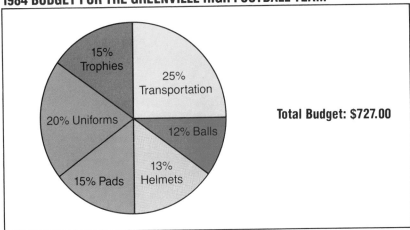

1984 BUDGET FOR THE GREENVILLE HIGH FOOTBALL TEAM

15% Trophies
25% Transportation
20% Uniforms
12% Balls
15% Pads
13% Helmets

Total Budget: $727.00

39. How much did the team spend on transportation? on pads?

40. How much did the team spend on uniforms? on helmets?

BONUS

There are 200 coins in Hal's change jar.

1. If 20 coins are quarters, what percent are quarters?

2. If 30% are dimes, how many dimes does he have?

3. If $\frac{2}{5}$ of the coins are pennies, how many pennies does he have?

4. If 40 coins are nickels, what percent are nickels?

RETEACHING

A. To write a percent as a decimal or a fraction, think of a percent as the ratio of a number to 100. So, 25% is 25:100, or $\frac{25}{100}$.

Write 73% as a decimal.
Think: $73\% = \frac{73}{100}$. Write 0.73.

Write 5% as a decimal.
Think: $5\% = \frac{5}{100}$. Write 0.05.

You can write a decimal as a percent.

Write 0.47 as a percent.
Think: $0.47 = \frac{47}{100}$. Write 47%.

Write 0.06 as a percent.
Think: $0.06 = \frac{6}{100}$. Write 6%.

B. You can also write a percent as a fraction or a fraction as a percent.

Write 65% as a fraction.

Percent   Fraction   Write in simplest form.

$65\% \longrightarrow \frac{65}{100} \longrightarrow \frac{13}{20}$

Write $\frac{11}{25}$ as a percent.

Find an equivalent fraction that has a denominator of 100.

$\frac{11}{25} = \frac{11 \times 4}{25 \times 4} = \frac{44}{100}$

Write the fraction as a percent.

$\frac{44}{100} = 44\%$

Write as a decimal.

1. 15%   **2.** 17%   **3.** 80%   **4.** 20%   **5.** 7%

Write as a percent.

6. 0.85   **7.** 0.54   **8.** 0.09   **9.** 0.08   **10.** 0.3

Write as a percent.

11. $\frac{15}{25}$ **12.** $\frac{48}{50}$ **13.** $\frac{4}{5}$ **14.** $\frac{1}{2}$ **15.** $\frac{20}{20}$

Write as a fraction in simplest form.

16. 85%   **17.** 30%   **18.** 9%   **19.** 42%   **20.** 116%

334

ENRICHMENT

Sales Tax and Discount

Some states and cities have **sales taxes.** The *sales tax* is a percentage of the cost of an item. It is added to that cost. When finding sales tax, round up. Pablo buys a book on the history of the railroad for $15.21. The sales tax is 6%. What is the total cost of the book?

Find 6% of $15.21	Round up to the nearest cent.	Add to the cost of the book.
$15.21 × 0.06 ——— $0.9126	$0.9126 → $0.92	$15.21 + 0.92 ——— $16.13

The total cost of the book is $16.13.

Some stores give **discounts.** A *discount* is also a percentage of the cost of an item. It is subtracted from that cost. When finding discounts, round down. Harry buys a book on airplanes for $7.38. The store gives a discount of 5%. What is the cost of the book?

Find 5% of $7.38	Round down to the nearest cent.	Subtract from the cost of the book.
$7.38 × 0.05 ——— $0.369	$0.369 → $0.36	$7.38 − 0.36 ——— $7.02

The cost of the book is $7.02.

Find the cost.

1. Price: $5.22
   Sales tax: 5%

2. Price: $6.22
   Sales tax: 6%

3. Price: $2.87
   Sales tax: 8%

4. Price: $9.47
   Sales tax: 4%

5. Price: $7.75
   Discount: 6%

6. Price: $5.33
   Discount: 7%

7. Price: $9.98
   Discount: 3%

8. Price: $1.28
   Discount: 5%

9. Price: $2.05
   Sales tax: 4%

10. Price: $8.12
    Discount: 9%

11. Price: $3.53
    Sales tax: 3%

12. Price: $4.07
    Discount: 10%

TECHNOLOGY

In BASIC, you can use parentheses to tell the computer the order in which to do the operations.

1. Complete this program so that when you RUN it, it prints this. 4 PLUS 30 IS 34

   ```
   10   LET A = (6 + ▒) / 4
   20   LET B = (12 − 6) * (4 + ▒)
   30   LET C = A + B
   40   PRINT A "PLUS" B "IS" C
   ```

2. You can use parentheses in PRINT statements, too. What does the computer type when you give these instructions?

   ```
   PRINT (6 * 4) / (2 + 2)
   PRINT (10 / 2) − (5 − 2)
   ```

   You can use a set of parentheses inside another set.

   PRINT 10 − (5 * (3 − 1))

   ```
                    5 *   2
                         10
           10 −   10
                  0
   ```

3. What will the computer type when you give these instructions?

   ```
   PRINT (17 − (5 − 3)) / 5
   PRINT 2 + ((2 * 3) − 4)
   ```

Notice that there are always the same number of left parentheses Ⅱ(as there are right parentheses)Ⅱ.

Here is a four-line program that stores the number 1 in variable Z.

```
10   LET X = 1 + 2 + 3
20   LET Y = X * 4
30   LET Z = 25 − Y
40   PRINT Z
```

Here is a program that does the same computation on one line.

```
10   LET Z = 25 − ((1 + 2 + 3) * 4)
40   PRINT Z
```

336

4. Rewrite this program so that it does the computation on one line.

```
10   LET I = 5 + 4
20   LET J = I / 9
30   LET K = J − 1
40   PRINT K
```

Even without parentheses, the computer will do some parts of the computation before other parts. You already know that the computer will first calculate inside parentheses, if there are any. After those calculations are done, the computer will do all the multiplication and division from left to right. Then all the addition and subtraction is done from left to right. For instance, this PRINT statement will multiply 7 and 3 to find 21, and then add 3 and 21. It will print 24.

PRINT 3 + 7 * 3

Here is another example.

PRINT 9 − (8 / 2 + 5)

This instruction has a set of parentheses. The computer will start to do the computation inside the parentheses first. Inside the parentheses, there are two operations. The division is done first: $8 \div 2 = 4$. Then $4 + 5 = 9$. Now the rest of the equation is easy: $9 - 9 = 0$.

5. What will each instruction print?

PRINT 6 + 6 / 2

PRINT (6 + 6) / 2

PRINT 10 * 10 − 3

PRINT 10 * (10 − 3)

PRINT 8 * 10 − (5 * 6)

PRINT (6 − (8 − 5)) + 7

PRINT 10 * (15 − (2 * 5)) / 5

PRINT 8 / 2 + 10 − 5

CUMULATIVE REVIEW

Write the letter of the correct answer.

1. $\frac{4}{9} \times \frac{5}{6}$

    a. $\frac{9}{54}$

    b. $\frac{10}{27}$

    c. $\frac{20}{15}$

    d. not given

2. $\frac{1}{3}$ mile = ▧ ft

    a. 1,760 ft

    b. 1,760 yd

    c. 5,280 ft

    d. not given

3. $15 \times \frac{1}{6}$

    a. $2\frac{1}{3}$

    b. $2\frac{1}{2}$

    c. $3\frac{1}{2}$

    d. not given

4. $\frac{1}{9} \times 81$

    a. $8\frac{1}{9}$

    b. 9

    c. $9\frac{1}{9}$

    d. not given

5. If the temperature falls from 55°F to ⁻55°F, by how many degrees has the temperature changed?

    a. 55°

    b. 100°

    c. 110°

    d. not given

6. 0.2×0.3

    a. 0.006

    b. 0.6

    c. 6

    d. not given

7. $\frac{5}{6} + \frac{7}{10}$

    a. $\frac{12}{16}$

    b. $\frac{13}{15}$

    c. $1\frac{8}{15}$

    d. not given

8. $6,736 \div 22$

    a. 36 R4

    b. 306

    c. 306 R4

    d. not given

9. $53 \times 5,697$

    a. 301,941

    b. 305,750

    c. 315,351

    d. not given

TRAIN SCHEDULE

Landstown to Cow Bluffs		
Landstown	**Poplar**	**Cow Bluffs**
12:10	12:24	12:55
1:30	1:44	2:15
4:55	5:09	5:40

Cow Bluffs to Landstown		
Cow Bluffs	**Poplar**	**Landstown**
6:14	6:19	6:53
6:38	6:44	7:12
6:54	6:59	7:34

Use the train schedule to answer the questions.

10. How long does it take to ride from Landstown to Cow Bluffs?

    a. 15 min

    b. 35 min

    c. 45 min

    d. not given

11. When does the last train arrive in Landstown?

    a. 4:55

    b. 6:53

    c. 7:34

    d. not given

Use as many vocabulary words about geometry as is possible to describe our solar system. Or, if you prefer, draw a diagram of our solar system, and then label all the geometric shapes.

11 GEOMETRY

Basic Ideas of Geometry

You use geometry whenever you ask questions about the size, shape, volume, or position of anything. The word *geometry* comes from two Greek words that mean "earth" and "to measure."

A **point** is an exact location in space. A point is named by a capital letter.	Point *P*, or *P* • *P*
A **line** is a straight path that goes on forever in two directions. A line is named by any two points on it.	Line *AB*, or $\overleftrightarrow{AB}$
A **ray** has one endpoint and goes on forever in one direction. A ray is named by an endpoint and any other point on it.	Ray *LM*, or $\overrightarrow{LM}$
A **line segment** is a part of a straight line. A line segment is named by its two endpoints.	Line segment *PQ*, or $\overline{PQ}$
A **plane** is a flat surface that goes on forever in all directions. A plane may be named by a small letter.	Plane *t*
Intersecting lines meet or cross each other. *N* is the point of intersection.	$\overleftrightarrow{HI}$ intersects $\overleftrightarrow{UV}$.
Parallel lines never intersect.	$\overleftrightarrow{RS}$ is parallel to $\overleftrightarrow{XY}$, or $\overleftrightarrow{RS} \parallel \overleftrightarrow{XY}$.

Identify and name each figure.

1.

2.

3.

4.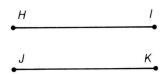

5.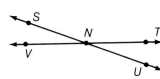

6.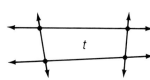

Name all the line segments.

7.

Copy the four points on a piece of paper, and then complete.

8. Draw $\overline{BC}$.

9. Draw $\overleftrightarrow{AC}$.

10. Draw $\overrightarrow{AD}$.

11. Draw intersecting lines $\overleftrightarrow{AC}$ and $\overleftrightarrow{BD}$.

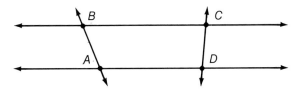

Draw the figure.

12. $\overleftrightarrow{AB} \parallel \overleftrightarrow{CD}$

13. $\overleftrightarrow{PQ}$ intersecting $\overleftrightarrow{RS}$ at point O

14. plane g

Complete. Use the diagram to answer.

15. Name the intersecting lines.

16. Name the parallel lines.

17. Name the points.

18. Name the line segments.

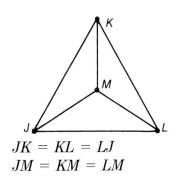

Draw a figure to show how you could arrange four points, J, K, L, and M, so that there are only two different distances between them. The figure on the right shows one way to do this. How many different ways can you solve this problem?

$$JK = KL = LJ$$
$$JM = KM = LM$$

Angles

Angles are measured in degrees.

 1 degree (1°)

The nose cone of the space shuttle has to be built at a certain angle. The builders measure the angle.

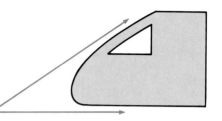

A. An angle, ∠BAC, is formed by rays AB and AC. AB and AC are the sides of the angle. A is the **vertex** of the angle.

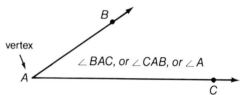

∠BAC, or ∠CAB, or ∠A

B. An angle that forms a square corner is a **right angle.** A right angle measures 90°.

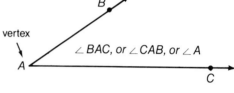

means this is a right angle

∠MLN

C. An angle that is less than a right angle is an **acute angle.** An acute angle measures less than 90°.

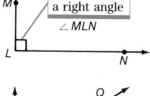

∠RPQ

D. An angle that is greater than a right angle is an **obtuse angle.** An obtuse angle measures more than 90°.

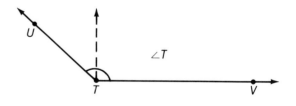

∠T

E. An angle whose rays point in exactly opposite directions is a **straight angle.** A straight angle measures 180°.

∠JIK

F. Perpendicular lines intersect to form right angles.

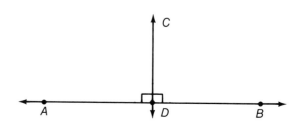

Give three different names for each angle.

1.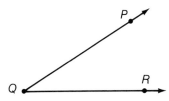

2.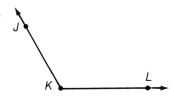

3.

Name the vertex, the sides, and the angle.

4.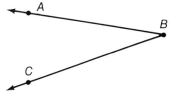

5.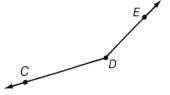

6.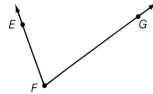

Identify the angle as *acute, obtuse, right,* or *straight.*

7.

8.

9.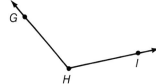

10.

11.

12.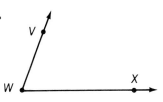

Are the lines perpendicular? Write *yes* or *no.*

13.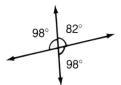

14.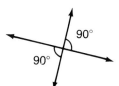

15.

16.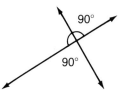

CHALLENGE

Trace the points at the right. Draw line segments from each point to Q. Label the angles that are formed as *acute, obtuse, right,* or *straight.*

1. ∠RQA   **2.** ∠RQB   **3.** ∠RQC   **4.** ∠RQD   **5.** ∠RQE

6. ∠FQR   **7.** ∠GQR   **8.** ∠HQR ★**9.** ∠IQJ   ★**10.** ∠KQE

Measuring Angles

A. A **protractor** can be used to measure angles. Place the center of the protractor on the vertex of the angle.

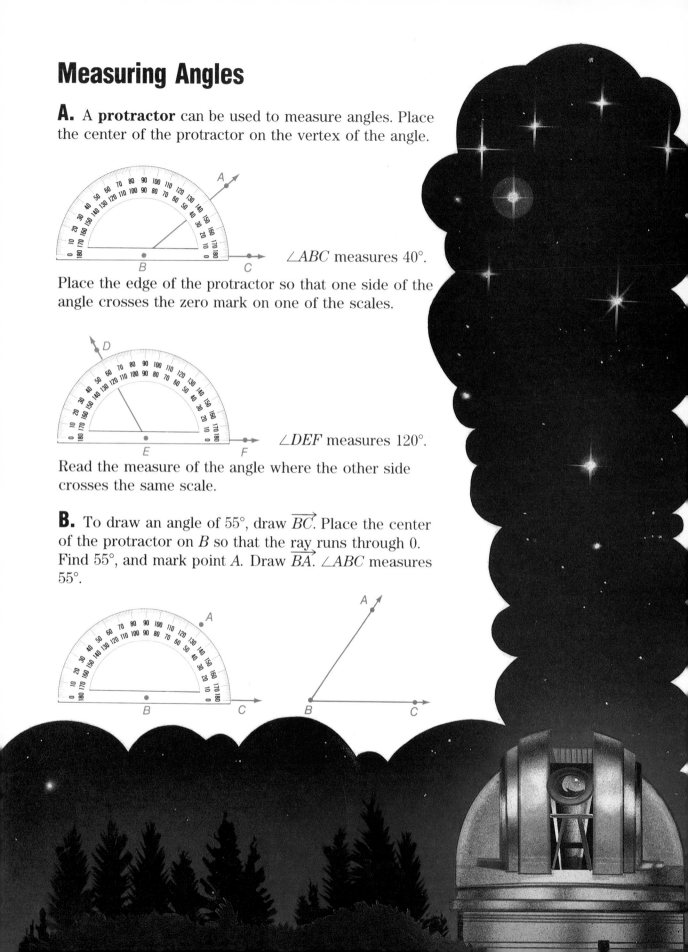

∠*ABC* measures 40°.

Place the edge of the protractor so that one side of the angle crosses the zero mark on one of the scales.

∠*DEF* measures 120°.

Read the measure of the angle where the other side crosses the same scale.

B. To draw an angle of 55°, draw $\overrightarrow{BC}$. Place the center of the protractor on *B* so that the ray runs through 0. Find 55°, and mark point *A*. Draw $\overrightarrow{BA}$. ∠*ABC* measures 55°.

Use a protractor to measure each angle.

1.
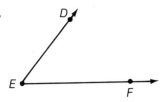

2.
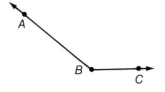

3.

4.

5.

6.
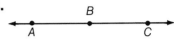

7.
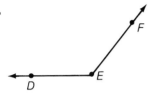

8.
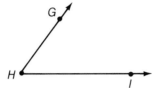

9.
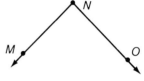

Draw each angle.

10. $135°$ **11.** $60°$ **12.** $30°$ **13.** $45°$ **14.** $127°$

15. $37°$ **16.** $100°$ **17.** $145°$ **18.** $166°$ **19.** $9°$

20. $15°$ **21.** $78°$ **22.** $132°$ **23.** $59°$ **24.** $5°$

25. $170°$ **26.** $40°$ **27.** $120°$ **28.** $50°$ **29.** $149°$

FOCUS: ESTIMATION

You can make a protractor and use it to sort angles.

To make a protractor:
1. Trace a circle.   **2.** Cut out the circle.
3. Fold it in half.   **4.** Label 0°, 90°, and 180°.

Copy and complete the chart.
Use your protractor to help sort these angles.

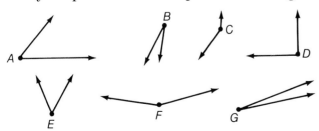

About 0°:	▣
Less than 90°:	▣
About 90°:	▣
More than 90°:	▣
About 180°:	▣

Triangles

A. Rockets traveling through space do not need wings. When a space shuttle reenters Earth's atmosphere, it uses triangular wings to glide to a landing.

A **triangle** is any figure that has three sides and three angles.

Read: triangle *ABC*.
Write: △ *ABC*.

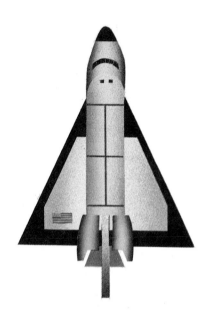

B. Some triangles have special names.

equilateral
All three sides
are of equal length.

isosceles
At least two sides are
of equal length.

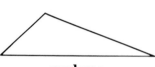

scalene
Each side is a
different length.

right
Has one
right angle.

acute
Has three
acute angles.

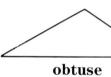

obtuse
Has one
obtuse angle.

C. The sum of the measures of the angles in any triangle is 180°.

Here is a way to show this.

Separate the angles
in △ *ABC*.

Place them along
a straight angle.

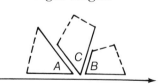

Since a straight angle measures 180°, the sum of the measures of the angles of △ *ABC* is 180°.

346

Name each triangle, and write if it is *equilateral, isosceles, scalene,* or *right.*

1.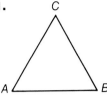

2.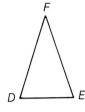

3.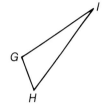

4.

Name each triangle, and write if it is *right, obtuse,* or *acute.*

5.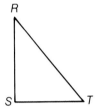

6.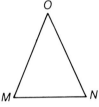

7.

8.

Find the measure of the missing angle.

9.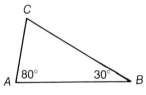

10.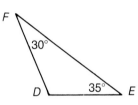

11.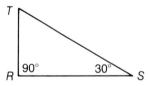

12.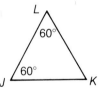

13.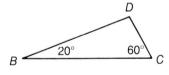

14.

Solve.

15. This isosceles triangle has two angles of equal measure, and one angle of 90°. What is the measure of each of the other two angles?

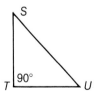

ANOTHER LOOK

Complete.

1. 20% of 50 =

2. 80% of 25 =

3. 60% of 30 =

4. 40% of 75 =

5. 90% of 60 =

6. 90% of 90 =

7. 70% of 70 =

8. 80% of 50 =

PROBLEM SOLVING
Making a Table To Find a Pattern

Many problems require that you find what comes next in a group of numbers, letters, or objects. To do this, you should study the information you are given to see whether there are any patterns that can be used as clues.

> As a space probe descends to the surface of a planet, it begins to slow its speed to make a soft landing. It begins its descent at 1,024 miles from the surface of the planet, at a speed of 486 miles per hour. At 256 miles, the probe is moving at 162 miles per hour. At an altitude of 64 miles, it has slowed to 54 miles per hour, and at 16 miles, the probe is moving at 18 miles per hour. At what speed will the probe be moving when it is at an altitude of 1 mile?

You can make a table to show the pattern of changes in altitude and speed.

SPACE-PROBE DESCENT

Speed	486	162	54	18
Altitude	1,024	256	64	16

You can see that there is a pattern to the changes in speed and altitude. Each time, the speed is $\frac{1}{3}$ of the previous speed. The altitude is $\frac{1}{4}$ the previous altitude. By dividing the speed by 3, we know the next number in the pattern. By dividing the altitude by 4, the next number in that pattern can be found.

SPACE-PROBE DESCENT

Speed	486	162	54	18	6	2
Altitude	1,024	256	64	16	4	1

Copy and complete the table to solve.

1. In a simulator, astronauts practice docking spacecraft with other spacecraft. To succeed, they must reduce their speed $\frac{1}{7}$ of their previous speed for every 10 km they come closer to the docking target. If they are traveling at 16,807 k/h at 50 km from the target, how fast will they be traveling at 20 km from the target?

DOCKING SIMULATION

Distance	60	50	40	30	20	10	0
Speed							

2. After docking, the astronauts increase their speed and reverse of the docking pattern. What is their speed at 60 km?

Solve.

3. While practicing a satellite launch from the simulator, the satellite is tracked by a computer. At 1 minute after release, the satellite is 100 yards from the launch vehicle. At 2 minutes, it has increased to 180 yards away. At 3 minutes the satellite is 324 yards from the launch vehicle, and at 4 minutes, 583.2 yards away. How far will the satellite have traveled at 6 minutes?

4. Space probes like the *Viking Lander* explore the surface of planets. One probe explores layers of soil. If it drills a hole that is 60 mm deep in 9 min, 50 mm deep in $7\frac{1}{2}$ min, and 40 mm deep at 6 min, how long had the probe been drilling when it reaches a depth of 20 mm?

★5. Astronauts need good reflexes to pilot spacecraft. The simulator flashes the pattern at right on its screen. The astronaut has to find the next figure as fast as possible. What would the next figure in the pattern be?

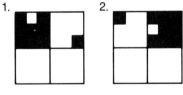

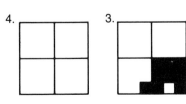

Polygons

A. **Polygons** are closed figures that consist of three or more line segments. They are closed figures because you can draw a line around their boundaries without ever coming to an end.

B. A **quadrilateral** is a polygon that has four sides.

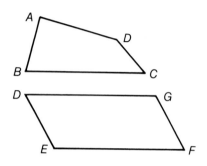

A quadrilateral whose opposite sides are the same length and are parallel to each other is a **parallelogram.** $\overline{DE} \parallel \overline{GF}; \overline{EF} \parallel \overline{DG}$.

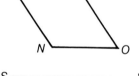

A parallelogram whose sides are all the same length is a **rhombus.**

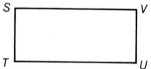

A parallelogram that has four right angles is a **rectangle.**

A rectangle whose sides are all the same length is a **square.**

C. Other polygons are also named according to the number of sides they have.

pentagon	**hexagon**	**octagon**	**decagon**
five sides	six sides	eight sides	ten sides

A polygon whose sides and angles are all equal is a **regular polygon.**

D. A line segment that connects two vertices of a polygon and is not a side is a **diagonal.** $\overline{BD}$ is a diagonal of square *ABCD*.

Name the polygon.

1.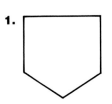

2.

3.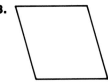

4.

5.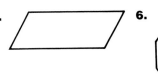

6.

7.

8.

9.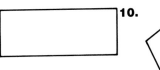

10.

11.

12.

Identify the polygon.

13. a quadrilateral whose opposite sides are parallel and are of equal length

14. a rectangle that has equal sides

15. a six-sided figure

16. a five-sided figure that has all sides and angles equal

17. a quadrilateral that has all sides equal but no right angles

Copy the figure. Draw its diagonals.

18.

19.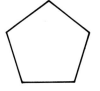

20.

Solve. Use the Infobank on page 420.

21. A constellation is a group of stars whose outline suggests a picture or a familiar shape. Name the shape of each highlighted constellation: Cetus, Orion, Pegasus, Camelopardalis, and Auriga.

Circles

A. The outline of a crater on the moon's surface suggests a circle. Scientists often use photographs to measure the size or the diameter of these circular craters.

Every point on a circle is the same distance from a point called the **center** of the circle.

A **chord** is a line segment that has its endpoints on the circle. $\overline{AD}$ and $\overline{BD}$ are chords.

A **diameter** is a chord that passes through the center of a circle. $\overline{BD}$ is a diameter. It is twice the length of the radius.

A **radius** is a line segment that has one endpoint on the circle and one endpoint on the center. $\overline{OC}$ is a radius.

B. You can use a compass to construct a circle that has a given radius.

Open the compass to the given radius.

Place the metal tip on a point, and turn the compass completely around the point.

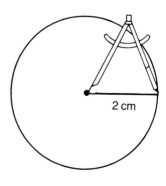

2 cm

Write *chord*, *radius*, or *diameter* for the given segment.

1.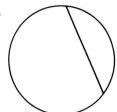

2.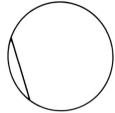

3.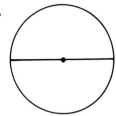

4.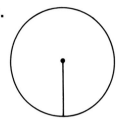

Copy and complete. Give the missing radius or diameter.

	5.	6.	7.	8.
Diameter	4 cm	▦	▦	26 cm
Radius	▦	5 cm	6 cm	▦

	9.	10.	11.	12.
Diameter	10 cm	▦	▦	44 cm
Radius	▦	18 cm	20 cm	▦

Use a compass to draw a circle that has the given radius.

13. 8 cm   **14.** 6 cm   **15.** 14 cm   **16.** 10 cm   **17.** 13 cm   **18.** 20 cm

Solve.

19. The Clavius Crater is one of the largest craters on the moon. The edge is about 130 km from the center of the crater. What is the diameter of the crater?

20. One crater has a diameter of 60 km. Another crater has a radius of 40 km. Which crater is wider?

MIDCHAPTER REVIEW

Use the figures at the right to answer.

1. Name all the line segments in Figure 1.

2. Name the parallel lines in Figure 1.

3. Use your protractor to find the measure of ∠B.

4. Find the measure of ∠C.

5. Is the triangle in Figure 2 right, equilateral, or isosceles?

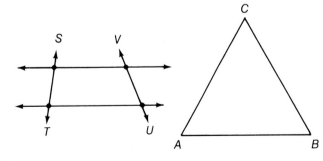

Fig. 1   **Fig. 2**

PROBLEM SOLVING
Practice

Write the letter of the most sensible answer.

1. An elevator takes people to the Moon Bus platform. A total of 25 people can fit in the elevator at one time. There are 279 people waiting for the elevator. How many trips must the elevator make to carry all the passengers to the platform?

   a. 11   **b.** 11 R4   **c.** 12

2. Each moon truck needs 12 wheels in order to run. Today, the Apex Truck Factory's wheel department made 173 wheels. How many moon trucks can be assembled with those wheels?

   a. 14   **b.** 14 R5   **c.** 15

Use the schedule to solve Exercises 3–8.

MOON BUS TRIPS TO LUNAR CITY San Francisco to Lunar City STARTING MAY 3830				
San Francisco	8:00 PM	11:00 PM	2:00 AM	5:00 AM
Dallas	9:35	———	———	6:30
Miami	10:50	1:55	———	7:40
New York	11:40	2:45	5:30	8:30
Chicago	12:10	3:15	6:00	———
Lunar City	8:15 AM	11:00 AM	2:05 PM	4:35 PM

3. You want to reach Lunar City by about 2:00 P.M. At what time would you have to leave New York?

4. Garth lives 25 minutes from the Miami bus stop. When must he leave home to catch the 7:40 A.M. bus?

5. You board the Moon Bus in San Francisco at 5:00 A.M. How many cities do you stop in before you arrive in Lunar City?

6. How long will it take you to go from Dallas to Lunar City if you take the bus that leaves San Francisco at 8:00 P.M.

7. At what time would you leave San Francisco in order to make the quickest possible trip to Lunar City?

8. If you leave Dallas at 9:35 P.M., how long will it take you to reach Chicago?

Make a plan. Then solve.

9. The moon travels around Earth at a rate of 2,287 miles per hour. At that rate, how many miles does the moon travel between noon and 6:00 P.M.

10. Mercury circles the sun at about 30 miles per second. Mars move at about half that speed. In 1 hour, how many more miles has Mercury traveled than Mars?

Write a number sentence to solve each problem.

11. The diameter of Venus is about 400 miles less than Earth's diameter. The diameter of Earth is about 7,900 miles. What is the diameter of Venus?

12. Jupiter is the largest planet. Its diameter is about 11.2 times that of Earth. If Earth's diameter is 7,900 miles, what is the diameter of Jupiter?

Solve.

13. A day is much longer on the moon than it is on Earth. The ratio of moon days to Earth days is 3:42. Use equal ratios to find the number of moon days that would equal 28 Earth days.

14. A rocket launch can burn 15 tons of fuel per second. At that rate, how many tons of fuel can it burn in 1 minute?

15. Gravity on the moon is only $\frac{1}{6}$ that of gravity on Earth. Someone who weighs 120 pounds on Earth would weigh only 20 pounds on the moon. How much would a person who weighs 32 pounds on the moon weigh on Earth?

16. The *Apollo 16 Rover* moon vehicle traveled 11.2 miles per second, a lunar record. At that rate, about how long would the vehicle take to travel a distance of 10 miles?

17. The moon is full of mountains and craters. Some lunar mountains rise to a height of 15,000 feet. That is about 75% of the height of North America's tallest peak, Mount McKinley. About how tall is Mount McKinley?

18. Halley's Comet was discovered in 1682. By 1986, it will have appeared 4 times. An equal number of years passes between appearances. How many years have there been between appearances? What fraction of a century (100 years) is that?

Congruent Polygons

A. The solar panels of the space lab are **congruent** to each other. They are the same size and shape.

Line segments that have the same length are **congruent segments.**	*A* ————————— *B* *C* ————————— *D* Read: line segment *AB* is congruent to line segment *CD*. Write: $\overline{AB} \cong \overline{CD}$.
Angles that have the same measure are **congruent angles.**	*H* ∠ *J* ∠ Read: Angle *H* is congruent to angle *J*. Write: $\angle H \cong \angle J$.
Polygons that have the same size and shape are **congruent polygons.**	*N* △ *T* △ *L* ——— *M* *R* ——— *S* Read: Triangle *LMN* is congruent to triangle *RST*. Write: $\triangle LMN \cong \triangle RST$.

B. If two polygons are congruent, the matching or **corresponding** parts are congruent. $ABCD \cong EFGH$

$\angle A \cong \angle E$ $\overline{AB} \cong \overline{EF}$
$\angle B \cong \angle F$ $\overline{BC} \cong \overline{FG}$
$\angle C \cong \angle G$ $\overline{CD} \cong \overline{GH}$
$\angle D \cong \angle H$ $\overline{DA} \cong \overline{HE}$

You can check whether two figures are congruent by tracing one and placing it over the other. If they match, the figures are congruent.

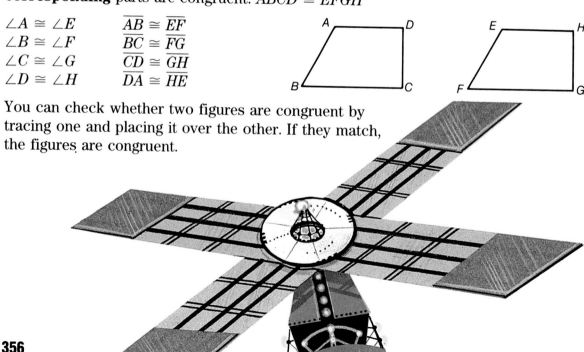

Is the line segment congruent to $\overline{AB}$? Write *yes* or *no*.
Use a ruler to measure. Trace to check.

1.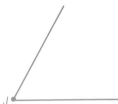

2.

3.

Is the angle congruent to $\angle J$? Write *yes* or *no*. Use a
protractor to measure. Trace to check.

4.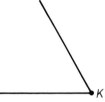

5.

6.

Is the figure congruent to $\triangle ABC$? Write *yes* or *no*. Use
a protractor and ruler to measure the angles and sides.
Trace to check.

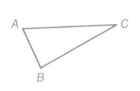

7.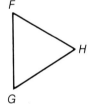

8.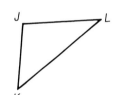

9.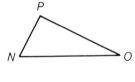

Figure $PQRS \cong$ Figure $WXYZ$
Use this information to answer Exercises 10 and 11.

10. Name the congruent sides.

11. Name the congruent angles.

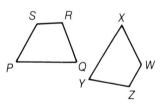

CHALLENGE

Long ago there was a great river. In the middle of the
river were two islands. The islands were connected to
the riverbanks and to each other by seven bridges.
Could a person cross all the bridges and never cross
the same bridge twice? If you think the answer is no,
copy the drawing with the least number of bridges
needed.

Symmetry

If you could fold this picture of the *Saturn 5* rocket along the red line, one half would fit exactly on top of the other half. The blue line is called a **line of symmetry.** The symmetrical shape of the *Saturn 5* helped it to fly to the edge of space.

Some figures have more than one line of symmetry.

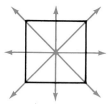 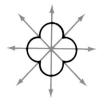 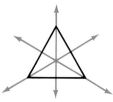

Some figures have no lines of symmetry.

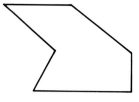 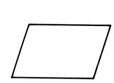 

Sometimes a picture that has two parts has a line of symmetry.

 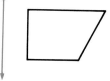 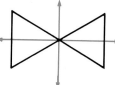

Is the blue line a line of symmetry? Write *yes* or *no*.

1.

2.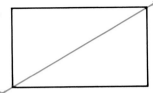

3.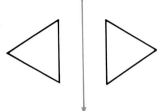

Copy each figure. Draw the line or lines of symmetry for each and count them.

4.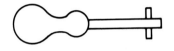

5.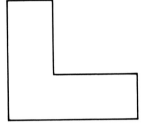

6.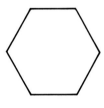

7.

8.

9.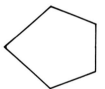

Copy and complete each figure so that the blue line is a line of symmetry.

10.

11.

12.

CHALLENGE

Copy and rearrange the circles in triangle *A* so that they match triangle *B*. You can only move three circles.

A

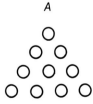

B

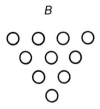

Similar Figures

A. An enlargement of a photograph allows small parts of the photo to be more clearly seen. The shapes of corresponding figures in the original and in the enlargement are the same.

Figures that have the same shape, but not the same size, are **similar.**

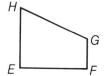

The two figures are similar.

Read: Quadrilateral *ABCD* is similar to quadrilateral *EFGH.*

Write: *ABCD ~ EFGH.*

B. In similar figures, the corresponding angles are congruent. $\triangle STU \sim \triangle KLM$

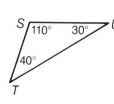
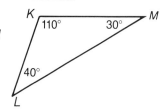

$\angle S \cong \angle K$
$\angle T \cong \angle L$
$\angle U \cong \angle M$

1. Name the figure that is similar to *ABCD.*

   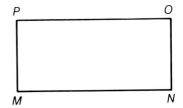

2. Name the triangle that is similar to $\triangle ABC$.

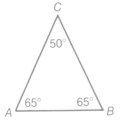 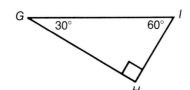 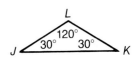

$\triangle TUV \sim \triangle PQR$ Use this information to answer Exercises 3 and 4.

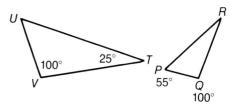

3. Find the measure of $\angle U$.

4. Find the measure of $\angle R$.

Write *true* or *false.*

5. All squares are similar.

6. All triangles are similar.

7. All parallelograms are similar.

8. All circles are similar.

9. Trace the triangle at the right. Then use your protractor to draw a triangle similar to the original one.

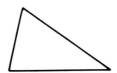

ANOTHER LOOK

Find the product. The answer must be in simplest form.

1. $\frac{1}{3} \times \frac{1}{3}$ **2.** $\frac{1}{3} \times \frac{1}{9}$ **3.** $\frac{4}{5} \times \frac{2}{3}$ **4.** $\frac{5}{8} \times \frac{1}{3}$ **5.** $\frac{1}{6} \times \frac{4}{7}$

6. $\frac{2}{7} \times \frac{1}{2}$ **7.** $\frac{4}{5} \times \frac{1}{6}$ **8.** $\frac{13}{14} \times \frac{2}{3}$ **9.** $\frac{8}{9} \times \frac{3}{4}$ **10.** $\frac{2}{5} \times \frac{3}{4}$

Points on a Grid

A. You can use a number pair, or **ordered pair,** to locate a point on a grid. The two numbers that you use are called **coordinates** of the point.

Find the location of point *A*.

Start at (0,0).
Move 3 spaces to the right.
Move 5 spaces up.
The ordered pair for point *A* is (3,5).

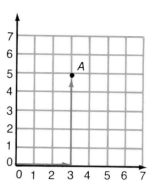

B. You can locate points on a grid and connect the points to form a geometric figure.

Locate (2,1), (1,3), (4,5), (7,3), and (6,1).

Connect the points in order to form a pentagon. Notice that the pentagon is a symmetrical figure.

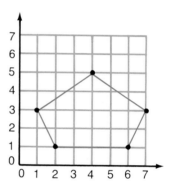

Checkpoint Write the letter of the correct answer.

1. Name the ordered pair for *M*.

    a. 0   **b.** 4   **c.** (0,4)   **d.** (4,0)

2. Name the point for (5,1).

    a. *N*   **b.** *P*   **c.** *Q*   **d.** *R*

3. Name the point for (6,0).

    a. *N*   **b.** *P*   **c.** *M*   **d.** *Q*

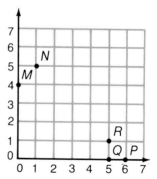

4. Name the ordered pair for *Q*.

    a. (1,5)   **b.** (0,5)   **c.** (5,0)   **d.** (5,1)

Name the ordered pair for each point.

1. *B*　**2.** *A*　**3.** *F*　**4.** *I*

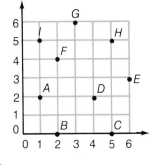

Name the point for each ordered pair.

5. (3,6) **6.** (4,2) **7.** (5,5) **8.** (6,3)

Graph each set of points on a grid. Then connect the points in order to form a geometric figure. Is the figure symmetrical? Write *yes* or *no*.

9. (1,1), (1,3), (6,3), (6,1), (1,1)

10. (1,2), (3,9), (5,2), (1,2)

11. (1,6), (5,10), (9,6), (7,1), (3,1), (1,6)

12. (7,1), (9,3), (9,6), (3,6), (3,3), (5,1), (7,1)

13. (3,5), (2,7), (7,7), (6,5), (3,5)

14. (3,9), (6,9), (8,7), (6,2), (3,2), (1,4), (1,7), (3,9)

CHALLENGE

On many maps, points are located by the regions or sectors that they lie in, rather than by the grid lines. On the map of the moon's features, the crater Tycho is located in sector C1. Name the sector for each of the following features.

1. Mare Frigoris

2. Altai Mountains

3. Schichard Crater

4. Copernicus Crater

5. Mare Tranquillitas

6. Aristarchus Crater

Name the feature found in each of the following sectors.

7. A3

8. E5

9. C4

10. D1

11. C2

12. E4

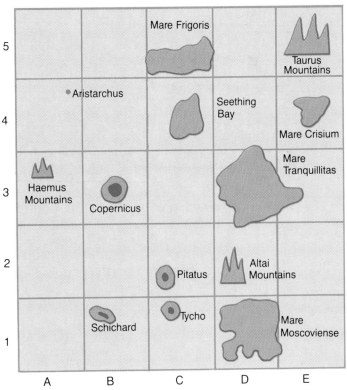

Perimeter

A. When a rocket is launched into space, it gives off great amounts of heat and gas. To protect the scientists and workers, a fence is put up around the launch area. To find how much fence is needed, you need to know the perimeter of the launch area.

The **perimeter** of a figure is the distance around it. To find the perimeter, add the measures of its sides.

$$\begin{array}{r} 75 \\ 60 \\ 75 \\ + 60 \\ \hline 270 \end{array}$$

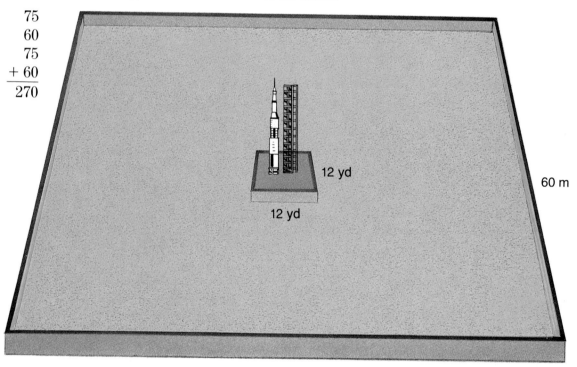

The perimeter of the launch area is 270 m. A total of 270 m of fence are needed.

B. The launchpad itself is a square at the center of the launch area. What is the perimeter of the launchpad?

You can find the perimeter of a regular polygon by multiplying the length of a side by the number of sides.

Find the measure of a side.	Find the number of sides.	Multiply.
12 yd	**4**	**4 × 12 = 48**

The perimeter of the launchpad is 48 yd.

Find the perimeter of each figure.

1.

5 m 9 m 6 m 4 m

2.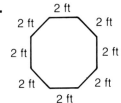

2 ft 2 ft 2 ft 2 ft 2 ft 2 ft 2 ft 2 ft

3.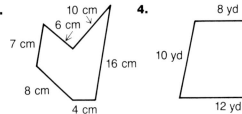

10 cm 6 cm 7 cm 16 cm 8 cm 4 cm

4.

8 yd 10 yd 10 yd 12 yd

Use a centimeter ruler to find the perimeter of each figure.

5.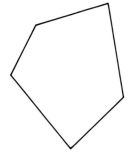

6.

7.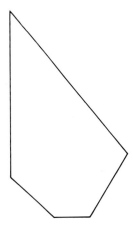

Solve.

8. A viewing platform is constructed near the launchpad of a weather satellite. Everyone watching from the platform has a clear view of the launchpad. The platform has sides that have lengths of 34 yd, 34 yd, 59 yd, and 59 yd. What is the perimeter of the platform?

★9. Rusty constructs a model rocket that is an exact replica of the *Apollo 11* spaceship which landed on the moon. His launchpad is a regular hexagon that has one side 23 in. long. What is the perimeter of Rusty's launchpad?

CHALLENGE

You can find the perimeter of circular objects. First wrap a piece of string around the outside of the object. Then use a ruler to measure the string. This is the perimeter of the object. Try this at home using different-sized cans.

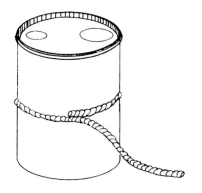

More Practice, page 441

Area of Rectangles

A. The *Viking Lander* was the first spacecraft to set down on the planet Mars. Many photographs of the surface of the planet were taken. Grids on the photographs allowed scientists to measure the precise areas that the photographs covered. What is the area of the grid on the photograph of Mars?

The **area** of a figure is the number of square units it contains. To find the area, you can count square units.

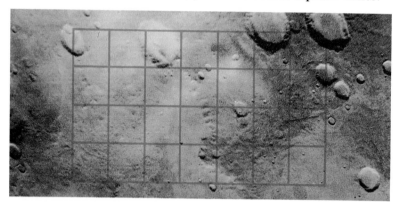

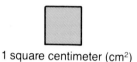

1 square centimeter (cm²)

There are 28 squares in the photograph.

Read: 28 square centimeters.
Write: 28 cm².

The units **km² (square kilometers), m² (square meters)**, and **mm² (square millimeters)** are some other metric units of area.

B. You can also find the area of a rectangle by multiplying the length times the width.

Area of rectangle = length × width
$A = l \times w$
$A = 8 \times 3$
$A = 24$
The area of the rectangle is 24 m².

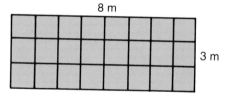

C. You can find the area of a square the same way you would for a rectangle.

There are 5 rows of 5 squares.
So, $5 + 5 + 5 + 5 + 5 = 25$, or $5 \times 5 = 25$.
The area of the square is 25 ft².

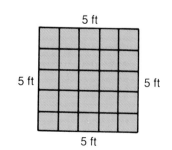

Count to find the area.

1.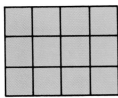

2.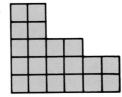

3.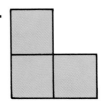

Multiply to find the area.

4.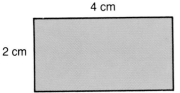

5.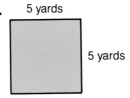

6.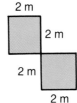

Multiply to find the area of the rectangle or square.

7. $l = 7$ m, $w = 4$ m

8. $l = 16$ ft, $w = 12$ ft

9. $l = 9$ mm, $w = 9$ mm

10. $l = 7$ km, $w = 6$ km

11. $l = 25$ miles, $w = 2$ miles

12. $l = 19$ in., $w = 5$ in.

13. $l = 15$ cm, $w = 15$ cm

14. $l = 42$ m, $w = 7$ m

15. $l = 28$ yd, $w = 28$ yd

16. $l = 9$ mm, $w = 5$ mm

17. $l = 50$ miles, $w = 8$ miles

18. $l = 30$ m, $w = 30$ m

19. $l = 13$ ft, $w = 5$ ft

20. $l = 21$ m, $w = 15$ m

★**21.** a square: one side = 23 yards

★**22.** a square: perimeter = 36 cm

Solve.

23. Scientists plan to track the space shuttle with a laser beam fired from Earth and bounced off a target on the shuttle. The target is a rectangle 8 m wide and 10 m long. What is the area of the shuttle target?

24. Astronauts returning from the moon land their capsule in the ocean. This is called *splashdown*. If the section of ocean they are to land in is 12 miles long and 14 miles wide, what is the area of the splashdown site?

CHALLENGE

Draw two lines to divide the shape into four congruent pieces.

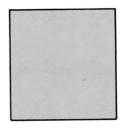

More Practice, page 442

Area of Triangles

Apollo astronauts have driven on the moon three times, in a vehicle called the *Lunar Rover*. The *Rover* has a special antenna which sends and receives radio signals to and from the *Lunar Lander*. The antenna looks like a sunflower whose petals resemble right triangles. The larger the area of the triangles, the better the antenna works. What is the area of the blue triangle on this antenna?

To find the area of a right triangle, you can think of it as part of a rectangle.

The height of the triangle is 4 m. It is the same as the length of the rectangle.
The base of the triangle is 3 m. It is the same as the width of the rectangle.

The area of the triangle is $\frac{1}{2}$ the area of a rectangle.

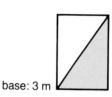

height: 4 m

base: 3 m

Area of rectangle = length × width

$$A = l \times w$$
$$A = 4 \times 3$$
$$A = 12$$

Area of triangle = $\frac{1}{2}$ (base × height)

$$A = \frac{1}{2} (b \times h)$$
$$A = \frac{1}{2} (3 \times 4)$$
$$A = \frac{1}{2} \times 12$$
$$A = 6$$

The area of each triangle on the antenna is 6 m².

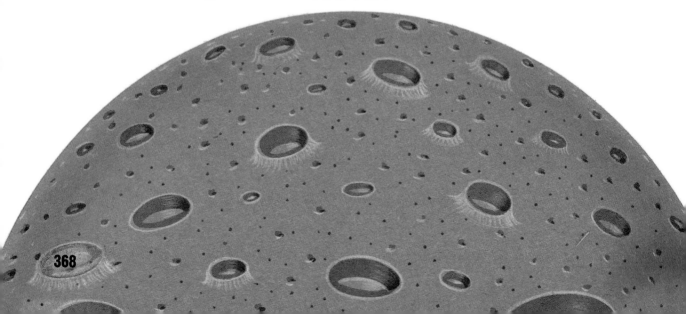

Count to find the area.

1.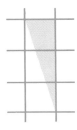

2.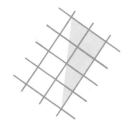

3.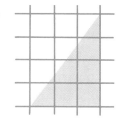

Multiply to find the area.

4.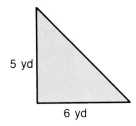

5 yd

6 yd

5.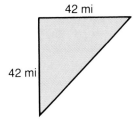

42 mi

42 mi

6.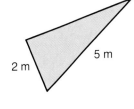

5 m

2 m

Copy and complete the chart.

A TRIANGLE WITH:

	Base (b)	Height (h)	Area (A)
7.	6 m	11 m	▦ m^2
8.	12 cm	9 cm	▦ cm^2
9.	2 mi	17 mi	▦ mi^2
10.	15 km	8 km	▦ km^2
11.	6 yd	7 yd	▦ yd^2
12.	17 in.	18 in.	▦ $in.^2$
★13.	▦ ft	9 ft	36 ft^2
★14.	▦ m	6 m	21 m^2

CHALLENGE

The drawing is made up of a number of overlapping triangles. Copy the drawing and find a systematic way to find and list all the triangles.

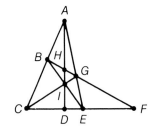

More Practice, page 442

PROBLEM SOLVING
Writing a Number Sentence

Writing a number sentence can help you solve a word problem. A number sentence can help you figure out the relationship between the numbers you know and the number you need to find.

Marcia learned that because the moon travels around Earth, a spacecraft must be aimed at a spot far ahead of the moon. The spacecraft's speed and the moon's speed must be carefully calculated so that they both arrive at the same spot at the same time. The moon travels 11,455 miles in the first 5 hours of a spaceflight. How far does the moon travel in the first hour?

1. List what you know and what you want to find.

The moon travels 11,455 miles in 5 hours.
How far does the moon travel in 1 hour?

2. Think about how to use this information to solve the problem.

You know that the number in each group is the same. You want to find how many in each group. So, divide.

3. Write a number sentence. Use n to stand for the number you want to find.

miles traveled $\div$ hours $= n$
$11{,}455 \div 5 = n$

4. Solve. Write the answer.

$11{,}455 \div 5 = 2{,}291$
$n = 2{,}291$

The moon travels 2,291 miles in the first hour.

Write the letter of the number sentence you would use
to solve the problem.

1. During lift-off, the first stage of the rocket-propelled *Apollo 11* traveled at 984 miles per hour for 0.042 hours. How far did the rocket climb during this stage?

  a. $984 + 0.042 = n$ **b.** $984 - 0.042 = n$
  c. $984 \times 0.042 = n$ **d.** $984 \div 0.042 = n$

2. The *Apollo* command module was 11 m long. The *Apollo* lunar module was 1.6 times shorter than the command module. What was the length of the lunar module?

  a. $11 + 1.6 = n$ **b.** $11 - 1.6 = n$
  c. $11 \times 1.6 = n$ **d.** $11 \div 1.6 = n$

Solve.

3. Technicians on Earth use instruments to monitor closely the astronauts' heartbeats during liftoff. One astronaut's heart rate registered 180. His normal heart rate is $\frac{1}{3}$ of that. What is his normal heart rate?

4. During the rocket's third stage, the spacecraft travels 2,025 miles in 5 minutes. What distance does the spacecraft travel in 3 minutes of third-stage flight?

5. Spaceflight *Apollo 11*, the first expedition to land on the moon, lasted 195 hours. Of that time, 13% actually was spent on the moon. How many hours were spent on the moon?

6. A spacecraft needs to travel at a speed of 7 miles per second to escape Earth's gravity. At that speed, how many seconds would it take to travel the 24,997 miles around Earth?

7. Between 1958 and 1983, 18 scientific satellites were sent into orbit above Earth. During the same period, 6 weather satellites were placed in orbit. What is the ratio of scientific satellites to weather satellites placed in orbit during that period?

8. One *Skylab* mission lasted for 84 days. To keep in shape, the astronauts spent 48 hours exercising. What is the ratio of the number of days spent exercising to the total number of days of the space mission?

Solid Figures

A. The nose of the *Atlas Centaur* rocket has the shape of a cone. A *cone* is an example of a solid figure.

B. Each flat surface of a prism is a **face.**
A prism has two congruent faces, called **bases.**
The rectangular prism in the figure has six faces.

Two faces intersect in an **edge.**
The rectangular prism has twelve edges.

The edges of a solid figure intersect at a
vertex. The rectangular prism has eight vertices.

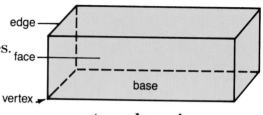

rectangular prism

A prism is often named by the shape of its base.

**triangular
prism**

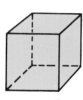

cube

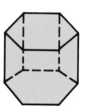

**hexagonal
prism**

C. A **pyramid** has one base. The other faces are in the shape of triangles. A pyramid can be named by the shape of its base.

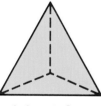

**triangular
pyramid**

**square
pyramid**

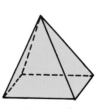

**rectangular
pyramid**

**hexagonal
pyramid**

Here are some other solid figures.

cylinder

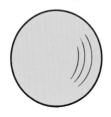

sphere

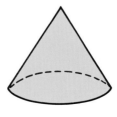

cone

Name the solid figure that is shaped like the object.

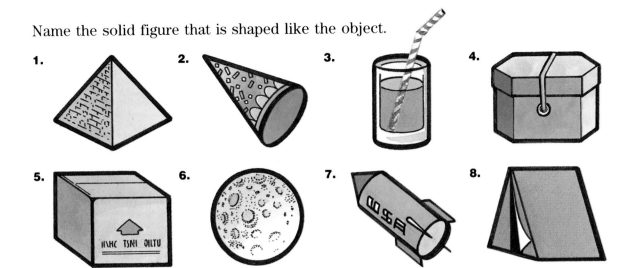

1.   **2.**   **3.**   **4.**

5.   **6.**   **7.**   **8.**

Count the number of faces, edges, and vertices each figure has.

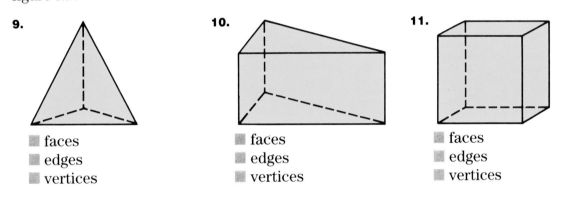

9.
- ▨ faces
- ▨ edges
- ▨ vertices

10.
- ▨ faces
- ▨ edges
- ▨ vertices

11.
- ▨ faces
- ▨ edges
- ▨ vertices

Copy and complete the table.

Figure	Number of faces	Number of edges	Number of vertices
square pyramid	**12.** ▨	**13.** ▨	**14.** ▨
hexagonal prism	**15.** ▨	**16.** ▨	**17.** ▨
cone	★**18.** ▨	★**19.** ▨	★**20.** ▨

CHALLENGE

The surface area of a figure is the sum of the areas of its faces.
Find the surface area of the rectangular prism shown at the right.

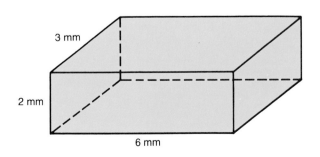

3 mm

2 mm

6 mm

Volume

A. The **volume** of a rectangular prism is the number of cubic units that are needed to fill it.

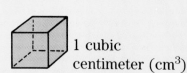 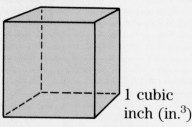

1 cubic centimeter (cm³)

1 cubic inch (in.³)

To find the volume of a rectangular prism, you can count the number of cubic units.

Each layer has six cubic centimeters.
There are 2 layers.
$6 + 6 = 12$

The volume is 12 cubic centimeters.

You can also multiply to find the volume of a rectangular prism.

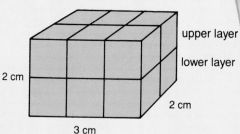

upper layer

lower layer

Volume of rectangular prism = length × width × height

$V = l \times w \times h$
$V = 3 \times 2 \times 2$
$V = 12$

B. Modular lockers on the space shuttle contain scientific equipment to be used for experiments. Each locker is 17 in. by 20 in. by 10 in. You can use the following formula to find the volume of the locker.

$V = l \times w \times h$
$V = 20 \times 17 \times 10$
$V = 3,400$

The volume of a modular locker is 3,400 in.³

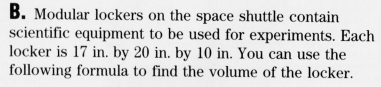
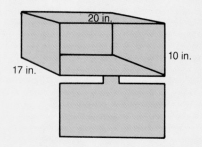

374

Find the volume of each.

1.
3 cm
3 cm
1 cm

2.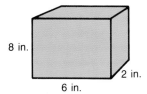
8 in.
2 in.
6 in.

3.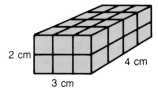
2 cm
4 cm
3 cm

4.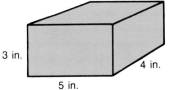
3 in.
4 in.
5 in.

5.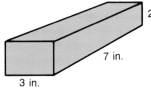
2 in.
7 in.
3 in.

6.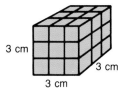
3 cm
3 cm
3 cm

Copy the table.
Use the formula $V = l \times w \times h$ to complete.

length (*l*)	width (*w*)	height (*h*)	Volume (*V*)
5 cm	3 cm	2 cm	**7.**
16 mm	12 mm	7 mm	**8.**
4 in.	1 in.	6 in.	**9.**
7 yd	5 yd	3 yd	**10.**
★**11.**	3 ft	5 ft	105 ft³

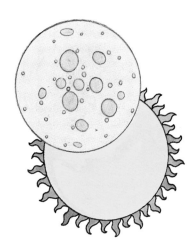

Solve.

12. An astronaut has a toolbox that is 24 in. long, 12 in. wide, and 8 in. high. What is the volume of the toolbox?

13. A cubic storage box has a volume of 1,000 cubic centimeters. How long is each edge of the storage box?

CHALLENGE

Use the fewest possible number of colors to find how many colors you would need to paint the cube at the right so that each face is only one color, and no face is the same color as the face next to it.

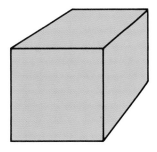

CALCULATOR

You can use a calculator to help you calculate perimeter and area. Remember you cannot enter fractions into a calculator. You must convert any fractions to decimals.

Example: Find the perimeter of a square that measures $8\frac{1}{2}$ ft. per side. You can write $8\frac{1}{2}$ as 8.5.
To find the perimeter:

Press: [4] [×] [8] [.] [5] [=]

The display should show 34. The perimeter is 34 ft.

Copy and complete the table.

Room	Perimeter	Area
1. Living Room		
2. Kitchen		
3. Bedroom 1		
4. Bath		
5. House		

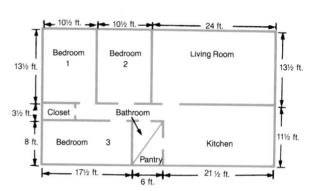

To the nearest 0.05 yd^2 how much carpet would it take to carpet these rooms?

What is the cost of each of the following if the carpet is $17 per square yard?

Room	Carpet Sq. Feet	Square Yards	Cost
Example: Living Room	324 ft^2	36 yd^2 (324 ÷ 9)	$ 612.00
6. Bedroom 2			
7. Bedroom 3			
8. Kitchen			
9. TOTAL			

GROUP PROJECT

Space or Bust

The problem: After visiting the planetarium, you and your classmates decide to build your own model of a spaceship. Think about the Key Questions before you draw your design.

Key Questions

- Will it be modeled after a real rocket, or will you design your own?
- What materials will you need?
- What will the skin of the spaceship be?
- What scale will you use?
- How many people will it be able to carry?
- Will the spaceship be designed for a long trip into space or for short trips to neighboring planets and the moon?
- How much storage space will be needed for fuel, food, water, air, and other supplies?

CHAPTER TEST

Identify and name each figure. (page 340)

1.   S T

2.   G H

3. 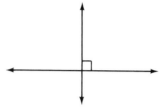 m

4.   L M
   A B

Are the lines perpendicular? Write *yes* or *no*. (page 340)

5.

6.

Use your protracter to draw the following angles.
Identify them as acute, obtuse, right, or straight. (page 344)

7.   32°   **8.**   90°   **9.**   180°   **10.**   100°

11. Is the triangle equilateral or isoceles? (page 346)

12. Is the triangle acute or obtuse? (page 346)

Name the polygon or the part of a circle. (pages 350, 352)

13.   **14.**   **15.**   **16.**

17. Are the two figures congruent?
Write *yes* or *no*. (page 356)

18. Is the line a line of symmetry?
Write *yes* or *no*. (page 358)

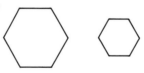

19. Find the measure of $\angle R$.

20. Find the measure of $\angle V$.

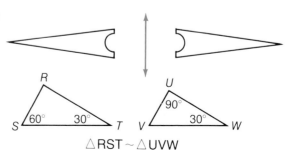

$\triangle RST \sim \triangle UVW$

Name the coordinates for each point. (page 362)

21. *A*   **22.** *D*

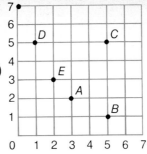

Name the point for each ordered pair. (page 362)

23. (2,3)   **24.** (5,1)

Find the perimeter. (page 364)

25. a square, one side = 9 in.

26. an equilateral triangle, one side = 6 ft

Find the area. (pages 366 and 368)

27. a triangle, $b = 6$ cm, $h = 9$ cm

28. a rectangle, $l = 7$ yd, $w = 6$ yd

Find the volume. (pages 372 and 374)

29. a rectangular prism, $l = 4$ ft, $w = 6$ ft, $h = 2$ ft

Solve. Make a table to find the pattern. (pages 348–349)

30. A *centrifuge* is used to test astronauts' ability to take high speeds. During one test, the centrifuge moved at 11 miles per hour in 0.5 seconds, 22 miles per hour in 1 second, 33 miles per hour after 1.5 seconds, and 44 miles per hour after 2 seconds. What was the machine's speed at 4, at 5.5, and at 8 seconds?

31. To test deceleration, the centrifuge was started at 264 miles per hour and slowed to 231 miles per hour after 4 seconds, 198 miles per hour after 8 seconds, and 165 miles per hour after 12 seconds. How fast was the machine moving after 20 seconds? How long did it take to stop?

Solve. (pages 370–371)

32. The distance from Earth to the moon and back is 496,000 miles. One spacecraft made the round trip in 8 days, traveling the same distance each day. How many miles did the spacecraft travel each day?

33. A cometary probe travels 32,704 kilometers each day as it moves towards Halley's comet. How far will the comet travel after 1 month? (HINT: a month is 30 days.)

379

RETEACHING

A. You can use a protractor to measure angles. Place the center of the protractor on the vertex of the angle.

Place the edge of the protractor so that one side of the angle crosses the zero mark on one of the scales.

Read the measure of the angle where the other side crosses the same scale.

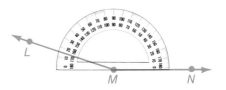

B. To draw an angle of 165°, draw $\overrightarrow{MN}$. Place the center of the protractor on M so that the rays run through O. Find the obtuse mark for 165°, and mark point L. Draw $\overrightarrow{ML}$.

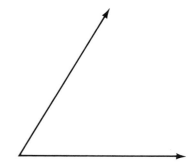

Use a protractor to measure each angle.

1.

2.

3.

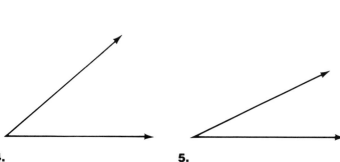

4.

5.

6.

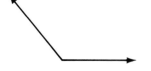

Use a protractor to draw each angle.

7. 32°   **8.** 80°   **9.** 35°   **10.** 12°

11. 90°   **12.** 125°   **13.** 165°   **14.** 147°

ENRICHMENT

Slides, Flips, and Turns

You can change the position of a figure by **sliding** it, **flipping** it, or **turning** it.

To *slide* a figure, you move it along a line.

To *flip* a figure, you "flip" it about a line of symmetry.

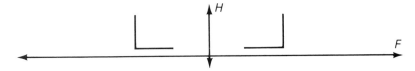

To *turn* a figure, you turn it on a curved path around a point.

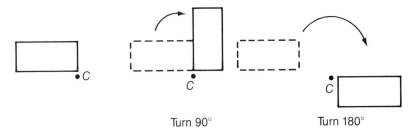

Turn 90°   Turn 180°

Compare the figure on the right to the figure on the left. Tell whether it was *turned, flipped,* or *slid*.

1.

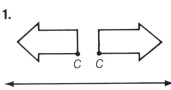

3.

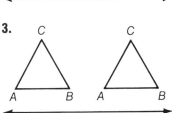

2.

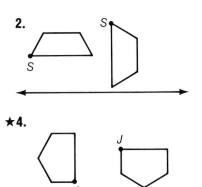

★4.

CUMULATIVE REVIEW

Write the letter of the correct answer.

1. Compute $\frac{7}{8} = \frac{21}{\blacksquare}$.

   a. 16   **b.** 20

   c. 24   **d.** not given

2. Write as a percent: $\frac{192}{100}$.

   a. 19.2%   **b.** 192%

   c. 1,920%   **d.** not given

3. Write 75% as a fraction.

   a. $\frac{2}{3}$ **b.** $\frac{3}{4}$

   c. $\frac{4}{5}$ **d.** not given

4. Write 58% as a decimal.

   a. 0.058   **b.** 0.508

   c. 0.58   **d.** not given

5. 37% of 1,500

   a. 540   **b.** 550

   c. 555   **d.** not given

6. What percent of 25 is 5?

   a. 20%   **b.** 125%

   c. 500%   **d.** not given

7. $9 \div \frac{1}{5}$

   a. $\frac{1}{45}$ **b.** 45

   c. 90   **d.** not given

8. Which unit would you use to measure the mass of a pen?

   a. milligrams   **b.** grams

   c. kilograms   **d.** not given

9. $4,056 \div 45$

   a. 9 R6   **b.** 90

   c. 90 R6   **d.** not given

10. $18\overline{)\$36.95}$

   a. $2.05   **b.** $2.0527

   c. $20.52   **d.** not given

11. 121×483

   a. 57,443   **b.** 65,975

   c. 72,323   **d.** not given

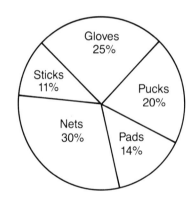

Total Costs: $2,500

12. How much was spent on pucks?

   a. $500.00   **b.** $1,000.00

   c. $1,500.00   **d.** not given

13. How much was spent on nets?

   a. $75.00   **b.** $750.00

   c. $1,250.00   **d.** not given

Some communities in the United States were planned more carefully than others. Have you looked at a map of your town or city? Are the streets arranged in any kind of pattern? Is the layout of the oldest part of the city different from the layout of the modern part? Draw a map of your town. Label areas of historic interest.

12 STATISTICS AND PROBABILITY

Making a Bar Graph

The population of the thirteen original colonies changed from 1660 to 1780. You can make a **bar graph** to show and compare population growth.

To make a bar graph:

1. Round the numbers in the data table to a convenient place.

2. Draw and label the vertical and horizontal axes. Choose intervals between the numbers that best display the data. Title the graph.

3. Draw the bars on the graph.

You can see from the graph how the population grew from 1660 to 1780.

POPULATION OF AMERICAN COLONIES: 1660–1780

Year	Actual population	Rounded to hundreds of thousands
1660	75,058	100,000
1690	210,372	200,000
1720	466,185	500,000
1750	1,170,760	1,200,000
1780	2,780,369	2,800,000

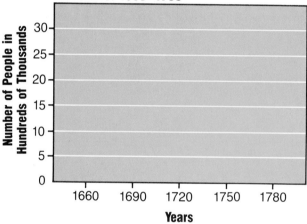

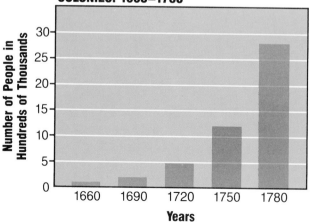

Solve.

1. The table lists the populations of California and of four other states in 1850. Copy and complete the bar graph. Use the data in the table. Round each population to the nearest tenth of a million.
3,097,394 ⟶ 3.1 million

POPULATIONS OF SELECTED STATES: 1850

State	Actual population	Rounded amounts
New York (N.Y.)	3,097,394	
California (Calif.)	92,597	
Michigan (Mich.)	397,654	
Virginia (Va.)	1,119,348	
Tennessee (Tenn.)	1,002,717	

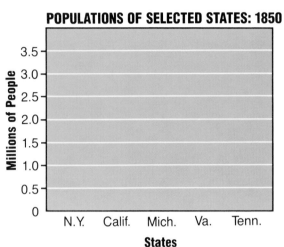

2. Which states have about double the population of Michigan?

3. The table lists some results of the 1980 census. Use the table to write and solve your own problem.

4. Make a bar graph that shows the number of students in each grade at your school. Use it to estimate the size of next year's classes.

PEOPLE PER SQUARE MILE: 1980

State	People per square mile
New York (N.Y.)	371
Florida (Fla.)	180
California (Calif.)	151
Texas (Tex.)	54

ANOTHER LOOK

Solve.

1. 6325 g = ▨ kg
2. 300 m = ▨ km
3. 15 kg = ▨ g
4. 932 km = ▨ m
5. 388 m = ▨ mm
6. 8,043 mL = ▨ L
7. 93 km = ▨ cm
8. 13 L = ▨ mL

More Practice, page 443

Making a Pictograph

Pictographs use symbols to represent large numbers. You can draw a pictograph to show the growth of United States cotton production from 1790 to 1810.

To make a pictograph:

1. Round the numbers in the data table to a convenient place.

U.S. COTTON PRODUCTION: 1790–1810

Year	Bales of cotton	Bales in five thousands
1790	3,000	5,000
1795	17,000	15,000
1800	73,000	75,000
1805	148,000	150,000
1810	178,000	180,000

2. List the years vertically as shown in the chart. Title the graph.

U.S. COTTON PRODUCTION: 1790–1810

Year	Bales of cotton
1790	
1795	
1800	
1805	
1810	

3. Choose a symbol to represent an approximate amount of cotton.

Let □ = 10,000 bales
Let ▽ = 5,000 bales

List the symbols. Replace the numbers with symbols. Complete the pictograph.

U.S. COTTON PRODUCTION: 1790–1810

Year	Bales of cotton
1790	▽
1795	□▽
1800	□□□□□□□▽
1805	□□□□□□□□□□□□□□□
1810	□□□□□□□□□□□□□□□□□□

□ = 10,000 bales   ▽ = 5,000 bales

You can see from the pictograph how cotton production increased from 1790 to 1810.

Solve.

1. Copy and complete the pictograph. Use the data in the table.

U.S. CORN PRODUCTION: 1940–1980

Year	Bushels of corn
1940	2,500,000,000
1950	3,000,000,000
1960	4,500,000,000
1970	4,000,000,000
1980	6,500,000,000

U.S. CORN PRODUCTION: 1940–1980

Year	Bushels of corn
1940	
1950	
1960	
1970	
1980	

◯ = 1 billion bushels

2. Use the information from the pictograph to tell whether corn production should rise or fall by 1990.

3. The table lists the average number of bushels harvested per acre for certain crops grown in the United States. Round the numbers to the nearest ten thousand. Let △ = 10,000 bushels. Then make a pictograph that uses all of this information.

HARVESTS OF SOME U.S. CROPS

Farm crop	Bushels per acre	Rounded amounts
Corn	91,000	
Sorghum	46,300	
Wheat	33,400	
Rye	24,400	

4. Native Americans taught the Pilgrims how to grow corn, one of America's favorite vegetables. Make a pictograph that gives your classmates' favorite vegetables.

ANOTHER LOOK

Find the area of

1. a triangle.

base = 4 in.
height = 19 in.
area = ▨

2. a square.

sides = 9 mm
area = ▨

3. a rectangle.

length = 17 km
width = 4 km
area = ▨

4. a triangle.

base = 12 yd
height = 13 yd
area = ▨

Making a Broken-Line Graph

Broken-line graphs are used to show a continuing trend during a period of time. You can make a broken-line graph to show the growth of railroads in the United States from 1840 to 1870.

To make a broken-line graph:

1. Round the numbers in the data table to a convenient place.

GROWTH OF U.S. RAILROADS: 1840–1870

Year	Miles of track	Miles in thousands
1840	2,818	3,000
1845	4,633	5,000
1850	9,021	9,000
1855	18,374	18,000
1860	30,626	31,000
1865	35,085	35,000
1870	52,933	53,000

2. Draw and label the vertical and horizontal axes. Choose intervals between the numbers that best display the data. Title the graph.

3. Place the points on the graph. Then connect all the points with line segments.

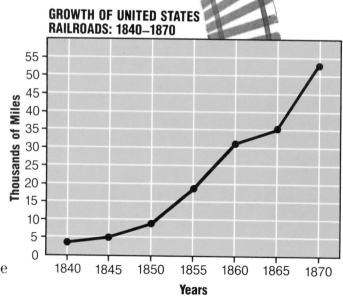

GROWTH OF UNITED STATES RAILROADS: 1840–1870

You can see from the graph how the railroads grew from 1840 to 1870.

Solve.

1. About 122 million cars use the roads in the United States. Copy and complete the broken-line graph to show car sales in the United States.

U.S. CAR SALES: 1979–1984

Year	Number of cars	Rounded amounts
1979	10,750,000	11
1980	9,000,000	9
1981	8,000,000	8
1982	8,200,000	8
1983	9,200,000	9
1984	10,500,000	11

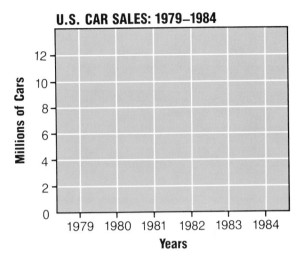

2. According to your broken-line graph, did car sales go up or down between 1979 and 1981? Do you think car sales will go up or down between 1984 and 1990?

3. Use the data in the table to make a broken-line graph to show car sales in the United States from 1946 to 1950.

4. What is different about the trend shown in the broken-line graph from 1979 to 1984 and the trend shown in the graph from 1946 to 1950?

U.S. CAR SALES: 1946–1950

Year	Number of cars
1946	2,100,000
1947	3,500,000
1948	3,900,000
1949	5,100,000
1950	6,600,000

ANOTHER LOOK

Divide.

1. $17.5 \div 4$
2. $89.1 \div 9$
3. $3 \div 6$
4. $25.3 \div 4$
5. $0.175 \div 5$
6. $36.3 \div 5$
7. $705.1 \div 4$
8. $0.304 \div 4$

PROBLEM SOLVING
Interpreting a Graph

Graphs are a good way to show and compare data. But you must be careful when comparing graphs. Sometimes they can be misleading.

> Use the two line graphs to answer the question. Which area had more miles of railroad track in 1880, the United States or Europe?

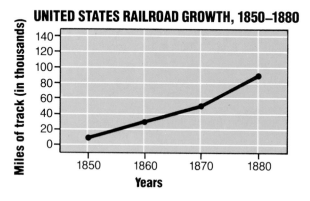

UNITED STATES RAILROAD GROWTH, 1850–1880

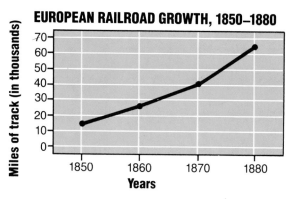

EUROPEAN RAILROAD GROWTH, 1850–1880

- The title and the labels at the bottom tell you that both graphs show railroad growth between 1850 and 1880. The graph at the left shows railroad growth in the United States. The graph at the right shows railroad growth in Europe.

- The labels at the left side of each graph mark the number of miles of railroad track in each area. The spaces are marked in 20,000-mile units on the first graph. The second graph is marked in 10,000-mile units.

- You have to read carefully to compare the information in these graphs. The broken line appears to be about the same, but the information is not presented in the same way in both graphs.

By using the scale at the left of each graph, you can see that Europe had 65,000 miles of track in 1880. The United States had 90,000 miles of track in 1880. In 1880, the United States had more miles of railroad track.

Use the information in the graphs to solve each problem.

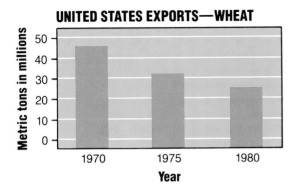

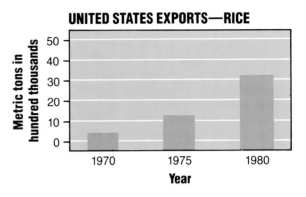

1. Did the United States export more wheat or more rice in 1975?

2. Did the United States export more wheat or more rice in 1980?

3. Which export shows the greatest change between 1970 and 1980?

4. About how many more metric tons of rice were exported in 1970 than in 1980?

5. Which export had the greatest decrease?

6. How many more metric tons of wheat were exported in 1970 than in 1975?

7. Did rice exports in 1980 equal wheat exports in 1975?

Mean, Median, Mode, and Range

A. A candidate for the office of President of the United States must be at least 35 years old. Find the mean age of our first five Presidents on Inauguration Day.

George Washington   age 57
John Adams   age 61
Thomas Jefferson   age 57
James Madison   age 57
James Monroe   age 58

To find the **mean,** or average, you add and then divide by the number of addends.

$57 + 61 + 57 + 57 + 58 = 290$
$290 \div 5 = 58$ The mean age was 58.

Means can also be written as decimals.

$65 + 42 + 71 + 82 + 41 = 301$
$301 \div 5 = 60.2$

B. If you put the numbers in order, the number in the middle is the **median.**

57, 57, 57, 58, 61   The median age is 57.

C. Look at the list of ages. The number that appears most often is the **mode.** Find the mode.

57, 61, 57, 57, 58   The mode is 57.

D. The difference between the greatest number and the least number on a list is called the **range.**

57, 61, 57, 57, 58
$61 - 57 = 4$ The range is 4.

Checkpoint   Write the letter of the correct answer.

1. Find the mean: 6, 5, 9, 22, 3.

   a. 6   **b.** 9   **c.** 19   **d.** 225

2. Find the median: 4, 12, 14, 19, 25.

   a. 4   **b.** 12   **c.** 14   **d.** 25

3. Find the mode: 1, 36, 45, 1, 17.

   a. 1   **b.** 20   **c.** 36   **d.** 44

4. Find the range: 9, 41, 7, 4, 4.

   a. 4   **b.** 7   **c.** 13   **d.** 37

Find the mean. Then find the median.

1. 9, 7, 8

2. 3, 9, 16, 19, 6

3. 14, 8, 6, 2, 5

4. 19, 6, 8, 16, 2

★**5.** 1, 9, 3, 7, 6

★**6.** 3, 7, 21, 8, 4

Find the mode. Then find the range.

7. 6, 6, 8, 14

8. 2, 23, 4, 9, 23, 10

9. 22, 32, 23, 16, 23

10. 117, 98, 102, 99, 98

11. 56, 42, 83, 42, 56, 83, 56

12. 34, 67, 85, 19, 34, 19, 34

Solve. For Problem 14, use the Infobank.

13. John Adams, the first Vice-President of the United States, was 54 years old when elected. The next three Vice-Presidents, Thomas Jefferson, Aaron Burr, and George Clinton, were 54, 45, and 66 years old when elected.
What was their mean age?
What was the mode?

14. Our largest President, William Howard Taft, weighed more than 300 pounds and was about 6 feet tall. Use the information on page 420 to write and solve your own word problem.

MIDCHAPTER REVIEW

Find the mean. Then find the median.

1. 6, 12, 9

2. 11, 5, 9, 3, 2

3. 23, 17, 31, 19, 15

4. 15, 9, 27, 17, 21

Find the mode. Then find the range.

5. 98, 84, 89, 91, 84

6. 73, 83, 74, 84, 39, 74

7. 94, 83, 59, 96, 38, 96

8. 104, 47, 39, 105, 39, 39, 47

PROBLEM SOLVING
Making a Diagram

Making a diagram can help you solve some problems. One kind of diagram that is often helpful is a tree diagram. A tree diagram is so named because it resembles the branches of a tree.

> The United States has honored leaders such as George Washington, Thomas Jefferson, and Abraham Lincoln. National holidays and monuments have been dedicated to them, and pictures of each appear on money and stamps. How can you best list the number of honors all three Presidents could have received?

You can use a tree diagram to solve this problem easily. First, list the possible combinations. Then, count them.

Person	Honor	Combination
G. Washington	Holiday	Washington–Holiday
	Monument	Washington–Monument
	Money	Washington–Money
	Stamp	Washington–Stamp
T. Jefferson	Holiday	Jefferson–Holiday
	Monument	Jefferson–Monument
	Money	Jefferson–Money
	Stamp	Jefferson–Stamp
A. Lincoln	Holiday	Lincoln–Holiday
	Monument	Lincoln–Monument
	Money	Lincoln–Money
	Stamp	Lincoln–Stamp

There are 12 possible combinations. Another way to find the number of possible combinations is to multiply the number of people by the number of honors.

people   honors   combinations
$$3 \times 4 = 12$$

Copy and complete each diagram.

1. Bruce, James, and Karen are each giving a report on a famous person from United States history. They can choose Thomas Paine, John Adams, or James Madison. How many combinations of students and reports can there be?

Student	Report	Combination
Bruce	Paine	Bruce–Paine
	Adams	Bruce–Adams

2. The students of Hill School are going to put up a lobby display as part of their study of United States history. They can display a copy of the Declaration of Independence, the Constitution, or the Bill of Rights. The document can be displayed in a wall case, a glass cabinet, or a pedestal case. How many combinations are there?

Document	Display	Combination
Constitution	wall case	Constitution–wall case
	cabinet	Constitution–cabinet

Solve. Use a tree diagram if needed.

3. Mrs. Jones, Ms. Martinez, and Mr. Zedd are taking their classes on a trip. They will be visiting historical sites. They can go to Boston, Philadelphia, or Washington, D.C. How many combinations can there be of classes and sites?

4. Chris, Sandy, Kim, and Rita must each portray a famous woman from the Revolutionary War period. They can be Abigail Adams, Dolley Madison, Betsy Ross, or Martha Washington. How many combinations can there be?

5. The 6 fifth-grade classes in Abbey School are each doing a presentation about a famous person in our country's history. The possible subjects are John Adams, Ben Franklin, Alexander Hamilton, and George Washington. How many combinations can be made?

Probability

A. To prove his strength, George Washington once threw a silver dollar across the Potomac River. What is the probability that it landed heads?

It is **equally likely** that the coin will land either heads or tails.

Probability can be shown as a fraction.

$$\frac{\text{Number of favorable outcomes (heads)}}{\text{Number of possible outcomes (heads + tails)}} = \frac{1}{2}$$

There is 1 chance out of 2 of heads. So, the probability of the coin landing heads is $\frac{1}{2}$.

B. Is it equally likely that the spinner will stop on red or on yellow?

Find the probability of each event.

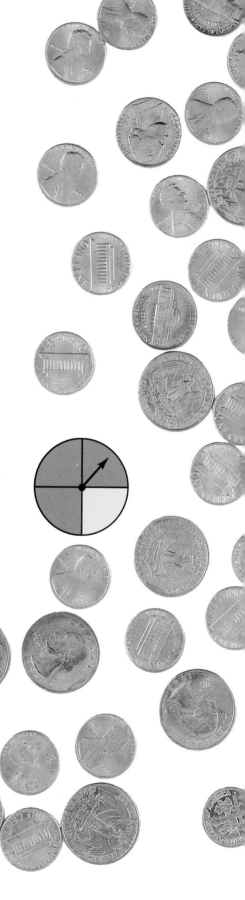

Stopping on red
Favorable outcomes: 3
Possible outcomes: 4
Probability: $\frac{3}{4}$

Stopping on yellow
Favorable outcomes: 1
Possible outcomes: 4
Probability: $\frac{1}{4}$

$$\frac{3}{4} > \frac{1}{4}$$

It is more probable that the spinner will stop on red. The two events are **not equally likely.**

Suppose you pick one marble from the jar with your eyes closed. Then you return it to the jar. Write a fraction for the probability of picking

1. a red marble.
2. a green marble.
3. a yellow marble.
4. a blue marble.
★5. a green or a blue marble.
★6. a blue or a yellow marble.

Suppose the cards are turned face down and rearranged. You pick one card. Then you return it. Write a fraction for the probability of picking

7. the letter *W*.
8. the letter *T*.
9. the letter *N*.
10. a vowel.
11. a consonant.

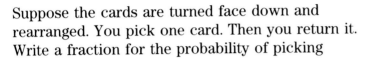

Solve.

Jamie is going to write a report about a President on this chart. Suppose he places all the names in a hat and picks one. What is the probability that

12. he will pick George Washington?

13. he will pick a President whose first name is John?

14. he will pick a President whose last name begins with *M*?

15. he will pick a President who was born in South Carolina?

16. he will pick a President who was born in New York or in Massachusetts?

★17. he will pick a President who was *not* born in Massachusetts?

President	Birthplace
George Washington	Virginia
John Adams	Massachusetts
Thomas Jefferson	Virginia
James Madison	Virginia
James Monroe	Virginia
John Quincy Adams	Massachusetts
Andrew Jackson	South Carolina
Martin Van Buren	New York

More Practice, page 444

More Probability

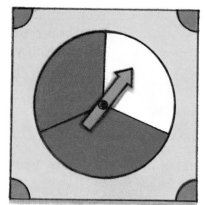

A. What is the probability that the spinner will stop on a color of the United States flag?

$$\frac{\text{Favorable outcome (red + white + blue)}}{\text{Possible outcomes (red + white + blue)}} = \frac{3}{3}, \text{ or } 1$$

It is **certain** that the spinner will stop on a color of the flag.

So, the probability of a certain event is 1.

What is the probability that the spinner will stop on green?

$$\frac{\text{Favorable outcomes (green)}}{\text{Possible outcomes (red + white + blue)}} = \frac{0}{3}, \text{ or } 0$$

It is **impossible** that the spinner will stop on green.

So, the probability of an impossible event is 0.

B. Andy plans to spin the spinner 30 times. He can predict about how many times the spinner will stop on red, on white, or on blue.

To make a prediction, multiply the probability of each separate event by the total number of spins.

Event	Total spins	Predicted outcome
$\frac{1}{3}$ (probability of red)	× 30	About 10 spins will stop on red.
$\frac{1}{3}$ (probability of white)	× 30	About 10 spins will stop on white.
$\frac{1}{3}$ (probability of blue)	× 30	About 10 spins will stop on blue.

Andy spins the spinner 30 times. He keeps a tally of his results. A **tally** is a way to record results.

Outcome	Prediction	Tally
Red	10	卌 IIII
White	10	卌 III
Blue	10	卌 卌 III

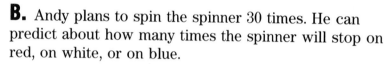

I = 1 spin

卌 = 5 spins

Andy's results are close to his predicted outcome.

Suppose you pick a marble from the bag. Predict the probability of each event below. Write *certain* or *impossible* for each event. What is the probability of picking

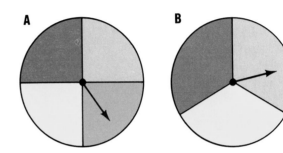

1. a yellow marble?
2. a red or a green marble?
3. a white marble?
4. a green or a red marble?

Use the spinners to answer the questions.

5. Is spinner A as likely to stop on green as on red?
6. What is the probability that spinner A will stop on either red or green?
7. Use spinner B to solve. What is the probability that it will stop on blue? on yellow?

8. As an experiment, predict how many times a coin will land either heads and how many times will it land tails in 50 tosses. Then toss a coin 50 times and record the results. What was your prediction? What were your results? How accurate was your prediction?

★9. Construct a six-sided number cube. Number each side from 1 to 6. How many times will side number 4 turn up in 24 tosses? How many times will either sides 2 or 5 turn up in 36 tosses? Toss the cube and compare your results with your predictions.

CHALLENGE

When there are an even number of numbers, the median is the average of the two middle numbers.
Find the median: 12, 15, 17, 25, 33, 36
                       middle numbers

$(17 + 25) ÷ 2 = 21$ The median is 21.

Find the median.

1. 11, 13, 23, 27, 28, 30
2. 3, 7, 9, 16
3. 8, 11, 13, 21, 26, 29
4. 23, 24, 27, 31, 36, 37
5. 4, 5, 7, 9, 15, 28, 29, 39
6. 18, 36, 41, 47, 49, 53, 55, 56

PROBLEM SOLVING
Practice

Solve. Use a table if needed.

1. Some early settlers traveled across the United States. After 3 days, they had gone 48 miles. After 6 days, they had gone 96 miles. After 9 days, they had gone 144 miles. At that rate, how long did it take them to go 432 miles?

2. Two mapmakers were canoeing down a river. After 15 minutes, they had gone 0.7 miles down the river. After 30 minutes, they had gone 0.14 miles, and after 45 minutes, 0.21 miles. Their destination was 11.2 miles away. For how long did they travel?

Make a list to solve.

3. A bookshop is having a sale on books about United States history. The owner decorates the store for the sale with red, white, and blue ribbons. If he uses 2 colors at a time, how many possible color combinations can he make?

4. At the "Americana Sale," you can choose 3 books for $25: 1 about agriculture, industry, or transportation; 1 about explorers, inventors, or Presidents; and 1 about literature or science. From how many combinations of subjects can you choose?

To solve, make a guess and then check your answer.

5. Grizzly bears were a danger faced by pioneers. Grizzly bears are very large. The weight of one particular grizzly is a 3-digit number. The sum of its digits is 14. The last digit is 0. If you drop that 0, the new number is not evenly divisible by 2, 5, or 7. What is the number?

6. Many pioneers died young, but Daniel Boone was an exception. The two digits of his age when he died have a product of 48. The 2-digit number is not evenly divisible by 17. What was Daniel Boone's age when he died?

Draw a picture to solve.

7. There are landmarks numbered from 1 to 9 along the westward trail. Explorers begin at number 3 and travel west 3 marks before going back 2. On the next day, they hike ahead 5. At which landmark are they?

8. Joe is judging a cooking contest. He has to judge the following dishes: stew, chicken, meat pie, baked beans, and ranch salad. If he begins with stew and finishes with salad, in how many orders can he taste the dishes?

Solve.

9. United States workers spend about 40 hours a week on the job. In the last century, the figure was different. It is also a 2-digit number. The difference between the digits is 6, and the sum of the digits is less than 10. If you multiply this number by 5, the product has a 0 in the ones and the tens places. How many hours did workers spend on the job in the last century?

10. The bookshop has another "Americana Sale." Customers can buy 1 hardcover or softcover book on discoveries, battles, or amazing feats; 1 hardcover or softcover book about art, music, or films; and 1 hardcover or softcover book on famous United States citizens. From how many combinations can the customers choose?

11. Books are displayed on 6 racks. Each rack holds books about a different subject. The subjects are history, science, literature, biography, politics, and art. If the racks are set up in pairs around the store, how many combinations of the 6 subjects are possible?

12. When explorers reached the new land, they hiked through the woods for $2\frac{1}{2}$ hours and stopped. Then they hiked 3 hours and made camp. On the next day, they hiked for a total of $6\frac{1}{2}$ hours, and on the third day, they hiked for $7\frac{1}{2}$ hours. At that rate, for how many hours altogether had they hiked after the fifth day?

13. Jack bought some books at the "Americana Sale." He donated $\frac{1}{2}$ of them to the library. He gave each of 3 friends the same number of books. He kept twice the number of books he gave to any of his friends. If he kept 4 books, how many books did he buy?

14. Traders traded beads and shells for Native American artifacts. One trader put 6 beads and 4 shells into a bag. He picked out 5 things at random to trade. How many combinations of beads and shells could he have picked out?

15. The town of Farmdale began with 200 people. After one year, there were 300 people. By the second year, the population had grown to 400, and in one more year, it had grown to 500 people. At that rate, what was the population in 7 years?

★16. From this lesson, choose one problem that you've already solved. Show how it can be solved by using a method that is different from the one you originally used.

LOGICAL REASONING

A. A statement that is true or false is a logical statement.

STATEMENT: It will snow on Thursday.

If it snows on Thursday, then the statement is *true*.
If it does not snow, then the statement is *false*.

Two statements that have exactly the same meaning
are called **equivalent** statements. Whenever one statement
is true, then the other statement is also true.

a. I am taller than my brother. ⎫
b. My brother is shorter than I. ⎬ Equivalent statements

a. The stoplight is green. ⎫ Not equivalent statements.
b. The stoplight is not red. ⎬ If the stoplight is yellow then only b is true

Polygon *ABCD* is a rectangle.
Write whether the statement is *true*
or *false*.

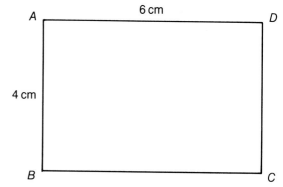

4. *ABCD* is a square.
5. *ABCD* is a parallelogram.
6. The perimeter of *ABCD* is 10 cm.
7. *ABCD* is a triangle.
8. *ABCD* is a quadrilateral.

Write whether statements *a* and *b* are *equivalent*
or *not equivalent*.

9. a. Yesterday was Tuesday.
   b. Tomorrow is Thursday.

10. a. Next month is January.
    b. Next month is not May.

11. a. Every day in May was rainy.
    b. May 15th was a sunny day.

12. a. Some of my friends are boys.
    b. None of my friends are girls.

13. a. Every visitor was unfriendly.
    b. None of the visitors were
       friendly.

14. a. My favorite number is less
       than 8.
    b. My favorite number is not
       greater than 8.

GROUP PROJECT

The Cold Old West

The problem: You and two friends are living in the days of the Old West. You have $100.00 to order supplies for the winter. How will you spend it?

Key Facts

- You have lived in the Old West for three years already.
- Winters can be very cold.
- The supplies you order must last through the winter.
- Your household is already furnished.
- Your transportation needs are already met.
- Your clothes are worn out.

General Equipment Company

Kettle	$ 1.50		Sleigh	$29.00
Leather boots	4.00		Blanket	3.50
Canning jar	0.25		Leather jacket	11.00
Chaps	12.00		Wool	0.05 per yard
Saddle	18.00		Cotton	0.05 per yard
Hat	2.25		Silk	0.10 per yard
Knife	1.75		Soap	0.15
Lantern	4.50		Socks	0.10 per pair
Rope	2.00		Shawl	3.00
Snowshoes	3.50 per pair		Candle	0.10

Order what you will need for the winter, but stay within your budget.

CHAPTER TEST

Solve. Copy and complete each graph using the information on the table to its right. (pages 384, 386, and 388)

1. Round each population to the nearest million.

U.S. FARM POPULATION

Year	Actual population	Rounded amounts
1900	29,875,000	
1925	31,190,000	
1950	23,048,000	
1975	8,253,000	

2. In which year was farm population about double the 1975 figure?

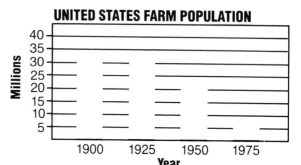

3. Round each number to the nearest half million.

NUMBER OF FARMS IN U.S.

Year	Actual number of farms	Rounded amounts
1900	5,737,000	
1925	6,471,000	
1950	5,648,000	
1975	2,314,000	

4. Were there more farms in 1900 or 1950?

NUMBER OF FARMS IN THE UNITED STATES

Year	Number of farms
1900	
1925	
1950	
1975	

◯ = 1,000,000   ◖ = 500,000

5. Round each number to the nearest ten million.

TOTAL ACRES OF FARMLAND

Year	Actual number of acres	Rounded amounts
1900	839,000,000	
1925	924,000,000	
1950	1,202,000,000	
1975	905,600,000	

6. Has the number of acres of farmland increased steadily since 1900?

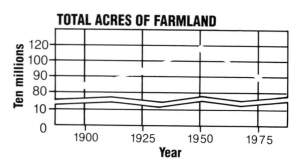

Find the mean, median, mode, and range. (page 392)

7. 109, 90, 91, 90

8. 83, 27, 83, 39, 42, 83, 42

A bowl has 14 beads. 3 are red, 2 are yellow, 4 are blue, and 5 are green. (page 396)

Write a fraction for the probability of picking a

9. red bead.　　**10.** blue bead.　　**11.** yellow bead.　　**12.** green bead.

A bag has 4 beads. 3 are brown, and 1 is orange.
Predict the probability of each event by writing *certain*
or *impossible*. (page 398)

What is the probability of picking a

13. red bead?　**14.** brown or orange bead?　**15.** orange or brown bead?　**16.** blue bead?

Solve. Use the graphs below. (pages 390–391)

POPULATION PER SQUARE KILOMETER OF THE CONTINENTS

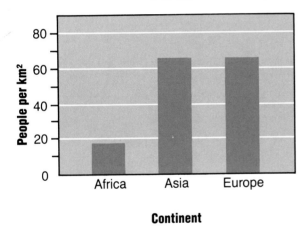

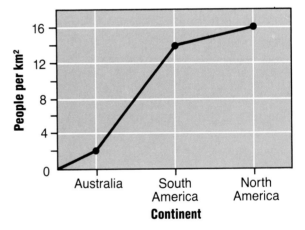

17. Does North America or Europe have more people per kilometer^2?

18. Does Asia or South America have more people per kilometer^2?

Solve. Use a tree diagram. (page 394–395)

19. Ron, Liz, and Clara are doing reports on philanthropists in New York history. They will include the Coopers, Houstons, Astors, and Rockefellers in their reports. How many combinations can be made?

20. Reports on six signers of the Declaration of Independence will be made by Sarah and Kyle. How many combinations can be made?

RETEACHING

A. Look at the spinner at the right. What is the probability that the spinner will stop on blue?

It is **equally likely** that the spinner will stop either on blue or on yellow.

Probability can be shown as a fraction.

$$\frac{\textbf{number of favorable outcomes (blue)}}{\textbf{number of possible outcomes (blue and yellow)}} = \frac{1}{2}$$

There is 1 chance in 2 that the spinner will stop on blue. So, the probability of blue is $\frac{1}{2}$.

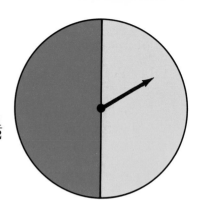

B. Find the probability of each event.

Stopping on A
Favorable outcomes: 1
Possible outcomes: 3
Probability: $\frac{1}{3}$

Stopping on B
Favorable outcomes: 2
Possible outcomes: 3
Probability: $\frac{2}{3}$

$$\frac{1}{3} < \frac{2}{3}$$

It is more probable that the spinner will stop on B. The two events are **not equally likely**.

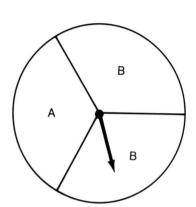

Write the probability as a fraction.
What is the probability that the first marble to be picked from the box will be

1. red?　　　**3.** white?

2. yellow?　　**4.** blue?

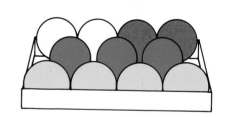

You have 26 cards each with a different letter of the alphabet written on it: 5 vowels and 21 consonants. What is the probability that you will pick

1. a **G** card?　　　**2.** an **S** card?

3. a **vowel** card?　　**4.** a **consonant** card?

ENRICHMENT

Double-Bar Graphs

Double-bar graphs are used to compare similar kinds of information. The bar graph below shows the population growth in the pioneer towns of Sage and Indian Bend between 1800 and 1860.

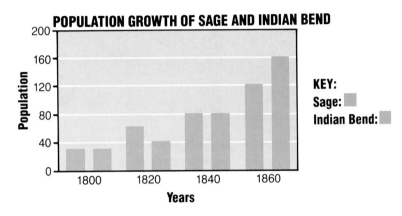

The key explains how to read the bars on the graph. The axis represents a different set of information. The titles of the axes identify this information.

Use the graph to answer the questions.

1. In what years did the population of Indian Bend double?

2. In which year was the population of Indian Bend greater than that of Sage? How many people lived in Sage that year?

Copy and complete the bar graph. Use the data in the chart.

NUMBER OF FARMS IN SAGE AND INDIAN BEND

Year	Sage	Indian Bend
1800	4	3
1820	6	7
1840	5	9
1860	8	15

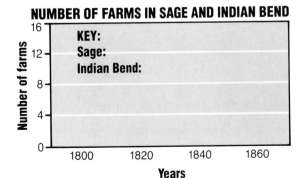

TECHNOLOGY

You can write BASIC programs to solve problems. Suppose you need to solve a problem such as this.

There are 340 children in a school. Of them, 25% are involved in sports. How many children are involved in sports?

You could write a computer program to give you the answer. Here's one way to do it.

```
10   LET NC = 340
20   LET P = .25
30   LET AN = NC * P
40   PRINT AN "CHILDREN ARE INVOLVED IN SPORTS"
```

This program stores the total number of children in a variable NC. It stores the percent in P. Remember that percents can be written as decimals. The answer is in AN. When you RUN this program, it will print this.

85 CHILDREN ARE INVOLVED IN SPORTS

1. Rewrite two lines of the program above so that you find the answer to this problem.

There are 270 children in another school. Of them, 20% are involved in sports. How many children are involved in sports?

2. What will the program print when it is RUN?

You can use parentheses to help solve multistep problems.

The fifth-grade class has a bake sale that lasts for three hours. They sell 60 baked goods in all. They sell 13 baked goods in the first hour and 30 in the second hour. How many baked goods do they sell in the third hour?

Here is one way to solve this problem.

```
10   LET AN = 60 − (13 + 30)
20   PRINT "THEY SOLD" AN "BAKED GOODS IN
THE THIRD HOUR."
```

408

3. Write a program to solve this problem.

The music teacher teaches four classes per day. She has 27 students in her first class, 14 students in her second class, and 20 students in her fourth class. She has 79 students in all. How many students does she have in her third class?

4. Write a program that has 20 print statements. The program should print the numbers 1 through 20. Each print statement should compute with the numbers 2, 3, and 4 only. You can add, subtract, multiply, or divide these three numbers. Here's a start.

```
10   PRINT 4 − 3
20   PRINT 4 / 2
30   PRINT 3 * (4 ÷ 4)
40   PRINT (4 / 2) * 2
```

When you use a comma in a PRINT statement, it acts like the tab key on your typewriter. It moves the cursor to the next printing area, or **print zone.**

Statement:	The computer will print this.
PRINT 1,2	1 2

Statement:	The computer will print this.
PRINT "HI","THERE"	HI THERE

5. Finish writing this program. It prints the number of students in each of three classes.

```
10   LET C1 = 15
20   LET C2 = 20
30   LET C3 = 23
40   LET T = C1 + C2 + C3
50   PRINT "CLASS 1",C1
```

When you RUN the program, it should print this.

```
CLASS 1   15
CLASS 2   20
CLASS 3   23
TOTAL     58
```

CUMULATIVE REVIEW

Write the letter of the correct answer.

1. Classify the angle.

90°

   a. acute   **b.** obtuse
   c. right   **d.** not given

2. Find the area of a triangle with base = 5 m; height = 8 m.

   a. 13   **b.** 20
   c. 40   **d.** not given

3. Find the area of a rectangle with length = 7 m; width = 4 m.

   a. 11 m   **b.** 11 m^2
   c. 14 m   **d.** not given

4. Find the volume of a cube with an edge of 9 in.

   a. 729 in.^{3} **b.** 729 in.^{2}
   c. 81 in.^{3} **d.** not given

5. Find the perimeter of a regular pentagon with a side 14 cm long.

   a. 70 cm^2 **b.** 70 cm
   c. 196 cm   **d.** not given

6. Give the name for *AB*.

   a. radius   **b.** diameter
   c. chord   **d.** not given

7. Choose the best unit for measuring the weight of a bowling ball.

   a. milligrams   **b.** grams
   c. kilograms   **d.** not given

8. 70% of 250

   a. 150   **b.** 175
   c. 200   **d.** not given

9. Write the answer in simplest form: $14\frac{5}{6} - 12\frac{7}{8}$.

   a. $1\frac{23}{24}$ **b.** $2\frac{1}{24}$
   c. $2\frac{23}{24}$ **d.** not given

10. Find the mean. 4, 7, 6, 10, 8

   a. 4   **b.** 6
   c. 7   **d.** not given

11. The Exotic Fruit Company delivers 6,783 coconuts in one week. The coconuts are packed 21 to a crate. How many crates do they deliver?

   a. 323   **b.** 325
   c. 142,443   **d.** not given

12. An office-supply company orders 3,575 pens. The pens are delivered in boxes, with no more than 43 to a box. How many boxes are delivered?

   a. 83   **b.** 84
   c. 85   **d.** not given

Help File

If you have trouble understanding the question, use one or more of these hints.

Organizing information
Look at the facts in the problem. Organize the facts by making a list. Use the list to help you solve the problem.

Find needed information
Some problems do not have all the facts you need to answer the question. Use resource books, periodicals and other outside sources to find the facts you need.

Crossing out extra information
Some problems contain too many facts. Make a list of the facts in the problem. Study the question. Cross out the facts you do not need.

Formulating a sensible question
Some problems do not ask a specific question. Look at the facts in the problem. Use the facts to write a sensible question.

Making a diagram
Sometimes it helps to draw a diagram that shows the facts in a problem. Study your picture, and use it to answer the question.

Choosing the operation
Look at the question. Study the numbers in the problem. Think about how the numbers could be used to answer the question. Choose the correct operation.

Being sure of the question
Be sure you know what you are being asked to do. Sometimes it helps to rewrite the question in your own words.

Finding information in the problem
Decide what the question is asking you to find. Write this down. Then find the facts in the problem that will help you answer the question.

Help File

You understand the question. To help you decide what to do, use one or more of these hints.

Writing a number sentence
Use a number sentence to decide how to use the numbers in the problem to answer the question.

Writing a simpler problem
Sometimes it helps to use simpler numbers. Use the simple numbers, then you use the actual numbers in the problem to find the answer.

Making a plan
Some problems require several steps. Make a plan. Follow it to be sure you have completed the steps needed to find the correct answer.

Estimating
Some problems can be answered with an estimate. Other problems require an exact answer.

Find the pattern
Look at the pattern of the items you are given. Use it to find the next item.

Making a table/list
Making a table or a list can often help you see the pattern that will enable you to name the next item.

Guessing an answer
Sometimes you can guess at an answer. Check your guess. Adjust it up or down to find the exact answer.

Converting measurement
Be sure all the units of measure in the problem are the same before you begin to solve the problem.

Using graphs
Some problems ask questions about graphs. Study the graphs to get the information you need.

Using maps, schedules, recipes and scale drawings
Some problems ask questions about maps, schedules, recipes or scale drawings. Study them to find the information you need.

Help File

If you have difficulty while you're solving the problem, use one or more of these hints.

Dividing
When dividing, remember to write zeros as needed in the quotient. See pages 152–155.

$$\begin{array}{r} 106 \\ 3\overline{)318} \end{array}$$

Solving problems with decimals
When multiplying, count the digits to the right of the decimal point in both factors. Their sum indicates how many decimal places must be in the product. Count from the right to place the decimal point. See pages 116–117.

$$\begin{array}{r} 0.35 \\ \times\ 0.2 \\ \hline 0.070 \end{array}$$

When dividing, you may have a remainder. Write zeros in the dividend, and continue to divide. Write zeros to the right of the decimal point as place holders. See pages 198–201.

$$\begin{array}{r} 0.01 \\ 8\overline{)0.12} \\ 8 \\ \hline 4 \end{array} \rightarrow \begin{array}{r} 0.015 \\ 8\overline{)0.120} \\ 8 \\ \hline 40 \\ 40 \\ \hline 0 \end{array}$$

Solving problems with fractions
To add fractions with unlike denominators, first find equivalent fractions with a common denominator. Then add the fractions, and write the sum in simplest form. See pages 250–251.

$$\frac{1}{6} + \frac{2}{4}$$
$$\downarrow \qquad \downarrow$$
$$\frac{2}{12} + \frac{6}{12} = \frac{8}{12} = \frac{2}{3}$$

Solving problems with percent
To write a fraction as a percent, first find an equivalent fraction with a denominator of 100. Then, write the fraction as a percent. See pages 322–323.

$$\frac{3}{4} = \frac{3 \times 25}{4 \times 25} = \frac{75}{100}$$
$$\frac{75}{100} = 75\%$$

To find what percent one number is of another, you need to compare the part to the whole. See pages 326–327.

$$\frac{60\ (part)}{120\ (whole)} \rightarrow \frac{1}{2} \rightarrow \frac{1 \times 50}{2 \times 50}$$
$$\frac{50}{100} = 50\%$$

Help File

When you want to know whether your answer is
correct, use these hints.

Checking for a reasonable answer
Does your answer make sense? If you aren't sure, try
estimating. If your estimated answer is very different
from your solution, compute the problem again.

Checking addition
Subtract one addend from the sum. The difference
should be the other addend.

$$
\begin{array}{r} 516 \\ +288 \\ \hline 804 \end{array}
\qquad
\begin{array}{r} {}^{7}\,{}^{9} \\ 8\,\cancel{0}\,4 \\ -2\,8\,8 \\ \hline 5\,1\,6 \ \checkmark \end{array}
$$

Checking subtraction
Add the difference to the smaller number in the
problem. The sum should be the larger number.

$$
\begin{array}{r} {}^{4}\,{}^{101} \\ \cancel{5}\,\cancel{1}\,6 \\ -2\,4\,8 \\ \hline 2\,6\,8 \end{array}
\qquad
\begin{array}{r} {}^{1}\,{}^{1} \\ 2\,6\,8 \\ +2\,4\,8 \\ \hline 5\,1\,6 \ \checkmark \end{array}
$$

Checking multiplication
Divide the product by one of the factors in the
problem. The quotient should be the other factor.

$$
\begin{array}{r} {}^{2} \\ 1\,4 \\ \times\ \ 6 \\ \hline 8\,4 \end{array}
\qquad
\begin{array}{r} 14 \\ 6\overline{)84} \ \checkmark \end{array}
$$

Checking division
Multiply the quotient by the divisor. The product should
be the dividend.

$$
\begin{array}{r} 13 \\ 4\overline{)52} \end{array}
\qquad
\begin{array}{r} 13 \\ \times\ \ 4 \\ \hline 52 \ \checkmark \end{array}
$$

When the quotient has a remainder, multiply as above.
Add the remainder to the product. There are several
ways to use the remainder in a division problem. Be
sure you have answered the question that is asked.

$$
\begin{array}{r} 13\ \text{R2} \\ 4\overline{)54} \end{array}
\qquad
\begin{array}{r} 13 \\ \times\ \ 4 \\ \hline 52 + 2 = 54 \ \checkmark \end{array}
$$

Checking fractions
To check fractions, use the same methods that you use
for whole numbers.

Checking decimals
Use the same methods that you use for whole numbers.
Make sure the decimal point is placed correctly.

Using outside sources
Sometimes you can use an outside source, such as a
reference book, to be sure your answer is correct.

Infobank

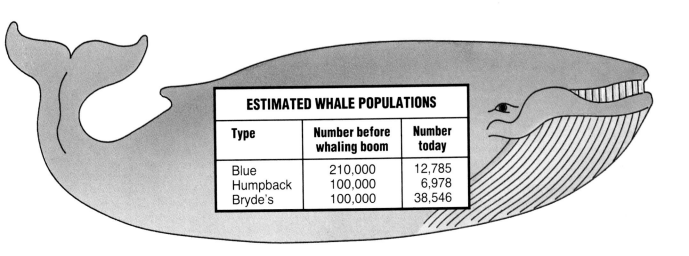

ESTIMATED WHALE POPULATIONS

Type	Number before whaling boom	Number today
Blue	210,000	12,785
Humpback	100,000	6,978
Bryde's	100,000	38,546

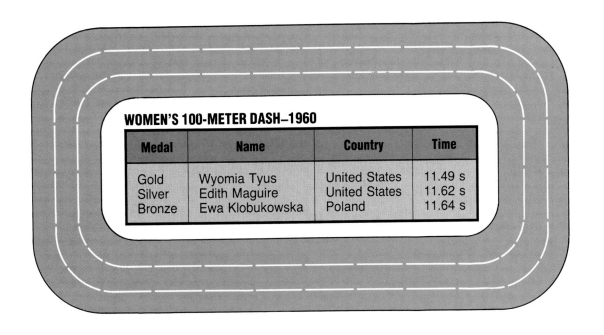

WOMEN'S 100-METER DASH–1960

Medal	Name	Country	Time
Gold	Wyomia Tyus	United States	11.49 s
Silver	Edith Maguire	United States	11.62 s
Bronze	Ewa Klobukowska	Poland	11.64 s

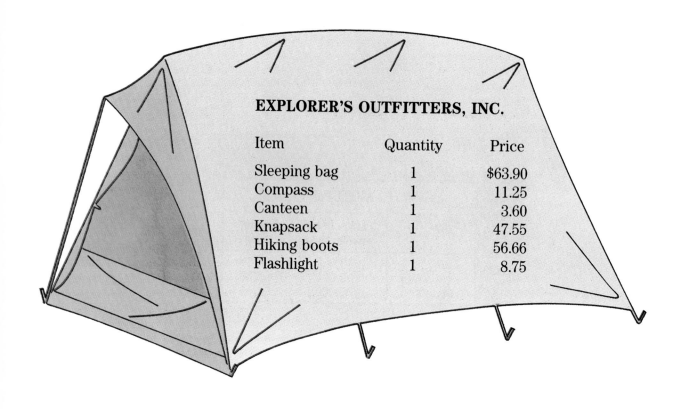

EXPLORER'S OUTFITTERS, INC.

Item	Quantity	Price
Sleeping bag	1	$63.90
Compass	1	11.25
Canteen	1	3.60
Knapsack	1	47.55
Hiking boots	1	56.66
Flashlight	1	8.75

THE AVERAGE COST OF A KILOWATT-HOUR OF ELECTRICITY

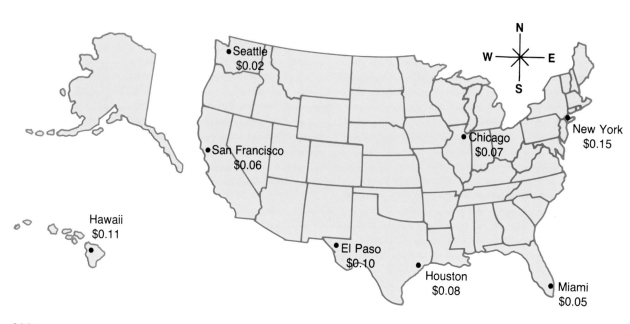

Seattle $0.02

San Francisco $0.06

Chicago $0.07

New York $0.15

Hawaii $0.11

El Paso $0.10

Houston $0.08

Miami $0.05

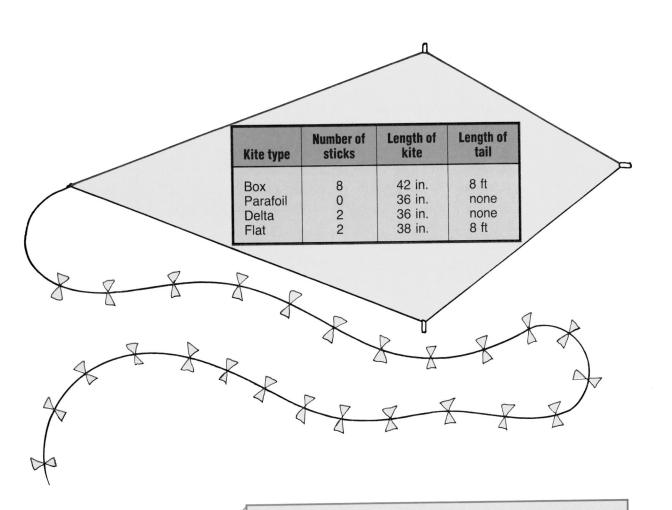

Kite type	Number of sticks	Length of kite	Length of tail
Box	8	42 in.	8 ft
Parafoil	0	36 in.	none
Delta	2	36 in.	none
Flat	2	38 in.	8 ft

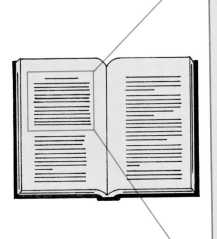

WORLD RECORDS

Longest Ferris Wheel Ride
The world record for the longest Ferris wheel ride is 768 hours, set in 1978 by Rena Clark and Jeff Block.

Most Participants In a Game of Musical Chairs
The world record for most participants in a game of musical chairs was set in 1982 by 4,514 Ohio State University students.

Most Hours Playing the Piano
The world record for most hours playing the piano is 1,218 hours, set in 1982 by David Scott.

Most Words Typed in One Hour
The world record for most words typed in one hour is 9,316 words, set in 1941 by Margaret Hanna.

BANDS OF THE ATMOSPHERE: LOWER BOUNDARIES

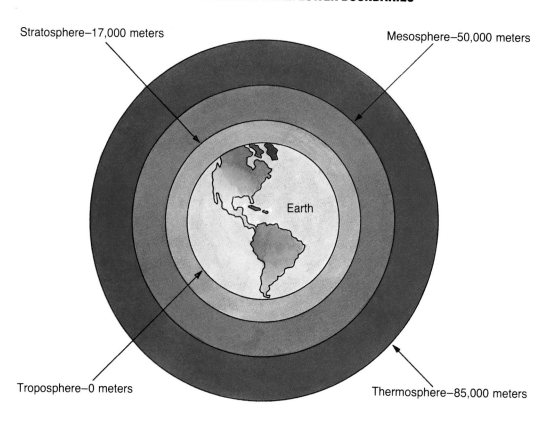

Stratosphere—17,000 meters

Mesosphere—50,000 meters

Earth

Troposphere—0 meters

Thermosphere—85,000 meters

LEADING WHEAT-GROWING STATES—1981

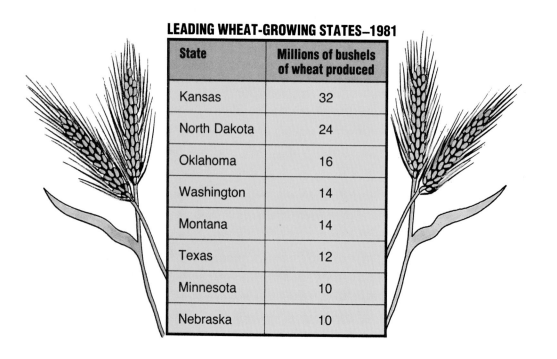

State	Millions of bushels of wheat produced
Kansas	32
North Dakota	24
Oklahoma	16
Washington	14
Montana	14
Texas	12
Minnesota	10
Nebraska	10

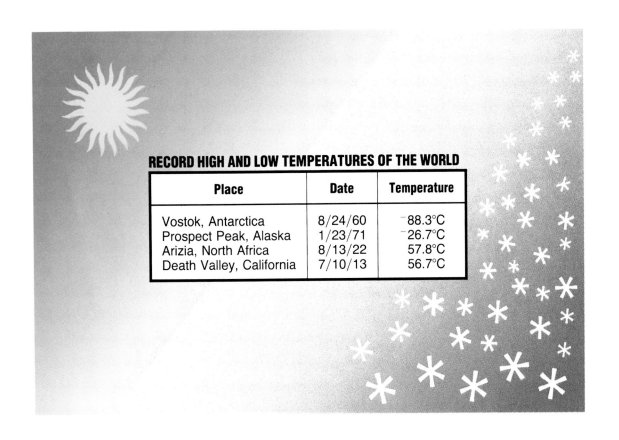

RECORD HIGH AND LOW TEMPERATURES OF THE WORLD

Place	Date	Temperature
Vostok, Antarctica	8/24/60	⁻88.3°C
Prospect Peak, Alaska	1/23/71	⁻26.7°C
Arizia, North Africa	8/13/22	57.8°C
Death Valley, California	7/10/13	56.7°C

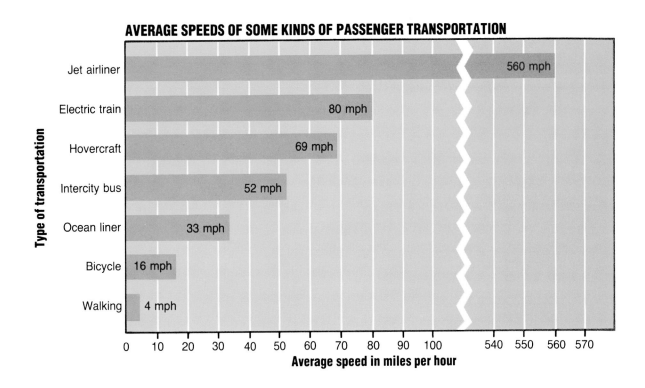

AVERAGE SPEEDS OF SOME KINDS OF PASSENGER TRANSPORTATION

Type of transportation

- Jet airliner — 560 mph
- Electric train — 80 mph
- Hovercraft — 69 mph
- Intercity bus — 52 mph
- Ocean liner — 33 mph
- Bicycle — 16 mph
- Walking — 4 mph

0 10 20 30 40 50 60 70 80 90 100 540 550 560 570

Average speed in miles per hour

SHAPES IN THE CONSTELLATIONS

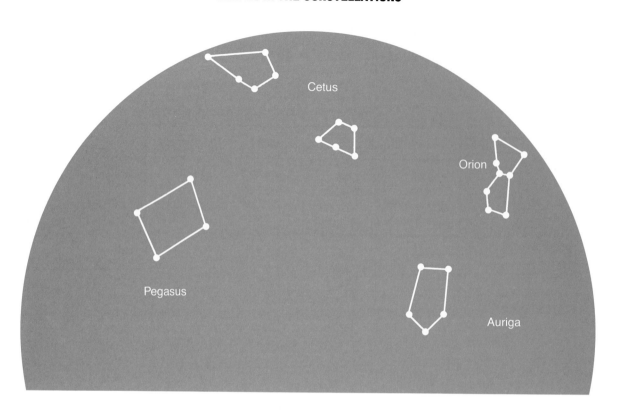

Cetus

Orion

Pegasus

Auriga

PRESIDENTIAL HEIGHT AND WEIGHT CHART

Name	Height	Weight in pounds
Calvin Coolidge	5 ft 8 in.	165
Herbert Hoover	5 ft 9 in.	200
Franklin Roosevelt	6 ft	200
Harry Truman	5 ft 7 in.	165
Dwight Eisenhower	5 ft 8 in.	180
John Kennedy	5 ft 10 in.	175
Lyndon Johnson	6 ft 1 in.	210
Richard Nixon	5 ft 9 in.	175
Gerald Ford	6 ft	200
Jimmy Carter	5 ft 9 in.	175
Ronald Reagan	5 ft 11 in.	185

More Practice

Chapter 1, page 7

Compare. Write >, <, or = for ●.

1. 639 ● 693 **2.** 536 ● 5,360 **3.** 3,777 ● 3,770 **4.** 821 ● 1,218

5. 579,038 ● 578,900 **6.** 11,010 ● 101 **7.** 357 ● 357 **8.** 770 ● 707

Order from the greatest to the least.

9. 3,275; 3,725; 375; 7,253 **10.** 3,297; 32,097; 2,397; 79 **11.** 923; 293; 329

12. 27,357; 27,573; 27,753; 27,735 **13.** 238; 328; 832 **14.** 6,750; 7,560; 6,570

Order from the least to the greatest.

15. 39,008; 39,080; 30,980; 39,800 **16.** 6,235; 6,523; 65,023; 65,325

17. 72,555; 75,525; 57,575; 75,557 **18.** 4,659; 46,000; 4,600; 460

Chapter 1, page 19

Estimate.

1. 189 − 52	**2.** 236 − 45	**3.** 568 − 327	**4.** $1.74 − 0.37	**5.** 483 − 269	**6.** 641 − 230

7. $3.68 − $1.54 **8.** 502 − 79 **9.** 876 − 637

10. 276 − 139 **11.** 710 − 335 **12.** 787 − 436

13. 875 − 693 **14.** 407 − 259 **15.** $6.72 − $3.59

Chapter 1, page 23

Add.

1. 356,081 + 7,974	**2.** 597,269 + 4,813	**3.** 918,597 + 24,987	**4.** $323.89 + 527.35
5. 87,405 + 62,567	**6.** 14,822 + 3,502	**7.** $63.52 + 35.98	**8.** 128,507 + 43,089

9. $65.73 + $32.98 **10.** 238,996 + 55,071 **11.** 329,554 + 817,111

12. $829.10 + $523.75 **13.** 134,431 + 58,111 **14.** 699,935 + 2,708

Chapter 1, page 25

Add.

1. 33,204
    3,224
+   9

2. 42,549
    35,240
    1,289
+   97

3. $15.29
    3.59
+   7.23

4. 87,853
    9,725
    14
+   12

5. 36,133
    3,411
+   3

6. $12.23 + $13.11

7. 47 + 3,116 + 542

8. 74,141 + 81,094 + 778

9. $37.89 + $19.98 + $3.23

10. 36,415 + 52,554 + 3,516 + 45

11. 27,055 + 38 + 2,913 + 516

Chapter 1, page 31

Subtract.

1. 237,155
− 12,791

2. $11.98
−   9.53

3. 84,695
−   499

4. 26,215
−   6,013

5. 114,458
− 73,817

6. $32.15
− 27.43

7. 26,337
−   633

8. 174,296
−   6,092

9. $95.18
− 59.37

10. 788,213
−   1,381

11. 385,883 − 56,867

12. 354,867 − 190,662

13. 54,923 − 4,712

14. 45,824 − 23,555

15. $46.44 − $38.56

16. 3,856 − 255

Chapter 1, page 33

Subtract.

1. $5.05
− 1.53

2. 708
− 72

3. 5,608
− 4,225

4. 90,009
− 5,270

5. 94,600
− 82,345

6. 9,005
−   254

7. 5,000
− 1,051

8. $4.03
− 3.29

9. 600,059
− 77,544

10. 230,009
− 127,122

11. 95,800 − 1,436

12. 865,005 − 552,244

13. 597,008 − 34,634

14. $6.08 − $5.05

15. $3,020.50 − $2,531.54

16. 3,670 − 647

Chapter 2, page 47

Write as a decimal.

1. twenty-two hundredths

2. nine and eleven hundredths

3. five hundredths

4. one hundred and seventy-nine hundredths **100.79**

Write the word name for each decimal. **See Answer Key.**

5. 6.47   **6.** 49.08   **7.** 2.76   **8.** 9.99   **9.** 0.82

10. 18.77   **11.** 44.89   **12.** 33.64   **13.** 80.05   **14.** 0.01

Chapter 2, page 49

Write as a decimal.

1. five and twenty-one thousandths

2. sixty-one thousandths

3. two hundred ninety-eight thousandths

4. six and four thousandths

Write the word name for each decimal.

5. 7.123   **6.** 8.365   **7.** 0.001   **8.** 0.021   **9.** 1.605

Write the value of the blue digit.

10. 713.307   **11.** 4.099   **12.** 0.263

13. 593.004   **14.** 427.446   **15.** 978.219

Chapter 2, page 53

Compare. Write $>$, $<$, or $=$ for ●.

1. 6.23 ● 6.32   **2.** 5.678 ● 5.6   **3.** 0.2 ● 0.20

4. 5.21 ● 5.021   **5.** 0.033 ● 0.33   **6.** 7.005 ● 5.707

Write in order from the least to the greatest.

7. 3.62, 36.2, 3.602   **8.** 0.018, 0.008, 0.801   **9.** 5.29, 5.2, 5.9

10. 6.098, 6.908, 6.089   **11.** 0.123, 1.013, 1.01   **12.** 32.9, 32.09, 3.29

Write in order from the greatest to the least.

13. 63.3, 6.33, 6.503   **14.** 0.012, 0.001, 0.011   **15.** 4.12, 0.412, 4.012

16. 0.81, 0.811, 0.085   **17.** 5.5, 5.55, 5.555   **18.** 28.13, 2.813, 28.3

Chapter 2, page 55

Round to the nearest whole number.

1. 3.52   **2.** 7.88   **3.** 2.39   **4.** 18.72   **5.** 135.26

6. 14.99   **7.** 12.05   **8.** 0.876   **9.** 5.23   **10.** 11.71

Round to the nearest tenth.

11. 14.29   **12.** 11.31   **13.** 15.65   **14.** 0.15   **15.** 2.54

16. 18.327   **17.** 72.275   **18.** 4.332   **19.** 4.37   **20.** 7.89

Round to the nearest hundredth.

21. 0.429   **22.** 0.215   **23.** 2.321   **24.** 8.588   **25.** 7.327

26. 0.813   **27.** 0.587   **28.** 25.631   **29.** 27.328   **30.** 52.551

Chapter 2, page 61

Add.

1. 2.27
    + 5.8

2. 3.653
    + 27.94

3. 7.2
    + 5.364

4. 8.53
    + 9.78

5. 6.038
    + 3.25

6. 4.037
    2.653
    + 8.215

7. 8.503
    5.675
    + 2.137

8. 6
    52.83
    + 0.2

9. 0.3
    0.5
    + 6.3

10. 7.87
    2.21
    + 0.54

11. 5.23 + 11.25   **12.** 16.32 + 25.238 + 2.1   **13.** 823.7 + 0.69

Chapter 2, page 63

Subtract.

1. 28.3
    − 0.28

2. $873.29
    − 565.38

3. 87.11
    − 60.42

4. $72.25
    − 61.15

5. 67.14
    − 28.09

6. 6.58
    − 5.39

7. 61.4
    − 0.04

8. 27.52
    − 16.45

9. 7.56
    − 4.87

10. 22.11
    − 19.67

11. 5 − 0.07   **12.** 28.3 − 0.28   **13.** 8.06 − 5.57   **14.** 32.3 − 29.5

Multiply.

1. $\begin{array}{r} 5 \\ \times\,0 \\ \hline \end{array}$
2. $\begin{array}{r} 9 \\ \times\,8 \\ \hline \end{array}$
3. $\begin{array}{r} 7 \\ \times\,6 \\ \hline \end{array}$
4. $\begin{array}{r} 4 \\ \times\,5 \\ \hline \end{array}$
5. $\begin{array}{r} 6 \\ \times\,3 \\ \hline \end{array}$

6. 3×8
7. $(4 \times 2) \times 5$
8. $(6 \times 0) \times 5$
9. $7 \times (1 \times 3)$

10. 8×0
11. 7×7
12. 9×9
13. 5×7

Complete. Write the name of the property.

14. $6 \times (7 + 4) = (6 \times 7) + (\blacksquare \times \blacksquare)$
15. $72 \times (63 \times 87) = (72 \times \blacksquare) \times 87$

16. $5 \times (3 + 4) = 15 + \blacksquare$
17. $7 \times \blacksquare = 7$
18. $9 \times \blacksquare = 3 \times 9$

19. $8 \times \blacksquare = 0$
20. $7 \times (5 + 2) = 35 + \blacksquare$
21. $5 \times \blacksquare = 5$

Multiply.

1. $\begin{array}{r} 80 \\ \times\,4 \\ \hline \end{array}$
2. $\begin{array}{r} 600 \\ \times\,5 \\ \hline \end{array}$
3. $\begin{array}{r} 5{,}000 \\ \times\,20 \\ \hline \end{array}$
4. $\begin{array}{r} 8{,}000 \\ \times\,7 \\ \hline \end{array}$
5. $\begin{array}{r} 6{,}000 \\ \times\,400 \\ \hline \end{array}$

6. $\begin{array}{r} 70 \\ \times\,10 \\ \hline \end{array}$
7. $\begin{array}{r} 400 \\ \times\,40 \\ \hline \end{array}$
8. $\begin{array}{r} 7{,}000 \\ \times\,40 \\ \hline \end{array}$
9. $\begin{array}{r} 7{,}000 \\ \times\,300 \\ \hline \end{array}$
10. $\begin{array}{r} 9{,}000 \\ \times\,5 \\ \hline \end{array}$

11. 70×900
12. $400 \times 6{,}000$
13. 10×20
14. 30×300

15. 70×800
16. $800 \times 4{,}000$
17. $60 \times 5{,}000$
18. $90 \times 3{,}000$

Multiply.

1. $\begin{array}{r} 635 \\ \times\,7 \\ \hline \end{array}$
2. $\begin{array}{r} 72 \\ \times\,5 \\ \hline \end{array}$
3. $\begin{array}{r} \$8.62 \\ \times\,6 \\ \hline \end{array}$
4. $\begin{array}{r} 23 \\ \times\,4 \\ \hline \end{array}$
5. $\begin{array}{r} 906 \\ \times\,7 \\ \hline \end{array}$

6. $4 \times \$3.28$
7. 9×17
8. $3 \times \$6.50$
9. 4×433

10. 6×133
11. 2×325
12. 5×115
13. 9×100

14. 5×141
15. 7×63
16. 8×35
17. 2×48

Chapter 3, page 89

Multiply.

1. $1,110 \times 7$

2. $5,731 \times 5$

3. $\$12.35 \times 6$

4. $6,742 \times 9$

5. $21,051 \times 8$

6. $42,323 \times 2$

7. $9,881 \times 7$

8. $1,441 \times 2$

9. $62,058 \times 4$

10. $\$29.88 \times 8$

11. $8 \times 31,223$

12. $5 \times 1,422$

13. $5 \times \$52.18$

14. $4 \times 41,526$

15. $9 \times \$75.30$

16. $7 \times 12,481$

17. $5 \times 11,010$

18. $3 \times 2,313$

19. $6 \times 1,000$

20. $8 \times 82,116$

21. $2 \times \$81.54$

22. $6 \times 3,647$

Chapter 3, page 93

Multiply.

1. $\$6.25 \times 91$

2. 711×75

3. $4,078 \times 57$

4. 23×42

5. $\$52.19 \times 83$

6. 556×20

7. $\$7.23 \times 78$

8. 12×11

9. 72×53

10. $2,456 \times 74$

11. $77 \times \$3.52$

12. $64 \times \$82.99$

13. $28 \times 4,329$

14. $33 \times 8,269$

15. 53×78

16. 23×827

17. 67×235

18. 14×12

Chapter 3, page 95

Multiply.

1. 477×177

2. 753×480

3. $\$3.96 \times 279$

4. 105×566

5. 704×639

6. $\$3.28 \times 215$

7. 450×347

8. 442×208

9. 603×578

10. $\$9.79 \times 281$

11. 647×745

12. $323 \times \$9.99$

13. 409×653

14. 231×828

Estimate. Write $>$ or $<$ for ●.

1. 3.4×7.6 ● 25

2. 4.2×6.5 ● 27

3. 2.6×4.1 ● 10

4. 6.3×24.2 ● 150

5. $7.62 \times \$3.29$ ● $\$25$

6. 2.38×25.45 ● 55

Estimate.

7.
$$\begin{array}{r} 8.7 \\ \times\ 3.1 \\ \hline \end{array}$$

8.
$$\begin{array}{r} 16.6 \\ \times\ 7.32 \\ \hline \end{array}$$

9.
$$\begin{array}{r} \$8.27 \\ \times\ 5.6 \\ \hline \end{array}$$

10.
$$\begin{array}{r} \$21.17 \\ \times\ 2.1 \\ \hline \end{array}$$

Multiply.

1.
$$\begin{array}{r} 4.8 \\ \times\ 0.68 \\ \hline \end{array}$$

2.
$$\begin{array}{r} 49.5 \\ \times\ 0.09 \\ \hline \end{array}$$

3.
$$\begin{array}{r} 0.76 \\ \times\ 0.8 \\ \hline \end{array}$$

4.
$$\begin{array}{r} 0.93 \\ \times\ 0.7 \\ \hline \end{array}$$

5.
$$\begin{array}{r} 6.28 \\ \times\ 0.3 \\ \hline \end{array}$$

6. 0.21×3.35

7. 0.3×31.23

8. 0.61×43.33

9. 0.2×27.27

Multiply. Round the product to the nearest cent.

10.
$$\begin{array}{r} \$36.78 \\ \times\ 0.13 \\ \hline \end{array}$$

11.
$$\begin{array}{r} \$7.28 \\ \times\ 0.6 \\ \hline \end{array}$$

12.
$$\begin{array}{r} \$33.59 \\ \times\ 0.39 \\ \hline \end{array}$$

13.
$$\begin{array}{r} \$2.15 \\ \times\ 0.3 \\ \hline \end{array}$$

14.
$$\begin{array}{r} \$6.23 \\ \times\ 0.5 \\ \hline \end{array}$$

Multiply.

1.
$$\begin{array}{r} 0.42 \\ \times\ 0.2 \\ \hline \end{array}$$

2.
$$\begin{array}{r} 0.4 \\ \times\ 0.1 \\ \hline \end{array}$$

3.
$$\begin{array}{r} 3.6 \\ \times\ 0.02 \\ \hline \end{array}$$

4.
$$\begin{array}{r} 2.2 \\ \times\ 0.03 \\ \hline \end{array}$$

5.
$$\begin{array}{r} 0.17 \\ \times\ 0.3 \\ \hline \end{array}$$

6. 2.2×0.04

7. 0.3×0.31

8. 1.1×0.07

9. 4.1×0.02

Multiply. Round the product to the nearest cent.

10.
$$\begin{array}{r} \$2.02 \\ \times\ 0.2 \\ \hline \end{array}$$

11.
$$\begin{array}{r} \$0.07 \\ \times\ 0.04 \\ \hline \end{array}$$

12.
$$\begin{array}{r} \$0.03 \\ \times\ 1.3 \\ \hline \end{array}$$

13.
$$\begin{array}{r} \$0.78 \\ \times\ 0.1 \\ \hline \end{array}$$

14.
$$\begin{array}{r} \$0.31 \\ \times\ 0.3 \\ \hline \end{array}$$

Chapter 5, page 135

Divide.

1. $5\overline{)38}$ 2. $7\overline{)44}$ 3. $3\overline{)25}$ 4. $8\overline{)49}$ 5. $9\overline{)53}$

6. $6\overline{)46}$ 7. $5\overline{)19}$ 8. $9\overline{)32}$ 9. $7\overline{)61}$ 10. $3\overline{)27}$

11. $35 \div 6$ 12. $59 \div 8$ 13. $13 \div 2$ 14. $87 \div 9$ 15. $15 \div 7$

16. $43 \div 7$ 17. $11 \div 3$ 18. $71 \div 8$ 19. $21 \div 4$ 20. $64 \div 9$

21. $\frac{68}{8}$ 22. $\frac{39}{7}$ 23. $\frac{26}{4}$ 24. $\frac{80}{9}$ 25. $\frac{77}{8}$

Chapter 5, page 145

Divide.

1. $3\overline{)111}$ 2. $3\overline{)87}$ 3. $2\overline{)106}$ 4. $6\overline{)76}$ 5. $4\overline{)221}$

6. $5\overline{)391}$ 7. $4\overline{)55}$ 8. $9\overline{)198}$ 9. $6\overline{)72}$ 10. $6\overline{)386}$

11. $87 \div 6$ 12. $740 \div 9$ 13. $105 \div 5$ 14. $145 \div 5$ 15. $429 \div 7$

16. $225 \div 9$ 17. $433 \div 5$ 18. $69 \div 3$ 19. $887 \div 9$ 20. $232 \div 9$

21. $\frac{488}{8}$ 22. $\frac{556}{9}$ 23. $\frac{644}{7}$ 24. $\frac{333}{8}$ 25. $\frac{265}{5}$

Chapter 5, page 147

Divide.

1. $3\overline{)1,782}$ 2. $6\overline{)5,046}$ 3. $9\overline{)991}$ 4. $7\overline{)917}$ 5. $2\overline{)858}$

6. $6\overline{)5,988}$ 7. $3\overline{)2,739}$ 8. $4\overline{)867}$ 9. $2\overline{)421}$ 10. $9\overline{)1,227}$

11. $874 \div 5$ 12. $884 \div 2$ 13. $968 \div 8$ 14. $495 \div 2$ 15. $818 \div 7$

16. $798 \div 7$ 17. $847 \div 7$ 18. $865 \div 4$ 19. $6,235 \div 5$ 20. $484 \div 4$

21. $\frac{1,870}{5}$ 22. $\frac{826}{2}$ 23. $\frac{963}{3}$ 24. $\frac{4,137}{7}$ 25. $\frac{757}{4}$

Chapter 5, page 149

Divide.

1. $8 \overline{)16,888}$
2. $7 \overline{)32,121}$
3. $4 \overline{)5,353}$
4. $3 \overline{)64,008}$

5. $5 \overline{)69,564}$
6. $4 \overline{)18,517}$
7. $2 \overline{)8,438}$
8. $8 \overline{)61,728}$

9. $20,686 \div 3$
10. $24,755 \div 6$
11. $37,180 \div 9$
12. $84,242 \div 9$

13. $67,255 \div 5$
14. $6,779 \div 2$
15. $48,793 \div 3$
16. $5,139 \div 4$

Chapter 5, page 153

Divide.

1. $2 \overline{)6,818}$
2. $5 \overline{)5,009}$
3. $2 \overline{)7,018}$
4. $2 \overline{)66,094}$
5. $3 \overline{)4,515}$

6. $8 \overline{)6,437}$
7. $8 \overline{)1,659}$
8. $7 \overline{)761}$
9. $6 \overline{)67,625}$
10. $9 \overline{)63,099}$

11. $6,358 \div 9$
12. $82,527 \div 3$
13. $19,511 \div 3$
14. $94,441 \div 2$

15. $6,808 \div 2$
16. $1,807 \div 2$
17. $1,859 \div 9$
18. $1,221 \div 4$

19. $758 \div 7$
20. $7,269 \div 9$
21. $79,266 \div 6$
22. $2,835 \div 4$

Chapter 5, page 155

Divide.

1. $3 \overline{)\$641.91}$
2. $8 \overline{)\$52.00}$
3. $2 \overline{)\$124.64}$
4. $2 \overline{)\$73.08}$

5. $3 \overline{)\$299.22}$
6. $2 \overline{)\$3.60}$
7. $6 \overline{)\$58.80}$
8. $6 \overline{)\$190.74}$

9. $\$506.87 \div 7$
10. $\$12.00 \div 3$
11. $\$15.00 \div 5$
12. $\$88.98 \div 2$

13. $\$396.48 \div 4$
14. $\$89.82 \div 6$
15. $\$259.04 \div 8$
16. $\$66.36 \div 4$

17. $\$106.35 \div 3$
18. $\$0.63 \div 7$
19. $\$8.10 \div 3$
20. $\$36.27 \div 3$

Chapter 6, page 167

Divide.

1. $30\overline{)180}$ 2. $40\overline{)320}$ 3. $20\overline{)160}$ 4. $10\overline{)60}$ 5. $70\overline{)210}$

6. $80\overline{)640}$ 7. $20\overline{)40}$ 8. $50\overline{)200}$ 9. $40\overline{)120}$ 10. $60\overline{)360}$

11. $490 \div 70$ 12. $540 \div 90$ 13. $480 \div 80$ 14. $180 \div 90$ 15. $250 \div 50$

16. $\frac{720}{80}$ 17. $\frac{320}{40}$ 18. $\frac{90}{30}$ 19. $\frac{550}{50}$ 20. $\frac{240}{60}$

Chapter 6, page 169

Estimate.

1. $22\overline{)364}$ 2. $36\overline{)576}$ 3. $32\overline{)838}$ 4. $86\overline{)2,759}$ 5. $42\overline{)964}$

6. $72\overline{)4,598}$ 7. $22\overline{)6,922}$ 8. $73\overline{)8,351}$ 9. $23\overline{)8,842}$ 10. $93\overline{)2,857}$

11. $457 \div 11$ 12. $2,614 \div 84$ 13. $3,234 \div 77$ 14. $7,174 \div 58$

15. $\frac{8,495}{55}$ 16. $\frac{1,917}{21}$ 17. $\frac{2,983}{58}$ 18. $\frac{3,553}{71}$ 19. $\frac{4,264}{44}$

Chapter 6, page 175

Divide.

1. $23\overline{)767}$ 2. $11\overline{)793}$ 3. $71\overline{)4,028}$ 4. $65\overline{)813}$ 5. $22\overline{)1,369}$

6. $33\overline{)431}$ 7. $21\overline{)869}$ 8. $83\overline{)6,343}$ 9. $72\overline{)6,565}$ 10. $31\overline{)2,932}$

11. $\$55.44 \div 88$ 12. $5,952 \div 27$ 13. $2,491 \div 53$ 14. $6,298 \div 77$ 15. $5,337 \div 62$

16. $794 \div 44$ 17. $\$80.51 \div 97$ 18. $2,983 \div 58$ 19. $7,115 \div 85$ 20. $1,752 \div 22$

Divide.

1. $13\overline{)793}$ 2. $12\overline{)\$7.32}$ 3. $63\overline{)1,384}$ 4. $55\overline{)3,740}$ 5. $78\overline{)3,539}$

6. $83\overline{)4,112}$ 7. $42\overline{)817}$ 8. $73\overline{)\$37.43}$ 9. $83\overline{)2,655}$ 10. $47\overline{)2,867}$

11. $829 \div 23$ 12. $1,517 \div 46$ 13. $\$16.10 \div 46$ 14. $864 \div 12$ 15. $979 \div 12$

16. $\frac{4,896}{68}$ 17. $\frac{769}{55}$ 18. $\frac{1,383}{22}$ 19. $\frac{1,659}{46}$ 20. $\frac{1,459}{21}$

Divide.

1. $79\overline{)9,218}$ 2. $33\overline{)16,938}$ 3. $35\overline{)4,480}$ 4. $81\overline{)57,964}$ 5. $11\overline{)6,751}$

6. $79\overline{)9,717}$ 7. $11\overline{)6,754}$ 8. $65\overline{)16,575}$ 9. $32\overline{)9,896}$ 10. $73\overline{)67,087}$

11. $19,078 \div 84$ 12. $27,653 \div 72$ 13. $8,567 \div 31$ 14. $9,362 \div 77$

15. $5,088 \div 24$ 16. $19,674 \div 31$ 17. $15,868 \div 41$ 18. $15,228 \div 42$

19. $\frac{8,625}{69}$ 20. $\frac{22,911}{81}$ 21. $\frac{26,157}{78}$ 22. $\frac{27,837}{42}$ 23. $\frac{7,819}{55}$

Divide.

1. $93\overline{)28,090}$ 2. $81\overline{)32,564}$ 3. $22\overline{)\$13.20}$ 4. $54\overline{)21,999}$ 5. $91\overline{)36,541}$

6. $42\overline{)8,509}$ 7. $91\overline{)9,454}$ 8. $76\overline{)8,134}$ 9. $58\overline{)40,865}$ 10. $52\overline{)26,439}$

11. $14,738 \div 71$ 12. $7,589 \div 36$ 13. $\$6.66 \div 74$ 14. $23,334 \div 33$

15. $6,615 \div 11$ 16. $5,762 \div 32$ 17. $11,548 \div 57$ 18. $14,849 \div 21$

19. $\frac{60,756}{86}$ 20. $\frac{20,012}{33}$ 21. $\frac{6,888}{34}$ 22. $\frac{29,766}{42}$ 23. $\frac{6,720}{32}$

Chapter 7, page 199

Divide.

1. $9\overline{)67.32}$ 2. $32\overline{)11.84}$ 3. $23\overline{)57.73}$ 4. $97\overline{)22.31}$ 5. $13\overline{)70.98}$

6. $21\overline{)6.51}$ 7. $8\overline{)50.4}$ 8. $33\overline{)36.63}$ 9. $23\overline{)55.43}$ 10. $14\overline{)11.746}$

11. $51\overline{)4.539}$ 12. $42\overline{)38.22}$ 13. $52\overline{)449.28}$ 14. $70\overline{)8.40}$ 15. $25\overline{)11.25}$

16. $2.282 \div 7$ 17. $69.72 \div 83$ 18. $9.152 \div 2$ 19. $118.56 \div 76$

Chapter 7, page 201

Divide.

1. $25\overline{)1.25}$ 2. $2\overline{)1.27}$ 3. $32\overline{)64.64}$ 4. $48\overline{)28.848}$ 5. $16\overline{)6.8}$

6. $32\overline{)6.496}$ 7. $79\overline{)3.397}$ 8. $25\overline{)12.3}$ 9. $22\overline{)9.13}$ 10. $41\overline{)8.241}$

11. $11.67 \div 15$ 12. $45.32 \div 44$ 13. $1.44 \div 36$ 14. $1.92 \div 96$

15. $0.32 \div 16$ 16. $46.09 \div 55$ 17. $8.505 \div 21$ 18. $7.84 \div 35$

Chapter 7, page 205

Measure the length of the piece of string to the nearest

1. ▦ cm. 2. ▦ mm. 3. ▦ cm ▦ mm.

Draw a line that measures

4. 4 cm. 5. 9.7 cm. 6. 15 cm. 7. 2.3 cm. 8. 12 cm.

Measure to find the distance around each shape.

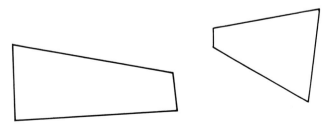

Which unit would you use to measure?
Write *millimeter, centimeter, meter,* or *kilometer.*

1. the length of a football field

2. the length of a pencil

3. the distance to the moon

4. the height of a flagpole

5. the length of a thumbtack

6. an airplane flight

Complete.

7. 23 km = ▒ m

8. 7 mm = ▒ cm

9. 0.03 m = ▒ cm

10. 0.427 cm = ▒ mm

11. 0.76 m = ▒ cm

12. 8,050 m = ▒ km

Which unit would you use to measure the capacity?
Write *milliter* or *liter.*

1. a bathtub

2. a swimming pool

3. a bottle cap

4. a pond

5. a straw

6. a cooler

Complete.

7. 4.5 L = ▒ mL

8. 0.57 L = ▒ mL

9. 352 mL = ▒ L

10. 3,232 L = ▒ mL

11. 0.565 L = ▒ mL

12. 3,252 mL = ▒ L

Which unit would you use to measure the mass?
Write *milligram, gram,* or *kilogram.*

1. a piece of paper

2. a dinosaur

3. a frog

4. an ocean liner

5. a feather

6. a paperweight

Complete.

7. 21,000 kg = ▒ mg

8. 63 g = ▒ mg

9. 56.4 g = ▒ kg

10. 320 mg = ▒ kg

11. 323 g = ▒ mg

12. 72.5 kg = ▒ g

Chapter 8, page 227

Find the least common multiple.

1. 7, 6
2. 2, 5, 4
3. 3, 9
4. 7, 8
5. 3, 4, 8
6. 6, 30
7. 4, 7
8. 5, 9, 3
9. 2, 11
10. 3, 13
11. 4, 6, 8
12. 2, 5, 8
13. 3, 4
14. 8, 9
15. 7, 14, 21
16. 3, 8
17. 4, 9
18. 6, 12, 8
19. 2, 7, 5
20. 2, 9
21. 4, 6, 9
22. 3, 6, 10
23. 4, 7
24. 2, 6
25. 2, 17

Chapter 8, page 229

Find the greatest common factor.

1. 16, 28
2. 8, 18
3. 8, 26
4. 14, 44
5. 56, 64
6. 72, 96
7. 24, 38
8. 6, 42
9. 22, 58
10. 21, 49
11. 33, 51
12. 9, 54
13. 24, 26
14. 5, 20
15. 34, 64
16. 30, 55
17. 9, 12
18. 26, 48
19. 27, 63
20. 77, 84
21. 56, 90
22. 60, 90
23. 45, 81
24. 14, 38
25. 21, 48
26. 28, 58
27. 40, 50
28. 12, 18, 24
29. 18, 36, 45
30. 16, 24, 32

Chapter 8, page 239

Complete.

1. $\frac{6}{8} = \frac{3}{\blacksquare}$
2. $\frac{3}{9} = \frac{\blacksquare}{3}$
3. $\frac{20}{25} = \frac{\blacksquare}{5}$
4. $\frac{27}{54} = \frac{1}{\blacksquare}$
5. $\frac{4}{6} = \frac{2}{\blacksquare}$
6. $\frac{18}{24} = \frac{3}{\blacksquare}$
7. $\frac{4}{34} = \frac{2}{\blacksquare}$
8. $\frac{20}{45} = \frac{\blacksquare}{9}$
9. $\frac{10}{16} = \frac{\blacksquare}{8}$
10. $\frac{6}{9} = \frac{2}{\blacksquare}$
11. $\frac{9}{27} = \frac{\blacksquare}{3}$
12. $\frac{16}{64} = \frac{1}{\blacksquare}$
13. $\frac{36}{90} = \frac{2}{\blacksquare}$
14. $\frac{72}{84} = \frac{\blacksquare}{7}$
15. $\frac{88}{99} = \frac{8}{\blacksquare}$

Write the fraction in simplest form.

16. $\frac{2}{38}$
17. $\frac{3}{6}$
18. $\frac{6}{72}$
19. $\frac{25}{125}$
20. $\frac{9}{12}$
21. $\frac{2}{8}$
22. $\frac{36}{144}$
23. $\frac{20}{50}$
24. $\frac{26}{39}$
25. $\frac{2}{4}$

Chapter 8, page 249

Estimate.

1. $\frac{1}{5} + \frac{7}{8}$ **2.** $\frac{2}{7} + \frac{6}{9}$ **3.** $\frac{5}{11} + \frac{3}{10}$ **4.** $\frac{4}{9} + \frac{6}{7}$ **5.** $\frac{7}{8} + \frac{7}{8} + \frac{3}{7}$

6. $\frac{1}{8} + \frac{2}{3}$ **7.** $\frac{4}{7} + \frac{2}{5}$ **8.** $\frac{7}{9} + \frac{7}{13}$ **9.** $\frac{4}{5} + \frac{3}{8}$ **10.** $\frac{5}{6} + \frac{1}{9} + \frac{2}{5}$

11. $\frac{6}{15} + \frac{3}{8}$ **12.** $\frac{7}{8} + \frac{17}{20}$ **13.** $\frac{5}{8} + \frac{4}{5}$ **14.** $\frac{2}{5} + \frac{6}{7}$ **15.** $\frac{7}{14} + \frac{2}{9} + \frac{8}{9}$

16. $\frac{5}{9} + \frac{7}{12}$ **17.** $\frac{9}{10} + \frac{3}{7}$ **18.** $\frac{6}{7} + \frac{11}{15}$ **19.** $\frac{1}{2} + \frac{8}{9}$ **20.** $\frac{1}{7} + \frac{16}{17} + \frac{4}{9}$

Chapter 8, page 251

Add. Write the answer in simplest form.

1. $\frac{3}{5} + \frac{1}{10}$ **2.** $\frac{3}{4} + \frac{1}{8}$ **3.** $\frac{4}{9} + \frac{3}{5}$ **4.** $\frac{1}{6} + \frac{2}{3}$ **5.** $\frac{1}{2} + \frac{6}{7}$

6. $\frac{7}{9} + \frac{3}{4}$ **7.** $\frac{3}{4} + \frac{5}{8}$ **8.** $\frac{5}{6} + \frac{1}{2}$ **9.** $\frac{2}{5} + \frac{7}{8}$ **10.** $\frac{2}{3} + \frac{8}{9}$

11. $\frac{1}{2} + \frac{3}{4}$ **12.** $\frac{9}{10} + \frac{4}{5}$ **13.** $\frac{2}{3} + \frac{4}{9}$ **14.** $\frac{5}{8} + \frac{1}{2}$ **15.** $\frac{1}{12} + \frac{5}{8}$

16. $\begin{array}{r} \frac{5}{9} \\ + \frac{5}{18} \\ \hline \end{array}$ **17.** $\begin{array}{r} \frac{4}{5} \\ + \frac{6}{9} \\ \hline \end{array}$ **18.** $\begin{array}{r} \frac{8}{15} \\ + \frac{7}{10} \\ \hline \end{array}$ **19.** $\begin{array}{r} \frac{5}{6} \\ + \frac{6}{7} \\ \hline \end{array}$ **20.** $\begin{array}{r} \frac{4}{9} \\ + \frac{8}{12} \\ \hline \end{array}$

Chapter 8, page 253

Subtract. Write the answer in simplest form.

1. $\frac{8}{9} - \frac{1}{9}$ **2.** $\frac{5}{7} - \frac{2}{3}$ **3.** $\frac{4}{6} - \frac{1}{6}$ **4.** $\frac{5}{6} - \frac{3}{4}$ **5.** $\frac{7}{8} - \frac{2}{5}$

6. $\frac{5}{7} - \frac{3}{7}$ **7.** $\frac{2}{4} - \frac{1}{4}$ **8.** $\frac{6}{10} - \frac{2}{7}$ **9.** $\frac{5}{9} - \frac{1}{2}$ **10.** $\frac{2}{3} - \frac{1}{3}$

11. $\frac{2}{3} - \frac{5}{9}$ **12.** $\frac{3}{5} - \frac{3}{10}$ **13.** $\frac{1}{2} - \frac{1}{4}$ **14.** $\frac{5}{9} - \frac{1}{3}$ **15.** $\frac{7}{10} - \frac{1}{5}$

16. $\begin{array}{r} \frac{3}{4} \\ - \frac{3}{10} \\ \hline \end{array}$ **17.** $\begin{array}{r} \frac{2}{5} \\ - \frac{1}{4} \\ \hline \end{array}$ **18.** $\begin{array}{r} \frac{8}{9} \\ - \frac{3}{4} \\ \hline \end{array}$ **19.** $\begin{array}{r} \frac{1}{2} \\ - \frac{1}{7} \\ \hline \end{array}$ **20.** $\begin{array}{r} \frac{2}{3} \\ - \frac{1}{5} \\ \hline \end{array}$

Estimate.

1. $1\frac{3}{5} + 4\frac{4}{9}$ **2.** $10\frac{1}{8} + 1\frac{4}{7}$ **3.** $5\frac{1}{8} + 7\frac{5}{6}$ **4.** $9\frac{3}{5} + 2\frac{5}{6}$ **5.** $8\frac{3}{10} + 5\frac{5}{6}$

Add. Write the answer in simplest form.

6. $7\frac{3}{4} + 3\frac{4}{5}$ **7.** $8\frac{5}{6} + 2\frac{2}{9}$ **8.** $5\frac{9}{10} + 1\frac{3}{4}$ **9.** $6\frac{1}{6} + 6\frac{9}{10}$ **10.** $3\frac{1}{2} + 5\frac{4}{7}$

11. $5\frac{1}{4} + 7\frac{5}{6}$ **12.** $6\frac{1}{2} + 10\frac{7}{9}$ **13.** $1\frac{2}{5} + 1\frac{2}{3}$ **14.** $4\frac{1}{3} + 1\frac{3}{4}$ **15.** $6\frac{5}{6} + 9\frac{1}{4}$

16. $\begin{array}{r} 8\frac{2}{5} \\ + 10\frac{3}{4} \\ \hline \end{array}$ **17.** $\begin{array}{r} 10\frac{3}{5} \\ + 2\frac{3}{4} \\ \hline \end{array}$ **18.** $\begin{array}{r} 6\frac{4}{7} \\ + 4\frac{1}{2} \\ \hline \end{array}$ **19.** $\begin{array}{r} 4\frac{3}{4} \\ + 5\frac{7}{10} \\ \hline \end{array}$ **20.** $\begin{array}{r} 1\frac{1}{2} \\ + 5\frac{5}{9} \\ \hline \end{array}$

Subtract. Write the answer in simplest form.

1. $8\frac{3}{8} - 2\frac{1}{8}$ **2.** $5\frac{6}{7} - 1\frac{3}{4}$ **3.** $3\frac{2}{5} - 1\frac{1}{5}$ **4.** $13\frac{7}{9} - 6$ **5.** $5\frac{5}{7} - 3\frac{2}{6}$

6. $12\frac{2}{3} - 5$ **7.** $8\frac{5}{9} - 3$ **8.** $7\frac{3}{4} - 4\frac{3}{8}$ **9.** $12\frac{2}{3} - 3\frac{1}{6}$ **10.** $14\frac{5}{6} - 12\frac{1}{2}$

11. $17\frac{2}{3} - 8\frac{1}{9}$ **12.** $15\frac{1}{2} - 14$ **13.** $16\frac{7}{10} - 10\frac{1}{2}$ **14.** $13\frac{7}{8} - 1\frac{3}{4}$ **15.** $3\frac{1}{3} - 1\frac{1}{6}$

16. $\begin{array}{r} 10\frac{7}{9} \\ - 9\frac{1}{3} \\ \hline \end{array}$ **17.** $\begin{array}{r} 11\frac{3}{4} \\ - 8\frac{1}{2} \\ \hline \end{array}$ **18.** $\begin{array}{r} 18\frac{7}{8} \\ - 1\frac{3}{4} \\ \hline \end{array}$ **19.** $\begin{array}{r} 14\frac{1}{2} \\ - 7 \\ \hline \end{array}$ **20.** $\begin{array}{r} 16\frac{2}{3} \\ - 3 \\ \hline \end{array}$

Subtract. Write the answer in simplest form.

1. $6\frac{1}{6} - 1\frac{3}{4}$ **2.** $4\frac{1}{4} - 1\frac{2}{5}$ **3.** $2\frac{1}{10} - 1\frac{1}{6}$ **4.** $20\frac{3}{5} - 4\frac{2}{3}$ **5.** $15\frac{2}{3} - 3\frac{3}{4}$

6. $8\frac{2}{9} - 2\frac{1}{2}$ **7.** $20 - 9\frac{1}{4}$ **8.** $10\frac{1}{3} - 1\frac{2}{3}$ **9.** $8\frac{3}{10} - 2\frac{7}{10}$ **10.** $16\frac{3}{10} - 13\frac{5}{6}$

11. $19\frac{1}{2} - 8\frac{2}{3}$ **12.** $13\frac{1}{3} - 6\frac{1}{2}$ **13.** $11\frac{3}{7} - 4\frac{5}{7}$ **14.** $18\frac{1}{2} - 15\frac{2}{3}$ **15.** $15\frac{1}{6} - 11\frac{7}{10}$

16. $\begin{array}{r} 8\frac{2}{10} \\ - 2\frac{3}{4} \\ \hline \end{array}$ **17.** $\begin{array}{r} 9 \\ - 8\frac{1}{2} \\ \hline \end{array}$ **18.** $\begin{array}{r} 19\frac{1}{9} \\ - 2\frac{1}{6} \\ \hline \end{array}$ **19.** $\begin{array}{r} 17\frac{3}{6} \\ - 13\frac{5}{6} \\ \hline \end{array}$ **20.** $\begin{array}{r} 9\frac{2}{5} \\ - 8\frac{1}{2} \\ \hline \end{array}$

Multiply. Write the product in simplest form.

1. $\frac{2}{3} \times \frac{1}{10}$ **2.** $\frac{1}{6} \times \frac{2}{3}$ **3.** $\frac{3}{7} \times \frac{1}{4}$ **4.** $\frac{1}{4} \times \frac{2}{5}$ **5.** $\frac{5}{6} \times \frac{1}{4}$

6. $\frac{1}{2} \times \frac{3}{4}$ **7.** $\frac{1}{8} \times \frac{1}{10}$ **8.** $\frac{2}{5} \times \frac{3}{4}$ **9.** $\frac{2}{3} \times \frac{3}{5}$ **10.** $\frac{1}{2} \times \frac{2}{9}$

11. $\frac{3}{4} \times \frac{4}{5}$ **12.** $\frac{4}{7} \times \frac{2}{5}$ **13.** $\frac{1}{3} \times \frac{5}{6}$ **14.** $\frac{2}{7} \times \frac{1}{3}$ **15.** $\frac{3}{4} \times \frac{5}{7}$

16. $\frac{1}{2} \times \frac{2}{3}$ **17.** $\frac{4}{7} \times \frac{3}{8}$ **18.** $\frac{1}{6} \times \frac{3}{5}$ **19.** $\frac{5}{8} \times \frac{2}{3}$ **20.** $\frac{1}{2} \times \frac{3}{5}$

21. $\frac{4}{5} \times \frac{7}{8}$ **22.** $\frac{2}{7} \times \frac{5}{9}$ **23.** $\frac{1}{3} \times \frac{1}{5}$ **24.** $\frac{4}{9} \times \frac{1}{6}$ **25.** $\frac{7}{9} \times \frac{1}{2}$

Multiply. Write the product in simplest form.

1. $\frac{3}{8} \times 4\frac{2}{3}$ **2.** $2\frac{5}{6} \times 8$ **3.** $\frac{4}{9} \times 1\frac{1}{8}$ **4.** $3\frac{1}{2} \times 2\frac{2}{3}$ **5.** $\frac{4}{7} \times 2\frac{2}{8}$

6. $8\frac{1}{4} \times \frac{5}{6}$ **7.** $7 \times 4\frac{2}{9}$ **8.** $6 \times 3\frac{4}{9}$ **9.** $8\frac{2}{5} \times \frac{1}{2}$ **10.** $\frac{2}{3} \times 4\frac{1}{8}$

11. $4\frac{4}{5} \times 3\frac{2}{3}$ **12.** $3\frac{1}{3} \times \frac{7}{8}$ **13.** $3\frac{2}{5} \times \frac{5}{6}$ **14.** $2\frac{4}{7} \times 4\frac{3}{4}$ **15.** $1\frac{1}{6} \times 4\frac{3}{4}$

16. $5\frac{2}{9} \times 7\frac{1}{2}$ **17.** $2\frac{1}{4} \times \frac{1}{6}$ **18.** $10\frac{2}{3} \times \frac{3}{8}$ **19.** $5 \times 3\frac{1}{5}$ **20.** $\frac{1}{2} \times 9\frac{1}{3}$

21. $2\frac{7}{10} \times \frac{2}{9}$ **22.** $\frac{3}{7} \times 2\frac{1}{4}$ **23.** $2\frac{1}{4} \times 3\frac{2}{3}$ **24.** $3\frac{3}{4} \times \frac{7}{10}$ **25.** $\frac{3}{7} \times 2\frac{1}{3}$

Divide.

1. $9 \div \frac{1}{7}$ **2.** $13 \div \frac{1}{2}$ **3.** $36 \div \frac{1}{6}$ **4.** $85 \div \frac{1}{2}$ **5.** $30 \div \frac{1}{5}$

6. $45 \div \frac{1}{9}$ **7.** $25 \div \frac{1}{10}$ **8.** $66 \div \frac{1}{3}$ **9.** $24 \div \frac{1}{8}$ **10.** $30 \div \frac{1}{8}$

11. $18 \div \frac{1}{9}$ **12.** $75 \div \frac{1}{10}$ **13.** $80 \div \frac{1}{6}$ **14.** $75 \div \frac{1}{5}$ **15.** $7 \div \frac{1}{7}$

16. $8 \div \frac{1}{8}$ **17.** $26 \div \frac{1}{4}$ **18.** $52 \div \frac{1}{3}$ **19.** $44 \div \frac{1}{4}$ **20.** $6 \div \frac{1}{8}$

21. $48 \div \frac{1}{3}$ **22.** $90 \div \frac{1}{4}$ **23.** $35 \div \frac{1}{10}$ **24.** $21 \div \frac{1}{7}$ **25.** $16 \div \frac{1}{6}$

Chapter 9, page 287

Add or subtract.

1. 3 h 14 min
   + 1 h 39 min

2. 12 min 15 s
   − 4 min 35 s

3. 6 h 5 min
   − 1 h 14 min

4. 2 h 35 min
   + 7 h 47 min

5. 2 min 4 s
   + 9 min 29 s

6. 45 h 12 min
   − 18 h 25 min

7. 8 h 30 min
   + 12 h 45 min

8. 24 h 6 min
   − 8 h 16 min

9. 16 h 21 min
   − 3 h 34 min

10. 47 min 12 s
    + 9 min 25 s

11. 12 h 39 min
    − 10 h 45 min

12. 55 min 12 s
    + 18 min 6 s

13. 4 min 55 s
    + 5 min 32 s

14. 8 h 45 min
    − 5 h 20 min

15. 29 min 7 s
    − 10 min 15 s

16. 6 h 25 min
    + 5 h 55 min

Chapter 9, page 293

Write *inches*, *feet*, or *yards*.

1. A car may be 10 ▨ long.

2. A book may be 10 ▨ wide.

3. A house may be 20 ▨ high.

4. A chalkboard may be 3 ▨ wide.

Complete.

5. 6 yd = ▨ in.

6. 7 ft = ▨ in.

7. 10,560 yd = ▨ mi

8. 48 in. = ▨ ft

9. $5\frac{1}{2}$ ft = ▨ in.

10. 12 yd = ▨ ft

11. 1 mi = ▨ ft

12. 54 in. = ▨ yd

13. 10 ft = ▨ in.

14. 24 ft = ▨ yd

15. 96 in. = ▨ ft

16. 6 mi = ▨ yd

Chapter 9, page 294

Write *cups*, *pints*, *quarts*, or *gallons*.

1. A coffee mug may hold 2 ▨.

2. A water tank may hold 200 ▨.

3. A bathtub may hold 25 ▨.

4. An oilcan may contain 1 ▨.

Complete.

5. $5\frac{1}{4}$ gal = ▨ qt

6. 2 pt = ▨ gal

7. 12 c = ▨ pt

8. 8 qt = ▨ gal

9. 48 c = ▨ gal

10. 2 gal = ▨ pt

11. 4 qt = ▨ c

12. 3 pt = ▨ c

13. 5 gal = ▨ c

14. 7 qt = ▨ c

15. 16 pt = ▨ gal

16. 7 c = ▨ pt

Find the next two equal ratios.

1. $\frac{1}{4} = \frac{2}{8} = \frac{\blacksquare}{\blacksquare} = \frac{\blacksquare}{\blacksquare}$

2. $\frac{2}{3} = \frac{6}{9} = \frac{\blacksquare}{\blacksquare} = \frac{\blacksquare}{\blacksquare}$

3. $\frac{4}{5} = \frac{8}{10} = \frac{\blacksquare}{\blacksquare} = \frac{\blacksquare}{\blacksquare}$

Find the missing number.

4. $\frac{84}{18} = \frac{14}{n}$

5. $\frac{n}{16} = \frac{15}{48}$

6. $\frac{16}{n} = \frac{96}{36}$

7. $\frac{99}{54} = \frac{n}{6}$

8. $\frac{n}{10} = \frac{33}{30}$

9. $\frac{18}{90} = \frac{1}{n}$

10. $\frac{50}{n} = \frac{100}{86}$

11. $\frac{n}{37} = \frac{12}{74}$

Write = or ≠ for ●.

12. $\frac{15}{8} ● \frac{90}{48}$

13. $\frac{14}{18} ● \frac{8}{2}$

14. $\frac{1}{13} ● \frac{2}{19}$

15. $\frac{18}{16} ● \frac{108}{96}$

16. $\frac{12}{21} ● \frac{5}{3}$

17. $\frac{2}{4} ● \frac{12}{24}$

18. $\frac{16}{13} ● \frac{17}{6}$

19. $\frac{4}{21} ● \frac{11}{14}$

Write as a percent.

1. $\frac{94}{100}$

2. $\frac{13}{100}$

3. $\frac{85}{100}$

4. $\frac{64}{100}$

5. $\frac{7}{100}$

6. $\frac{76}{100}$

7. $\frac{3}{100}$

8. $\frac{91}{100}$

9. $\frac{1}{100}$

10. $\frac{42}{100}$

11. $\frac{11}{100}$

12. $\frac{57}{100}$

13. $\frac{37}{100}$

14. $\frac{99}{100}$

15. $\frac{12}{100}$

16. $4:100$

17. 18 out of 100

18. $29:100$

19. 48 per 100

20. $95:100$

21. 42 to 100

22. $6:100$

23. 17 per 100

24. 32 per 100

25. 48 out of 100

26. $11:100$

27. 55 to 100

Write as a percent.

1. 0.62

2. 0.83

3. 0.32

4. 0.92

5. 0.66

6. 0.79

7. 0.78

8. 0.23

9. 0.07

10. 0.48

11. 0.19

12. 0.05

13. 0.09

14. 0.34

15. 0.17

Write as a decimal.

16. 72%

17. 15%

18. 4%

19. 86%

20. 49%

21. 97%

22. 6%

23. 51%

24. 63%

25. 8%

26. 13%

27. 25%

28. 7%

29. 35%

30. 14%

Write each percent as a fraction in simplest form.

1. 90%   **2.** 76%   **3.** 68%   **4.** 31%   **5.** 48%

6. 60%   **7.** 26%   **8.** 74%   **9.** 35%   **10.** 2%

Write each fraction as a percent.

11. $\frac{9}{100}$ **12.** $\frac{3}{4}$ **13.** $\frac{2}{5}$ **14.** $\frac{3}{10}$ **15.** $\frac{6}{25}$

16. $\frac{1}{20}$ **17.** $\frac{7}{28}$ **18.** $\frac{7}{10}$ **19.** $\frac{3}{25}$ **20.** $\frac{6}{20}$

Find the percent of each number.

1. 20% of 100   **2.** 6% of 250   **3.** 74% of 2,000   **4.** 6% of 350

5. 70% of 30   **6.** 60% of 470   **7.** 10% of 7   **8.** 80% of 65

9. 55% of 500   **10.** 35% of 700   **11.** 40% of 600   **12.** 15% of 900

13. 2% of 650   **14.** 4% of 950   **15.** 75% of 500   **16.** 5% of 240

17. 8% of 850   **18.** 50% of 660   **19.** 25% of 1,000   **20.** 44% of 500

Compute.

1. What percent of 15 is 12?   **2.** 91 is what percent of 910?   **3.** What percent of 500 is 90?   **4.** 7 is what percent of 35?

5. What percent of 20 is 6?   **6.** What percent of 700 is 63?   **7.** 3 is what percent of 15?   **8.** 9 is what percent of 300?

Find the percent that each part is of the whole.

9. 2 blue cars
10 cars
■% are blue.

10. 7 white houses
14 houses
■% are white.

11. 2 black mice
5 mice
■% are black.

Identify and name each figure.

1.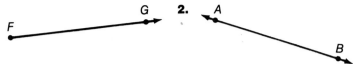

2.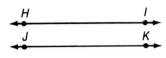

3.

Draw each figure.

4. line segment $\overline{XY}$

5. plane t

6. intersecting lines $\overleftrightarrow{TU}$ and $\overleftrightarrow{VW}$ at point Z

Is the figure congruent to $\triangle ABC$? Write *yes* or *no*.

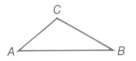

1.

2.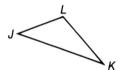

3.

Is the figure congruent to $\triangle STUV$? Write *yes* or *no*.

4.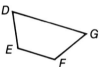

5.

6.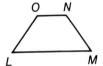

Is the figure congruent to $\square VWXY$? Write *yes* or *no*.

7.

8.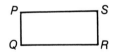

9.

Find the perimeter of each figure.

1.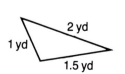

2.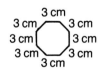

3.

4.

5. a pentagon
    each side = 2 ft

6. a square
    each side = 3 cm

7. a rectangle
    2 sides = 6 yd
    2 sides = 2 yd

Chapter 11, page 367

Find the area.

1.

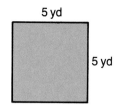
5 yd
5 yd

2.

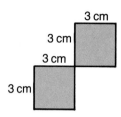
3 cm
3 cm
3 cm
3 cm

3.

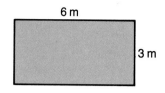
6 m
3 m

4. $l = 9$ cm $w = 4$ cm

5. $l = 12$ ft $w = 6$ ft

6. $l = 19$ in. $w = 7$ in.

7. $l = 4$ m $w = 4$ m

8. $l = 17$ km $w = 10$ km

9. $l = 8$ mi $w = 5$ mi

10. $l = 20$ yd $w = 7$ yd

11. $l = 9$ mm $w = 8$ mm

12. $l = 16$ ft $w = 11$ ft

Chapter 11, page 369

Find the area.

1.

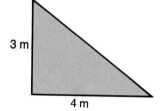
3 m
4 m

2.

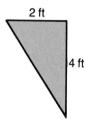
2 ft
4 ft

3.

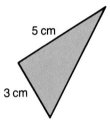
5 cm
3 cm

4. $b = 9$ m $h = 10$ m

5. $b = 6$ yd $h = 14$ yd

6. $b = 7$ in. $h = 7$ in.

7. $b = 11$ ft $h = 13$ ft

8. $b = 4$ mm $h = 7$ mm

9. $b = 10$ cm $h = 13$ cm

10. $b = 7$ km $h = 12$ km

11. $b = 5$ ft $h = 8$ ft

12. $b = 8$ mi $h = 9$ mi

Chapter 11, page 375

Find the volume.

1.

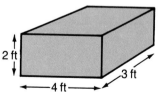
2 ft
4 ft
3 ft

2.

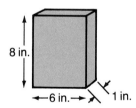
8 in.
6 in.
1 in.

3.

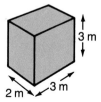
3 m
2 m
3 m

4. $l = 9$ in. $w = 3$ in. $h = 4$ in.

5. $l = 10$ mm $w = 6$ mm $h = 6$ mm

6. $l = 8$ km $w = 8$ km $h = 8$ km

7. $l = 32$ yd $w = 1$ yd $h = 2$ yd

8. $l = 16$ ft $w = 5$ ft $h = 9$ ft

9. $l = 20$ m $w = 7$ m $h = 7$ m

Copy and complete the bar graph. Use the data in the
table. Round each number to the nearest tenth of a
million. Then answer each question.

AREA OF THE UNITED STATES—1790–1970

Year	Area in square miles	Rounded
1790	888,811	
1820	1,788,006	
1850	2,992,747	
1880	3,022,387	
1970	3,618,467	

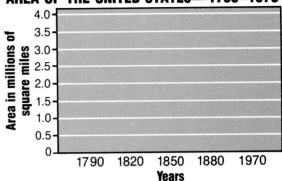

1. During which period did the area of
the United States double in size?

2. About how much did the area of the
United States increase between 1880
and 1970?

Copy and complete the broken-line graph. Use the data
in the table. Round each number to the nearest tenth of
a billion. Then answer each question.

POSTAL INCOME IN FIVE CITIES—1980

City	Income	Rounded
Boston	$224,428,760	
Chicago	$528,233,991	
Los Angeles	$271,136,828	
New York	$666,377,778	
St. Louis	$127,427,555	

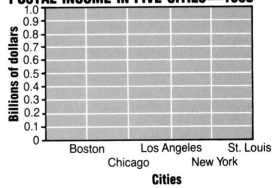

1. Which city had the most postal
income? Which city had the least
postal income?

2. Which two cities had almost the
same postal income?

Find the mean and the median.

1. 5, 4, 3

2. 25, 20, 15, 10, 18

3. 12, 8, 6, 2, 4

4. 9, 10, 11, 14, 19

5. 32, 18, 24, 42, 14

6. 7, 5, 9, 1, 4

Find the mode and the range.

7. 68, 52, 44, 68

8. 102, 34, 76, 34

9. 3, 9, 21, 9, 18

10. 29, 17, 51, 47, 17

11. 54, 26, 32, 54

12. 4, 1, 5, 11, 5

Write a fraction for the probability of picking

1. a blue marble.

2. a red marble.

3. a green marble.

4. a yellow marble.

5. a pink marble.

6. a black marble.

Write a fraction for the probability of picking

7. a *1* card.

8. a *2* card.

9. a *3* card.

10. a *4* card.

11. a *5* card.

12. a *7* card.

Predict the probability of each event.
Write *certain* or *impossible* for each event.

What is the probability of picking

1. a green marble?

2. a yellow or black marble?

3. a black or yellow marble?

4. a red marble?

What is the probability of landing on

5. blue?

6. red or white?

7. green?

8. green or yellow?

444

TABLE OF MEASURES

TIME

1 minute (min) = 60 seconds (s)
1 hour (h) = 60 minutes
1 day (d) = 24 hours
1 week (wk) = 7 days

1 year (y) = 12 months
1 year = 52 weeks
1 year = 365 days

METRIC UNITS

Length

1 centimeter (cm) = 10 millimeters (mm)
1 meter (m) = 100 centimeters
1 kilometer (km) = 1,000 meters

Capacity

1 liter (L) = 1,000 milliliters (mL)

Mass

1 gram (g) = 1,000 milligrams (mg)
1 kilogram (kg) = 1,000 grams

Temperature

0° Celsius (°C) . . . . Water freezes
100° Celsius (°C) . . . . Water boils

CUSTOMARY UNITS

Length

1 foot (ft) = 12 inches (in.)
1 yard (yd) = 3 feet
1 mile (mi) = 5,280 feet
1 mile = 1,760 yards

Weight

1 pound (lb) = 16 ounces (oz)
1 ton (T) = 2,000 pounds

Capacity

1 pint (pt) = 2 cups (c)
1 quart (qt) = 2 pints
1 gallon (gal) = 4 quarts

Temperature

32° Fahrenheit (°F) . . . . Water freezes
212° Fahrenheit (°F) . . . . Water boils

FORMULAS

AREA	Rectangle	$A = l \times w$
	Triangle	$A = \frac{1}{2}(b \times h)$
VOLUME	Rectangular Prism	$V = l \times w \times h$

SYMBOLS

$<$ is less than
$>$ is greater than
$\neq$ is not equal to
$\approx$ is approximately equal to
$4 \div 2$ 4 divided by 2
% percent

3:5 the ratio 3 to 5
° degree
A point A
$\overleftrightarrow{AB}$ line AB
$\overrightarrow{AB}$ ray AB
$\overline{AB}$ line segment AB

$\angle ABC$ angle ABC
$\triangle ABC$ triangle ABC
$\parallel$ is parallel to
$\cong$ is congruent to
$\sim$ is similar to
(5,3) the ordered pair 5,3

445

Glossary

Acute Angle An angle that measures less than 90°.

Acute triangle A triangle that has three acute angles.

Addends Numbers that are added.
Example: $5 + 7 = 12$
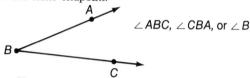
addends

Angle A figure formed by two different rays that have the same endpoint.

$\angle ABC$, $\angle CBA$, or $\angle B$

Area The number of square units needed to cover a surface.

Associative Property of Addition If the grouping of addends is changed, the sum remains the same.
Example: $(2 + 3) + 6 = 2 + (3 + 6)$

Associative Property of Multiplication If the grouping of factors is changed, the product remains the same.
Example: $(2 \times 2) \times 4 = 2 \times (2 \times 4)$

Average The average, or mean, of a set of numbers is the sum of the numbers divided by the number of addends.

BASIC *BASIC* stands for "Beginner's All-purpose Symbolic Instructional Code," a computer language.

Chord A line segment that has endpoints on a circle.

Circle A circle consists of all points in one plane that are the same distance from one point, called the *center.*

Circumference The distance around a circle.

Commutative Property of Addition If the order of two addends is changed, the sum remains the same.
Example: $6 + 3 = 3 + 6$

Commutative Property of Multiplication If the order of the factors is changed, the product remains the same.
Example: $7 \times 4 = 4 \times 7$

Composite number A number that has more than two factors.
Example: 24 is a composite number because it has 8 factors: 1, 2, 3, 4, 6, 8, 12, and 24.

Cone A solid figure that has a circular base.

Congruent Figures that are exactly the same shape and size are congruent.

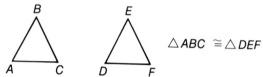

$\triangle ABC \cong \triangle DEF$

Cube A solid figure that has six square faces.

Cylinder A solid figure that has two congruent circular bases.

Debug To fix the problems in a computer program so that it will do what you want it to do.

Decimal A number that uses place value and a decimal point to show tenths, hundredths, thousandths, etc.
Examples: 0.09, 37.1486

Degree A unit of measure for circles, angles, and temperature.

Denominator In $\frac{5}{8}$, 8 is the denominator. It tells the total number of parts or groups.

Diagonal A line segment that is not a side and that joins two vertices of a polygon.

Diameter A chord that passes through the center of a circle.

Digit Any of the individual numerals 0, 1, 2, 3, 4, 5, 6, 7, 8, or 9 that are used to build the base-ten name for a number.

Distributive Property To find the product of a number times the sum of two addends, you can multiply each addend by the number and then add the products.
Example: $4 \times (3 + 8) = (4 \times 3) + (4 \times 8)$

Dividend The number that is divided.
Example: $10 \div 2 = 5$ $\quad$ $2\overline{)10}^{\,5}$
$\uparrow$
dividend ————↑

Divisible A number is divisible by another if it can be divided by that number with no remainder.
Example: 9 is divisible by 3, but not by 4.

Divisor The number that divides the dividend.
Example: $16 \div 8 = 2$ $\quad$ $8\overline{)16}^{\,2}$
$\uparrow$
divisor ↗

Edge Two faces of a prism intersect at an edge.

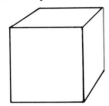

Equation A mathematical sentence that has an equals sign.

Equilateral triangle A triangle that has three sides of equal length.

Equivalent fractions Two or more fractions that name the same number.
Example: $\frac{6}{8} = \frac{3}{4}$

Even number A number that has the digit 0, 2, 4, 6, or 8 in the ones place. They are divisible by 2.

Expanded numeral A numeral expanded to show the value of each digit.
Example: $57,305 = 50,000 + 7,000 + 300 + 5$

Face The flat surfaces of a prism.

Factors Numbers that are multiplied.
Example: $3 \times 5 = 15$
$\quad\uparrow\quad\uparrow$
$\quad$ Factors

Flowchart A diagram that shows the steps to do something.

FOR/STEP/NEXT A three-part command that makes a computer repeat a step a given number of times.
```
10   FOR N = 2 TO 10 STEP 2
20   PRINT N
30   NEXT N
```
makes a computer print the numbers 2, 4, 6, 8, and 10.

Fraction A fraction is used to name parts of a whole, or parts of a group.

Gram A unit of mass in the metric system.

Greatest Common Factor The greatest common factor of two or more numbers is the greatest number that is a factor of each number.
Example: 9 is the greatest common factor of 18 and 27.

Hexagon A six-sided polygon.

IF/THEN A command that tells a computer to make a decision.
Example: If N < 10 THEN 40 tells a computer to go to line 40 if the number in the storage place N is less than 10.

Inequality A number sentence that contains <, >, or =.

Intersecting lines Lines that meet or cross at one point.

Isosceles triangle A triangle that has at least two congruent sides.

Least common denominator The least common multiple of the denominators of two or more fractions.
Example: For $\frac{1}{5}$ and $\frac{2}{3}$, 15 is the least common denominator.

Least common multiple The least common multiple of two or more numbers is the smallest number other than 0 that is a common multiple.
Example: For 4 and 7, 28 is the least common multiple.

447

Line A line is a straight path that goes on forever in two directions.

line *AB* or $\overleftrightarrow{AB}$

Line of symmetry The line along which a symmetrical figure can be folded so that the two halves match exactly.

Line segment A line segment is a part of a straight line. It is named by its endpoints.

line segment *AB* or $\overline{AB}$

LIST A command that tells a computer to show the program lines that are stored in its memory.

Liter A unit of liquid capacity in the metric system.

Mean The mean, or average, of a set of numbers is the sum of the numbers divided by the number of addends in the set.

Median The median is the middle number in an ordered set of numbers.

Meter A unit of length in the metric system.

Mixed Number A mixed number has a whole number part and a fraction part, such as $4\frac{2}{3}$.

Mode The mode is the number that occurs most often in a set of numbers.

Multiple A multiple of a number is the product of that number and any other whole number.
Example: 10, 25, and 40 are multiples of 5

Number line A line used to show numbers in order.

Number sentence An equation or inequality.
Examples: $13 - 5 = 8$ $n \times 4 = 32$
    $3 \times 4 < 16$ $8 + 6 > 11$

Numeral A name for a number.

Numerator In $\frac{6}{7}$, 6 is the numerator. It tells how many parts you are talking about.

Obtuse Angle An angle that measures more than 90 but less than 180°.

Obtuse triangle A triangle that has an obtuse angle.

Odd number A number that has 1, 3, 5, 7, or 9 in the ones place. They are not divisible by 2.

Ordered pair A pair of numbers, used to locate a point on a grid.

Parallel lines Lines in a plane that never intersect.

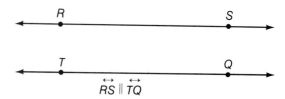

$\overleftrightarrow{RS} \parallel \overleftrightarrow{TQ}$

Parallelogram A quadrilateral whose opposite sides are the same length and are parallel.

Pentagon A five-sided polygon.

Perimeter The distance around a polygon—the sum of the lengths of its sides.

Period A group of three digits in a numeral set off by commas.

Perpendicular lines Lines that intersect and form right angles.

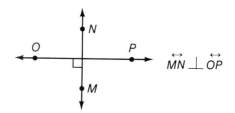

$\overleftrightarrow{MN} \perp \overleftrightarrow{OP}$

Plane A flat surface that goes on forever in all directions.

Point A point is an exact location in space.

Polygon A closed figure formed by line segments.

Prime factorization A composite number can be shown to be the product of its prime factors.
Example: The prime factorization of 18 is
    $2 \times 3 \times 3$.

Prime number A prime number has exactly two factors, itself and 1.

PRINT A command that tells a computer to output information on a screen or on paper.

Prism A solid with congruent polygons as two parallel faces.

Probability A comparison of the number of favorable outcomes to the total number of possible outcomes.

Property of One for Multiplication If one factor is 1, then the product is always the other factor.

Property of Zero for Addition If one of the addends is zero, the sum is equal to the other addend.
Example: $7 + 0 = 7, 0 + 9 = 9$

Property of Zero for Multiplication If one factor is 0, the product is always 0.
Example: $8 \times 0 = 0, 9,672 \times 0 = 0$

Protractor An instrument used to measure angles.

Pyramid A solid that has three or more faces that are triangles that have a common vertex and one face that is a polygon.

Quadrilateral A polygon with four sides.
Example: $36 \div 4 = 9 \leftarrow$ quotient $\rightarrow 6\overline{)42}^{\,7}$

Radius A line segment that has one endpoint on the circle and one endpoint on the center.

Range The difference between the greatest and the least number in a set of numbers.

Ratio A comparison between two numbers.

Ray A part of a line that begins at an endpoint and goes on forever in one direction.

ray AB or $\overrightarrow{AB}$

Rectangle A parallelogram that has four right angles.

Rectangular prism A three-dimensional figure that has six faces and eight corners. Its bases are rectangular.

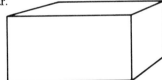

Rhombus A parallelogram whose sides are all the same length.

Right angle An angle that measures 90°.

Right triangle A triangle that has one right angle.

RND(1) In a computer program, RND(1) makes a computer pick a random 9-place decimal between 0 and 1.

Scalene triangle A triangle with no congruent sides.

Similar figures Figures that have the same shape but not necessarily the same size.

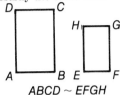

$ABCD \sim EFGH$

Simplest form A fraction is in simplest form if its numerator and denominator have no common factors other than 1.

Sphere A solid figure that has a surface that has all points the same distance from its center.

Square A rectangle whose sides are all the same length.

Standard numeral The usual way to name a number.
Example: The standard numeral for thirty-four is 34.

Symmetry A figure is symmetrical when there is a line about which the figure can be folded. The resulting figure matches the original.

Triangle A three-sided polygon.

Vertex The common endpoint of the sides of an angle or two sides of a polygon.

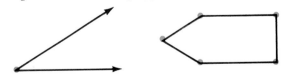

Volume The interior space of a solid figure.

Whole numbers Any of these numbers:
0, 1, 2, 3, . . .

INDEX

percent, 328–329, 371
ratios, 316–317, 355, 371
statistics and probability,
390–391, 394–395
time, 80–81, 288–289, 298–299,
354–355
Problem Solving—Data Sources
bar graph, 50, 391
broken-line graph, 50, 171, 390
catalog, 403
circle graph, 328–329
fact list, 37, 121, 159, 301, 331
Infobank, 415–420
map, 298–299
recipe, 316
schedule, 288, 354
table, 59, 279
Problem Solving—Estimation,
26–27, 58–59, 84–85, 176–177,
278–279
Problem Solving—Practice, 91,
170–171, 244–245, 354–355, 401
Problem Solving—Strategies
checking for a reasonable
answer, 80–81
choosing the operation, 34–35,
136–137, 234–235
checking that the solution
answers the question,
184–185
choosing/writing a sensible
question, 150–151
guessing and checking,
202–203
identifying needed
information, 118–119, 208–209
identifying extra information,
208–209
interpreting graphs, 390–391
interpreting the quotient and
the remainder, 156–157,
262–263
making change, 91
making a diagram, 394–395
making an organized list,
112–113
making a table to find a
pattern, 282–283, 348–349
solving twostep/multistep
problems/making a plan,
96–97, 214–215
using the Help File, 8–9,
108–109
using outside sources, 26–27
using a recipe, 316–317
using a schedule, 288–289
using a time-zone map, 298–299
writing a number sentence,
64–65, 142–143, 370–371

writing a simpler problem,
312–313
Products, 76–77
Programming languages. (See
BASIC; LOGO
Projects. (See Group projects)
Protractor, 344–345
Property
Associative
for addition, 10–11
for multiplication, 76–77
Commutative, 10–11
Distributive, 76–77
Grouping (See Associative)
Identity, 76–77
Order (See Commutative)
Zero
in addition, 10–11
in multiplication, 76–77
Proportions, 269, 314–315,
316–317
Pyramid(s), 372–373

Q _____

Quadrilateral, 350–351
Quart, 294

R _____

Radius, 352–353
Range, 392–393
Ratio(s), 308–309
and fractions, 308–311
and percents, 318–319
and scale drawings, 314–315
Ray, 340–341
Reading math, 120, 216, 330
Recipes, 294, 316–317
Rectangle, 350–351
area, 366–367
perimeter, 364–365
Rectangular prism, 372–373
volume of, 374–375
Regular polygon, 350–351
Remainder, 136–137
Rhombus, 350–351
RT (right), 72–73
Right angle, 342–343
Right triangle
area of, 368–369
Road map, 305
Roman numeral, 41
Rounding
with decimals, 54–55
with fractions, 248–250
whole numbers, 16–17

RUN, 192–193

S _____

Sales tax, 355
Scale drawing, 314–315
Scalene triangle, 346–347
Scientific notation, 125
Short division, 163
Skills Applications—Data
Sources
advertisement, 277, 319
charts, 7, 11, 13, 21, 89, 147, 155,
213, 227, 385, 387, 389, 397
grids, 362–363
pictures, 309
recipes, 294
scale drawings, 315
spinners, 399
Slides, 381
Solid figure(s), 372–373
Sphere, 372–373
Square
geometric figure
area, 366–367
perimeter, 364–365
Square units, 366–367
Standard form, 2–3
Statistics, 390–391, 396–399
data sources
graphs (making)
bar, 384–385
broken line, 388–389
double bar, 407
pictograph, 386–387
measures of central tendency
mean, 392–393
median, 392–393
mode, 392–393
range, 392–393
tally, 398–399
Subtraction
with decimals, 62–63
annexing zeros, 62–63
estimation in, 18, 23, 56–57
facts, 12–13
with fractions, 252–253
with larger numbers, 30–31
with mixed numbers, 258–261
with money, 28–29
rules, 12–13
with units of time, 286–287
of whole numbers, 12–13,
28–33
with zeros, 32–33
Sum(s), 10
Surface area, 373
Symmetry, line of, 358–359